Paul J. Bowman
April 22, 1970.

The Quest for Plants

The gods, that mortal beauty chase
Still in a tree did end their race;
Apollo hunted Daphne so
Only that she might laurel grow,
And Pan did after Syrinx speed
Not as a nymph, but for a reed.

ANDREW MARVELL *Thoughts in a Garden*

The Quest for Plants

A HISTORY OF

THE HORTICULTURAL EXPLORERS

Alice M Coats

STUDIO VISTA LONDON

Published in London 1969 by Studio Vista Limited
Blue Star House, Highgate Hill, London N19
Set in Bembo 12pt solid
Printed in Great Britain by
Staples Printers Limited at their Rochester, Kent, establishment

SBN 289 27985 2

Contents

Acknowledgments

The author wishes to express her gratitude to the following helpers: Mrs M. Archer, Mr R. Desmond, Miss S. Raphael, Mr P. Stageman, Mrs S. Stone and Miss I. Wilson, librarians respectively at the India Office, Kew, the Linnean Society, the Royal Horticultural Society, the Birmingham Library and the Birmingham Reference Library; to Dr W. T. Stearn and Mr J. S. L. Gilmour; to foreign correspondents Mrs Z. Artjushenko, Professor J. Ewan, Mrs M. Lawder, Dr G. H. M. Lawrence, Mrs E. du Plessis and Mr A. Probert; to Mrs H. Harrison, a collateral descendant of Thomas Coulter, for her obliging loan of family letters and papers; and to a number of personal friends who have supplied references, looked up references, translated passages in alien tongues, and tackled a mountain of typing.

She is also obliged to those who supplied material for the illustrations: the British Museum (pls 6, 14; pp 52, 152, 261); the Linnean Society (pls 3, 19, 20, 24); the National Portrait Gallery (pls 2, 23); the Royal Botanic Gardens, Kew (pls 1, 12, 16, 17, 18, 22; pp 21, 202); the Royal Horticultural Society (pl 8; p 9); Verwaltung der Staatliche Schlösser und Gärten, Berlin (pl 25); the Swedish Institute for Cultural Relations (pl 5); the Henry Herbarium, Glasnevin (pl 10); Mr Joe Elliott (pl 26); the Ashmolean Museum, Oxford (pl 4); the Royal College of Physicians (pl 7); Condé Nast Publications (pl 9); Mr F. Nigel Hepper (p 9). She is also obliged to Mrs Maureen Verity, who drew the map on p 159, and to the various photographers: Kenneth Collier, Twickenham (pls 1, 11, 12, 17, 18; pp 21, 52, 202); John R. Freeman Ltd, London (pls 3, 19, 20, 24); Svenska Porträttarkivet, Stockholm (pl 5); J. E. Downward, Essex (pl 10); Logan, Birmingham (pl 21); P. Siviter Smith (pl 26); Elsa Postel (pl 25).

Foreword

. . . Many shall run to and fro, and knowledge shall be increased.

DANIEL XII, 4

This book is chiefly, though not exclusively, concerned with the exploits of professional gardeners who travelled abroad expressly for the purpose of collecting hardy horticultural plants. There are many kinds of plant-hunters and it has been necessary to limit the enormous field by the exclusion (for the time being) of the pioneer botanists who sailed in naval vessels engaged on exploration or coastal surveys, but, being based on shipboard, were unable to penetrate far inland. Owing to limitations of space, only brief accounts are given of those who worked in the tropics, and the biographies of collectors before and after the period of their travels have been reduced to a minimum.

It is impossible, however, to make any firm distinction between the 'horticultural' collectors of ornamental plants, and the 'botanical' ones whose primary aim was the enlargement of the science; for the botanists introduced plants and the gardeners botanized. The most strictly scientific collectors were encouraged to send seeds to the botanic gardens that had been founded in order that the living plant might be studied; and from these gardens the more ornamental species spread into general cultivation. Moreover, the botanist often preceded the gardener, as the pilot-fish the whale; his travels aroused interest and stimulated demand, as well as showing where good plants were to be found. On the other hand it is perhaps not realized how much time even the most horticultural of collectors spent on the preparation of botanical specimens, nor how important a part the value of such collections played in the financing of plant-introduction. Very few expeditions, past or present, could be supported on the profits of gardening alone; botany, ornithology or some kindred science had to cover part of the cost.

The predominantly horticultural collectors have, on the whole, been badly served by the chroniclers, in comparison with the botanical ones. On occasions when scientific expeditions were accompanied by both a botanist and a gardener, the latter is hardly mentioned in the official reports, and any plants introduced are credited to others. In Aiton's *Hortus Kewensis*, for example, it is the patron who is almost invariably named as the introducer, and not the collector he employed. (An exception is made for some of Kew's own collectors; Aiton seems to have felt some diffidence about crediting their plants either to himself or to George III.) It was only when a gardener was himself capable of writing letters, journals or articles about his travels and his finds, that we know anything about his activities; and many such records must have been lost, especially before the days of the gardening and botanical periodicals. The taciturn ones, though perhaps no less meritorious, are in danger of being forgotten.

It is easy to over-emphasize the importance of 'first introductions'. It would be absurd, for example, to imply that if Busbecq had not introduced the tulip and the lilac to Vienna in 1562 we should still be without them today. There is a certain inevitability about the entry of the really first-class plants to cultivation; if one man does not bring them, another will. There are a few instances where the first introduction was also the only one; but gardeners probably owe more to the person who established a plant firmly in cultivation, than to its first discoverer. The pioneers, however, deserve the recognition, because it was they who endured the hardships.

Besides a good knowledge of botany and gardening, a collector frequently had some skill in ornithology, zoology, geology, surveying or medicine – the latter particularly desirable. He had to be adaptable and able to get on with natives, and his life often depended on his being a good shot and fisherman. He had also to have great tenacity and endurance, the conditions of travel being often such that only curiosity, the greatest human motive-power next to love and hunger, could enable him to support them. It follows that the successful collectors were very remarkable men, and their lives and characters well worth recording.

That the task of recording them was rather beyond my ability will be evident by the mistakes and omissions which with my greatest care and enthusiasm I have not been able to avoid; but 'None of us all can publish anything, but there may bee slippes and errours in many places thereof.'[1]

BIRMINGHAM 1969

[1] John Parkinson *Theatrum Botanicum* 1640

Queen Hatshepsut's Plant-Collectors (see p. 243)
By courtesy of the Royal Horticultural Society

The Mediterranean and Near East

The dates are those of the collector's first journey in the area

THE TURKISH EMPIRE

1546	Belon	1787	Sibthorp
1573	Rauwolf	1874	Elwes
1675	Wheler	1877	Maw
1700	Tournefort	1911	Ball
1782	Michaux	1932	Balls

THE MEDITERRANEAN CIRCUIT

North Africa

1749	Hasselquist	1852	Cosson
1761	Forsskӧl	1869	Maw
1798	Delile	1871	Hooker and Ball
1827	Webb	1936	Balls

The Iberian Peninsula

1607	Boel	1783	Masson
1611	Tradescant	1825	Webb
1751	Loefling	1869	Maw
1760	Alstroemer	c. 1880	Barr

'It would be quite lost labour to give any detailed account of this kind of scientific life, if it can be so-called, this dull occupation of plant-collectors.' JULIUS VON SACHS *History of Botany* 1890

'A tribute to botanical explorers is, we must say, well-earned. Of all deadly occupations this is surely the most fatal.' BOOK REVIEW *The Gardener's Chronicle* 1881

The Mediterranean and Near East

The Turkish empire

It is an axiom that the territory available for profitable plant-hunting shrinks as agriculture expands and civilization advances. The greater part of Europe was in a high state of cultivation long before botanical exploration began, and the well-established towns and means of communication made it easy for the remaining plants to be studied by knowledgeable local enthusiasts, without the necessity for deliberate collecting expeditions, except in the wildest parts.

Botanical exploration, as such, began with the Renaissance, and was at first largely concerned with the identification of plants mentioned in the Greek and Latin classics. It was inevitable, therefore, that the travellers should be drawn to the land of the sunrising – the Levant; and this, for more than 300 years, meant the Kingdom of Turkey. By 1526 the Ottoman Empire extended over Asia Minor, Egypt, Greece, the Balkans and Hungary; in 1529 Turkish armies were battering at the gates of Vienna; Algeria was conquered in 1541 and Tunisia in 1574. The slow process of liberation began in the Mediterranean with the War of Greek Independence in 1821–9, and was not completed till 1913. The Turks were exceedingly resistant to Western ideas of 'progress', and conditions of travel in their dominions did not change very much from the time of Belon to that of Sibthorp.

Pierre Belon (1518–1563) took his medical degree in Paris, and for some years afterwards supervised the making of an arboretum for the Bishop of Mans; his special interest in trees and in medicinal plants is apparent throughout his book. His journey to the East was undertaken under the patronage of François, Cardinal de Tournon, at whose expense he had already travelled in England, Germany, France and Italy; and its purpose was to see at first hand the 'curiosities' – plants, animals and antiquities – mentioned by classical authors. His three-year tour established a pattern which was followed by scientific travellers for nearly three centuries.

It began in 1546 with a prolonged stay in Crete, renowned in Roman times for the richness of its flora. Belon devotes two chapters to the plants he found there, including the white form of the oleander, and *Paeonia clusii*. He climbed Mt Ida three times, but was unable to find the expected wild raspberries, already named *Rubus idaeus*;[1] and he gives the first modern account of the methods by which ladanum was obtained from *Cistus ladaniferus*.

Having 'much haunted and travelled the yland of Crete or Candie',[2] Belon proceeded to Constantinople, where he obtained a safe-conduct through the French Ambassador, M. de Fumet, who was in good standing with the Turks. Then he doubled back to the islands of Imbros and Lemnos. On the latter – an island 'all humped with little hills' – his medical services were requested both by Greeks and Turks, including one of the principal men of the island. He announced, therefore, that he wanted to see all the products of the land, in order to choose the best medicine; and by this means got many plants brought to him that he would not have been able to seek out for himself.

From Lemnos, Belon sailed to Thassos (though blown out of course to Skyro on the way) and after a short stay crossed to Mt Athos, where he was pleased to find that the monks still employed the old plant-names used by Theophrastus, Dioscorides and Galen. Then he went to Salonika, and by land through Kavalla and Rodosto to Constantinople, with side-trips on the way to see the ruins of Philippi and the mines of Siderocaps and Ipsala.

Ships for Alexandria left Constantinople about the end of August, when a favourable north wind might be expected; and in August 1547 Belon embarked on a vessel carrying more than 100 Turkish passengers for Egypt. With stops at Gallipoli,[3] Scio and Rhodes they reached Alexandria safely, and after a stay of a few days Belon went to Rosetta and up the Nile to Cairo. There he again met M. de Fumet, and made his subsequent travels in the Ambassador's suite, the lawlessness of the country rendering a large party necessary. In this good company he went to see the Pyramids, and afterwards on a twenty-day tour via Suez to Mt Sinai – a journey so hard, dangerous and arid that it was considered nothing unusual that seven out of ten horses and more than half the camels died.

On 29 October 1547, the party left to return to Constantinople via Palestine. Hard travel brought them through Gaza and Rama to Jerusalem on 8 November. From there they made excursions to Jericho, the Jordan and Bethlehem before continuing north to Damascus, Baalbek, Hamah and Aleppo. At Aleppo the physician Belon noted that rhubarb-root was brought from Mesopotamia at the rate of twelve camel-loads at a time,

[1] Perhaps from the Ide mountains (Kaz Dag) in N.W. Asia Minor
[2] Lyte *A Nievve Herbal* 1578.
[3] Which he compared to 'Rie [Rye?] en Angleterre'

but he could not find out from what sort of plant it came. They went a little out of their way to visit Antioch; then by Adana, Mt Taurus and Iconium (at Christmas) to Carachara, where Belon spent the rest of the winter. He does not say whether he was still with the Ambassador's party when he crossed Mt Olympus to Brousa (Bursa) and so back to Constantinople in 1548; nor does he tell us anything about the return journey, though he is said to have rejoined the Cardinal de Tournon in Rome.

Belon entitled his subsequent book *Les Observations* (1554), and he was an observer only – he did not attempt to bring home seeds, plants or specimens, so far as is known. He relied only on descriptions and his own lively drawings, quite recognizable even when translated into the crude woodcuts of the time. He understood the importance of botanical geography; his method, he tells us, was to put 'some little branch or leaf' of everything he found into a bag and at the end of the day or during the noonday rest to go through them and write them all down, including the commonest – 'just to show that they are found in those places as well as with us'.[1] He noticed that the plane and the acacia do not shed their bark at the same season, and that the acacia folds up its leaves so closely that 350 leaflets can be covered with a thumb. In his book we find the first mention of the cherry-laurel – 'un arbre de Trapisonde qui porte des cerises' – and of the lilac, as a shrub cultivated by the Turks. He did not confine his attention to botany, but reported on animals – including a giraffe in captivity in Cairo – birds, snakes and insects, antiquities and local customs, and was a keen student of ichthyology. Fourteen years after escaping all the perils of his oriental journey, this 'man of high attempts in natural science'[2] was murdered by robbers in the Bois de Boulogne.

At the time of Belon's journey the herbarium had not been invented; the earliest recorded collection of dried plant-specimens was made three years later, in 1551. Within thirty years the practice had become so well established that our next collector carried paper for preserving plants as a matter of course. This was the German physician **Leonhardt Rauwolf** (d. 1596) who left Augsburg on 18 May 1573, 'chiefly to gain a clear and distinct knowledge of those delicate herbs, described by *Theophrastus*, *Dioscorides*, *Avicenna*, *Serapio*, *etc.* by viewing them in their proper and native places'[3] and to encourage the apothecaries to procure the right sorts for their shops.

After a three-month stay at Marseilles, Rauwolf sailed on 1 September for Tripoli (where he saw and described the banana) and presently 'propounded to myself to travel to Aleppo'. He arrived in mid-November and made a long stay, botanizing in the vicinity in company with a friend, Hans Ulrich Krafft. These excursions were made 'not without great pain and danger of being knock'd on the head', and on one occasion Rauwolf was attacked by a drunken Turk with a scimitar, and had to keep dodging round the trunk of an olive-tree until his assailant could be bribed to go away. His finds were 'preserved and glued to some paper, with great and

[1] *Les Observations*, trans. A.M.C.
[2] Gerard *Herball*
[3] The quotations are from Rauwolf's *Travels*, trans. by John Ray

peculiar care, so that they are to be seen with their natural colours so exact, as if they were green'. Like Belon, he notes that the Turks 'love to raise all sorts of flowers, in which they take great delight, and use to put them on their turbant, so I could see the fine plants that blow one after another dially [sic]without trouble'. Among the wild flowers he found near Aleppo was the hyacinth, already known in Europe as a cultivated plant.

On 13 August 1574, Rauwolf and a Dutchman from Aleppo disguised themselves as merchants in Turkish dress, and joined a caravan of Armenians going to Bir (Birejik) on the Euphrates. After waiting there for about a fortnight they procured a passage on a barge, and proceeded slowly down the river towards Baghdad. Like many subsequent botanists, Rauwolf found river-travel tedious. 'When we did land, and had time to spare, I used to look about me for some strange plants' but more frequently he saw bushes and trees on the banks and 'would fain have been at them to discern what they were . . . but I was forced to stay on the ship, and so I missed them'. At Deir, where they were held up by the Customs, he presented to the authorities 'some sheets of white paper, which they willingly received, and were so pleased with it, that some of them (as children do in our country, when we give them something that is strange or pleasing to them) smiled as often as they look'd on it'.

They reached Baghdad towards the end of October, and Rauwolf was tempted to go on to 'the Indies', but received letters which decided him to return to Aleppo. He set out on 16 December by a different route, across Mesopotamia. 'After we had lived for several days very hardly in the desarts, and spent our lives in misery', the small party with which he travelled joined a large caravan at Harpel, before crossing a dangerous stretch of country involving forced marches by day and a careful watch at night. In spite of all hazards and the winter season, Rauwolf still managed to collect, and crossed a deep ford on the River Caprus (Khabur?) 'not without detriment to my plants, which I carry'd on horseback before me'. Shortly afterwards the camp was attacked at night by a large band of robbers; Rauwolf was 'the left-hand man in the first rank again, with my scymeter drawn, and had before armed my breast with several sheets of paper, that I had brought with me to dry my plants in' – but at the sight of such determination, the enemy prudently withdrew. Zibin, and Urfa in the mountains, brought them again to Bir on the Euphrates, and so back to Aleppo on 10 February 1575.

At Aleppo Rauwolf lodged with some French merchants whom he had previously cured of 'several distempers', and got a much-needed rest and change of clothes; his others, he says, 'never came from my back in half-a-year's time'. In May he returned to Tripoli, just in time to escape a plot to arrest him for spying – i.e. botanizing in the hills – as an excuse to mulct him of a large fine; as he says of the Turks, 'so unjust, malicious and infidel a people are they, that one would hardly believe it'. At Tripoli he stayed with the French consul and resumed the practice of medicine with such success that his services were requested by the Patriarch of the Maronites, who invited him to return with him to his establishment on

Mt Lebanon. This gave Rauwolf a splendid opportunity for botanizing on the famous mountain; he climbed almost to the snowline to view the cedars, but could find only twenty-six aged trees and no young ones. More fortunate than Belon, he discovered a species of rhubarb,[1] 'the true *Ribes* of the Arabians', whose stalks were 'full of a pleasant sourish juice . . . whereof chiefly the true *Rob Ribes* is prepared, as I have seen it myself, and Serapio testifieth'. He also found many other plants new to him, including 'a pretty sort of Tulips with yellow stripes'.

He did not wish to return home without visiting the Holy Land, so took ship from Tripoli to Joppa on 7 September 1575, and thence to Rama, Jerusalem and other places, returning to Tripoli on 1 October. The homeward voyage began on 6 November, but his adventures were not yet at an end. Adverse winds drove them to shelter at the island of Calderon near Crete, where he found 'a kind of mandrake with blew flowers in great quantity'[2] and they nearly foundered in a hurricane off Cerigo, when the store holding the cannon-balls broke open and the balls 'ran up and down all over the ship according as she rolled' and 'the seamen prayed three several times'. After rest and repair at Argostoli in Cephalonia they proceeded slowly up the Adriatic, the ship so heavily loaded with merchandise that the passengers had to brave all weathers on deck, there being no room for them below. They reached Venice on 15 January 1576, and Rauwolf posted safely back to Augsburg, 'to the great rejoicing of my dear parents and relations, whom I found all in indifferent good health'.

Rauwolf only once refers to an attempt to introduce plants; seeds of a thorny shrub from Mt Lebanon were raised in an Augsberg garden, but died before flowering. There were probably other introductions, which will never be known. In enthusiasm, toughness and enterprise he was the epitome of the botanical and horticultural collector of all subsequent ages.

The traveller who set out some hundred years later, however, was of a different type – the gentleman-amateur, travelling for his own pleasure and at his own expense, but nevertheless making valuable contributions to scientific knowledge. Reading **George Wheler's** sober and learned *Journey into Greece* (1682) one hardly realizes that at the time he was a young man of twenty-five, and his journey only an extension of the Grand Tour on which he embarked with a tutor after leaving Oxford and before taking his degree. He tells us that having resolved on a voyage to the Levant, he 'hasten'd to Venice' in early June 1675, and sought out his friend Dr James Spon of Lyons; and hearing that there would soon be an opportunity of joining the party of an ambassador who was going to Constantinople, they obtained a passage in one of his galleys, and were off before the end of the month. Both were interested in coins and antiquities, but only Wheler collected plants; Spon, he tells us, 'did not at all concern himself about them'. Wheler began his 'simpling' on the first day of the voyage, when he went ashore on the tiny island of St André off the coast of Istria and found sixteen plants, including *Convolvulus wheleri*;

[1] *Rheum ribes*
[2] *Mandragora autumnale*

and continued to the very last moment, when, held up by adverse winds at Aspra Spitia Bay in Greece, he went once more ashore and got 'very many curious Plants' including three that were new to him.

They went from port to port down the Dalmatian coast to Spalato, whence the Ambassador, already weary of the sea, continued his journey overland, leaving the galleys to complete the sea-passage with the baggage; thence to Corfu, Zante and Cerigo, and without further stop through the Archipelago to Tenos. Here the ship was expected to remain some days; so Wheler and Spon hired a four-oared barque to go and inspect the ruins on the island of Delos. No sooner had they arrived than the weather worsened, and they were stranded for two nights, almost destitute of provisions and water; 'for the next meal', says Wheler, 'we were very sollicitous, not knowing whose turn it might first be, to have his haunches cut out, to serve for Venison to the rest'. They had also the anguish of seeing their ship sail from Tenos without them; but the same bad weather held her at up Micone, and when the wind and sea had a little abated they were able to cross the four-mile strait and rejoin her. And so by Tenedos, Troas, Imbros and Gallipoli to Constantinople on 13 September. Plague was present in the last two places, but that had to be accepted, the cities of Turkey being seldom free of it; 'so we thought we had as good begin here to accustom ourselves to its company'.

Several excursions were made in the neighbourhood of the capital, including one to the pleasant resort of Belgrade, where they saw many a 'stately Chiosque' paved with 'Purcelaine Tiles', and where Wheler found *Hypericum calycinum*, which he introduced to Britain and which was still known a century later[1] as 'Sir George Wheler's Tutsan'. On 6 October they joined a party of English merchants bound for Smyrna. The city of Brousa was their first objective; it had 'one of the pleasantest Comings to it imagineable' and here in a garden Wheler saw a weeping willow, a tree then unknown in England. The illness of a member of the party caused a delay which enabled Wheler and a travelling companion (not the unbotanical Spon) to visit Mt Olympus; they climbed to the snowline and in two hours got 'more curious Plants than I could ever since find Names for', though some were only in seed.[2] Their enthusiasm kept them on the mountain till nightfall, and they had difficulty in getting down in the dark. The party set out again on 13 October and reached Smyrna towards the end of the month.

After visiting Ephesus, Wheler and Spon sailed from Smyrna on 17 November, but encountered storms and did not reach Zante till Christmas Eve. At Patras they were again detained, first by trouble with the Customs and then by adverse winds, and finally left on 19 January, crossing the Gulf of Patras and travelling overland by way of Salone, Mt Parnassus and Livadia to Athens. Here they spent a month sight-seeing and making local excursions, though the weather was consistently bad – 'our Acquaintances would say, That if they wanted Rain for their Olive-Yards, they need but send to us to go abroad, and they should have it'.

[1] To Gilbert White of Selborne
[2] He is said to have introduced *Hypericum olympicum*

Nevertheless Wheler found about sixty plants, and gives an interesting account of methods of beekeeping practised on Mt Hymettus. Their longest excursion was to Eleusis, Megara and Corinth.

On 29 February 1676, they left Athens, meaning to go overland to Germany, but at Turcochorio they found the mountain-passes were still blocked by snow. Spon was anxious to get home, and decided to return by sea via Zante and Venice; but Wheler 'could not be so soon reconciled to the sea, in such bad Weather, and at that season of the Year', so he parted from his friend and made a detour that took him to Rimocastri and Mt Helicon. Returning by way of Aspra Spitia, Zante, Italy and France, he reached Canterbury with great joy and relief on 15 November.

Spon published an account of their joint travels in 1678, but Wheler's book was longer in preparation and did not appear till 1682. It was an immediate success; he was knighted the following September, and received an honorary degree the next year. He took orders – it is said, in fulfilment of a vow made at some moment of peril in his travels – and became Prebendary of Durham. In his book he names between 180 and 200 plants that he found – not necessarily new, even in his day – and he sent specimens home to the botanists, Ray, Plukenet and Morison; he also sent Morison seeds.

Spon's book was among those familiar to Tournefort, whose visit to the East was so well prepared that he was, it was said, 'naturalised by his learning' wherever he landed. **Joseph Pitton de Tournefort** (1656–1708) was destined for the church, but turned to botany as soon as his father's death freed him to do so, and had his first experiences of the joys and hazards of a plant-hunter in the mountains of Dauphiné and Savoy in 1678, and in the Pyrenees in 1681. On the latter journey he was robbed and stripped by the Miquelets,[1] who in consideration of his distress and the extreme cold, gave him back his coat – including, fortunately, some money tied in a handkerchief which had slipped down inside the lining. Having walked, bare-legged, to the nearest town, he was therefore able to buy 'a Thrum-Cap, Linen Trowsers and a Pair of Wooden Shoes'. Later, he had a narrow escape when a house in which he was lodging fell down in the night.

In 1683 Tournefort was appointed Professor of Botany at the Jardin du Roi, and did some rather more decorous plant-collecting in the botanic gardens of Spain, Portugal, Holland and England. He also devised one of the most important early systems of plant-classification, and was altogether an ideal person to be despatched by Louis XIV in 1700 on a voyage of scientific exploration, the object of which was primarily 'to discover the Plants of the Ancients, and others, which perhaps escaped their Knowledge'.[2]

Tournefort chose as his companions two colleagues of some years' standing; a German doctor called Andreas Gundelscheimer, whom he had sent to England in 1698 to see the collections brought back from Jamaica by his former pupil Sir Hans Sloane, and a young artist, Claude Aubriet,

[1] Spanish bandits

[2] Tournefort *Compleat Herbal* 1730 (Foreword)

who had drawn the plates for his *Elémens de Botanique* (1694). A physician was essential, Tournefort thought, when travelling in remote parts: '. . . it frets a man, too, to see fine Objects, and not be able to take Draughts of them'. Aubriet worked very hard, sketching landscapes, antiquities and local costumes as well as plants, and well deserved his commemoration in the aubrieta. Tournefort's account of the voyage takes the form of a series of letters to M. de Pontchartrain, and is written with a remarkable freedom of expression, as though the Secretary of State was the author's oldest friend. The result is one of the most engaging of travel-books, as though Pepys himself had gone a-roving.

The footing of intimacy is at once established. After a voyage of nine days from Marseilles – a quick passage, but it seemed long – the travellers landed at Canea in Crete, and started out with great expectations to explore the vicinity of the port. But alas! 'Discontent return'd at every step we took. . . . We ever and anon look'd at one another without opening our mouths, shruggling up our shoulders and sighing as if our very Hearts would break, especially as we follow'd those pretty Rivulets, which water the beauteous Plain of Canea, beset with Rushes and Plants so very common, that we would not have vouchsafed them a look at *Paris*, we whose Imagination was then full of Plants with silver Leaves, or cover'd with some rich Down as soft as Velvet, and who fancied that *Candia* could produce nothing that was not extraordinary.'[1] They spent three months (May–July 1700) on a thorough exploration of the island, which to some extent retrieved its reputation, save for Mt Ida, which they found 'nothing but a huge, overgrown, ugly, sharp-rais'd, bald-pated Eminence; not the least shadow of a Landskip, no delightful Grotto, no bubbling Spring or purling Rivulet to be seen . . . we found nothing but flint-stones, and but a few uncommon Plants, being scarce able to draw one Leg after the other'.

Tournefort had hoped to visit every island of the Greek archipelago; and though frustrated in this design by adverse winds and the advancing season, he actually landed on thirty-three of them – twenty-eight on the outward and five on the homeward journey. They delayed leaving Crete until they could do so in a French vessel, the local craft being too small and unsafe for the 100-mile crossing; and eventually embarked, as he coolly informs Pontchartrain, on one of the prohibited Corsairs – 'one of those your Lordship has forbidden pickeering from Iland to Iland for plunder. I promis'd the Master not to inform against him, and so he convey'd us to Argentière, the first of August.' This tiny island, where, we are told, 'the Women have no other Employment but making Love and Cotton Stockings', was a resort of the Corsairs, and being 'encumbered with our baggage, and reposing no great Confidence in the People of the Place' the travellers moved on next day to Milo, and thence to Antiparos, to see the famous grotto, where they descended into 'three or four frightful Abysses, one under another'.

And so from island to island; Paros of the monasteries, including St

[1] Quotations from Tournefort's *Voyage into the Levant*, trans. John Ozell, 1718

George of the Gooseberries ('a Fruit pretty rare in the East'); Naxia, birthplace of Bacchus; uninhabited Stenosa; volcanic Santorin; Delos with its antiquities; Thermia, where they were taken for banditti, and would have been attacked if it had not been noticed in time that four of the party wore hats, and must, therefore, be respectable; barren Joura, where they spent the night 'in a ruinated Chappel, where we durst not sleep for fear the Field-Mice should come and gnaw our Ears,' and finally Mycone, where they withdrew to wait for Spring, as the weather was getting very cold and the sea 'rougher every day than other'. Early in March 1701, they embarked on a Turkish vessel, and after calling at Lesbos, arrived in a few days at Constantinople.

Here they stayed for about three weeks, enjoying the hospitality of the French consul, the Marquis de Ferriol, and preparing for the next stage of the journey, which involved procuring Turkish costume, as it was unsafe to travel further in European dress. Actually, it was the Armenian habit that they assumed, as more convenient for riding; but they kept their Spanish leather boots, for the thin Turkish slippers were 'by no means fit for Persons who love to go a-simpling'. Their camping-equipment consisted of a tent, four leather sacks for baggage, 'and some Osier Baskets covered with a Skin to preserve our Plants, and the Papers which serv'd to dry them'; six plates, two bowls, kettles and cups of tinned copper, two leather water-bottles, a lantern and some wooden ladles. A good cloak was invaluable, and they also furnished themselves 'with Callicoe Drawers, which serve instead of Bed-clothes in this sort of Roads'.[1] They had a Greek as general servant and interpreter.

They might have stayed longer in Constantinople had not an excellent opportunity occurred of travelling with the party of the Turkish Bassa Cuperli, who was returning to Erzerum. This official already had one French physician in his suite, but was not averse to adding another; and as all the members of Bassa's household – wife, mother, daughter, major-domo and other officers, were ill on the journey, Tournefort and M. de St Lambert were both kept busy, It was advisable to travel in as large a party as possible, for fear of robbers, and the Bassa's troop was joined by various caravans of merchants, to the number eventually of some 600 strong.

It was a relatively small group, however, which left Constantinople on 13 April and entered the Black Sea in calm weather, only to be held up for ten days at the R. Riva by adverse winds and rain. Then came a leisurely progress in a convoy of boats from port to port along the coast, stopping at the least sign of bad weather, or even when it was pronounced by the soothsayers to be an unlucky day. On one occasion Tournefort himself advised a halt 'for the benefit of the sick', but actually to afford further opportunities to botanize. They got to Trebizond on 23 May – and found the rhododendrons in bloom.

Tournefort was the first botanist to see and describe the common rhododendron and azalea (*R. ponticum* and *R. luteum*) and to identify them

[1] Compare this modest outfit with the twenty-two mule-loads thought necessary by Ruiz and Pavon.

with the *chamaerhododendros* of Pliny, the source of the honey by which Xenophon's soldiers were poisoned in this very region. The tradition of its deleterious properties was still maintained. 'As beautiful as the Flower is', Tournefort said of the rhododendron, 'I did not judge it convenient to present it to the Bassa . . . but as to the flower of the preceding species [the azalea] I thought it so very fine, that I made up great Nosegays of it, to put in his Tent; but was told by his Chiara, that this Flower caus'd Vapours and Dizziness.'

From Trebizond the caravan proceeded by the easiest rather than the shortest route; its slow and devious progress suited the botanists, who were thus enabled to see more of the country. 'The Merchants laught heartily to see us mount and remount every moment, only to pick a few Herbs . . . At the next Lodging we described our Plants while our Meat was in our Mouths, and M. Aubriet drew all he could.' On 6 or 7 June they crossed the mountains of the Pontic Range, and descended on the 8th into a region where the flora was quite new and strange, so that 'we knew not which to fall on first'. The weather now grew oppressively hot, and marches were made by night, enabling the indefatigable naturalists to botanize in the daytime; when it was moonlight, they also collected along the route, the merchants 'laughing all the time, to see us three groping about in a Country dry and burnt up in appearance, but notwithstanding enrich'd with very fine Plants. When it was Morning, we review'd our Harvest, and found ourselves rich enough'.

Erzerum was reached on 15 June, and the Frenchmen stayed there for three weeks, making excursions in the neighbourhood. The first was to the source of the Euphrates, where they ran considerable risk of an attack by the Kurds; a party of the marauders was sighted, but an Armenian bishop who accompanied the botanists on this occasion acted as mediator and interpreter, and the encounter ended with an amicable exchange of gifts. It was near Erzerum that Tournefort found the Oriental Poppy (*Papaver orientale*), which he successfully introduced; as he puts it: 'This fine Species of Poppy is mightily pleas'd with the King's Garden, nay and with Holland too, where we have communicated it to our Friends.' Another find, which he thought 'one of the most beautiful Genus's of Plants that is in all the Levant' he named after Dr Morin of the Royal Academy of Sciences, who was the only one to succeed in raising the plant (*Morina persica*) from the seeds Tournefort sent home.

On 6 July the little party joined a caravan bound for Tiflis, and after some difficulty in getting passports at Cars, entered Persian territory on 14 July, and found the smiling friendly people a pleasant change from the severe and suspicious Turks. After five days in Tiflis they set out on 26 July for Three Churches, through the region in which Tournefort believed the garden of Eden to have been situated, though he was rather shaken to find common European weeds – burdock, plantain and nettles – growing by the wayside. At Erivan (8 August) they kissed the hand of the Patriarch of the Armenians, which 'pleas'd him much, for many Franks don't show him that Respect; but we would have kiss'd his Feet, if we had ever so little suspected that he requir'd it, we had so great need

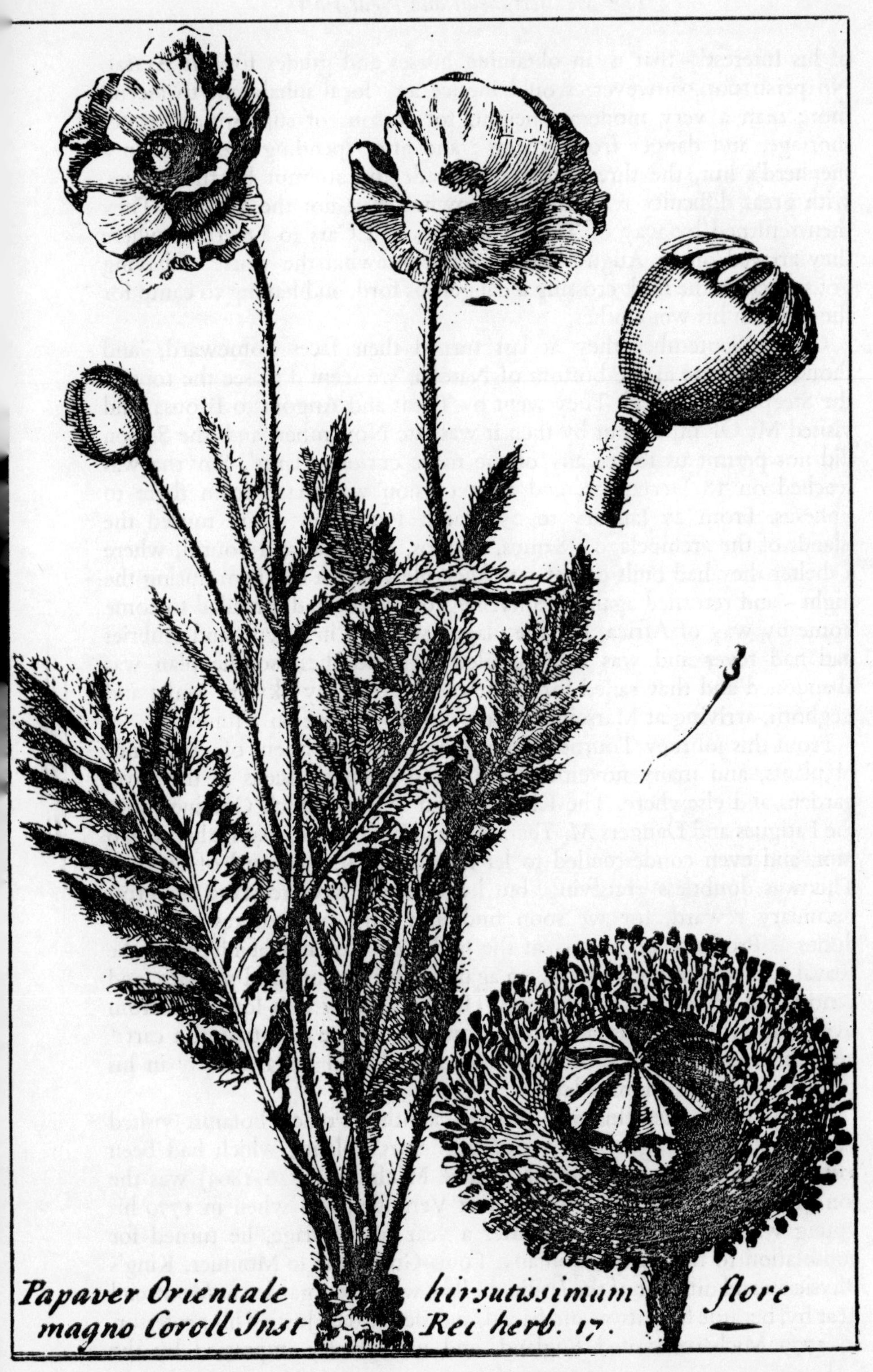

Aubriet's drawing of the Oriental Poppy. From Tournefort *A Voyage into the Levant* 1718

By courtesy of the Royal Botanic Gardens, Kew *Photo by Kenneth Collier*

of his Interest' – that is, in obtaining horses and guides for Mt Ararat. No persuasion, however, would induce any local inhabitant to ascend more than a very moderate height, for reasons of superstition, water shortage, and danger from 'tygers'; and after spending the night in a shepherd's hut, the three Europeans made the attempt by themselves, with great difficulty reaching the snowline, but not the summit. They then returned by way of Three Churches and Cars to Erzerum, where they arrived on 24 August, Tournefort somewhat the worse of having got soaked to the neck crossing a dangerous ford, and having to camp for the night in his wet clothes.

On 12 September they at last turned their faces homeward, 'and though we were at the bottom of Natolia, we seem'd to see the tops of the Steeples in France'. They went by Tocat and Angora to Brousa, and visited Mt Olympus, but by then it was late November, and 'the Season did not permit us to see any of the more curious Plants'. Smyrna was reached on 18 December, and an excursion was made from there to Ephesus. From 27 January to 25 March 1702, they again toured the islands of the archipelago – Samos, Patmos, St Minas and Fourni, where a shelter they had built out of wreckage blew down on them during the night – and returned again to Smyrna. Tournefort had intended to come home by way of Africa, but the plague was bad in Egypt, and Aubriet had had fever and was still troubled by headaches; so the plan was abandoned and they sailed for home on 13 April, by Skyros, Malta and Leghorn, arriving at Marseilles after a stormy passage on 3 June.

From this journey Tournefort brought home specimens of 1356 kinds of plants, and many novelties were raised from his seeds in the royal gardens and elsewhere. The King 'enter'd with so much Goodness into the Fatigues and Dangers *M. Tournefort* had undergone, that he bemoan'd him, and even condescended to let him know it, by word of mouth'. This was doubtless gratifying, but he seems to have received no great pecuniary reward, for we soon find him hard pressed, resuming his duties as Professor of Botany at the Jardin du Roi and of Physic at the Royal College, trying to work up again his private medical practice, and writing his travels in the evenings. He was already in a low state from overwork when he was accidentally crushed by the axle-tree of a cart;[1] he was 'seized with a spitting of blood' and died 'of a Dropsy in his Breast' some four months later.

Eighty years after this epic voyage, another French botanist visited Persia, and found again many of Tournefort's plants, which had been lost to sight during the interval. **André Michaux** (1746–1803) was the son of a farmer and estate-manager at Versailles, and when in 1770 his young wife died in childbirth after a year of marriage, he turned for consolation to the study of botany. Louis-Guillaume le Monnier, King's Physician and director of the Jardin du Roi, who had a garden at Montreuil near by, became his patron and friend, and Bernard de Jussieu his professor. In 1779 Michaux visited England, and was greatly impressed by the

[1] In what is now the Rue Tournefort

variety of foreign plants in cultivation, many of which he brought back to the gardens of France.

Impatient to travel, and not greatly caring where, Michaux at first obtained permission from the Spanish ambassador to botanize in Mexico; but the outbreak of war between England and Spain placed too many difficulties in the way. Just at this time (December 1780) Xavier Rousseau[1] returned from Baghdad where he had been acting-consul, to have his appointment confirmed and to obtain compensation for losses sustained in the local wars. He and his wife created a sensation by appearing at Court in oriental dress; and Michaux petitioned Marie Antoinette to be allowed to accompany them on their return to the East. Permission was granted; Monsieur, the King's brother, contributed a pension of 1200 livres and a commission to make natural history collections, and Michaux bought his equipment, including scientific instruments, from his own modest funds. The party sailed from Marseilles on 4 March 1782, and reached Alexandrietta (Iskenderun) on the 29th of the same month.

Syria was then in a state of turmoil, with Turk fighting against Arab and tribe against tribe, and their landing was delayed till 2 April because a battle with the local Payas was in progress in the town. It was a miserable little place, surrounded by stagnant marshes, but Michaux was wildly excited, and could not sleep at night, so eager was he for day. To be in Asia, and able to roam at will over plains covered with new flowers! He could not roam very far, however, as the lurking Payas rendered the mountains dangerous.

After a week the party got under way for Antioch and Aleppo, where they arrived on 14 April. Here they were obliged to stay for more than six months, while waiting for the cooler season in which to continue their journey (one feels this might have been better organized) and also while bargaining with a local Arab chief who was asking an exorbitant sum for a safe-conduct through his territory – and had 2000 men camped only three leagues away. During this time Michaux made three local excursions – as he philosophically remarked, if he wished to be always in safety, he would never get anywhere. Before leaving he sent home his first collections – birds, minerals, shells, medals and inscriptions, plant-specimens, bulbs and seeds, including those of *Michauxia campanuloides*.

They left Aleppo on 10 October in a caravan some 130 strong, including about a dozen Europeans; the desert crossing to the Euphrates and then down to Baghdad took forty-six days. All vegetation had been shrivelled by the heat, but Michaux managed to collect something every day, either at stopping-places or on risky detours along the route, digging at random in the hot soil and occasionally turning up bulbs, of species unknown; searching for seeds round the bases of withered plants or in places where he thought they might have been blown by the wind. By the time they reached Baghdad he had got seeds of 100 species of desert plants.

After spending the winter in Baghdad, Michaux parted with regret from M. Rousseau, and on 1 March 1783 set out again on his travels.

[1] Second cousin of the celebrated philosopher

He had first to seek out the Pasha of Baghdad, who was then in the field against the Arabs, and obtain a *tchoukada* or permit, which entitled him to the very necessary escort of a sheikh and four men. With these he got safely to Semaouah, and thence by boat to Korna and Basra (12 March). It was difficult to proceed further, for the Basra Turks were at war with the local Arab Kaibs, who lay in wait by the mouth of the Chotte el Arab and stopped all traffic going up and down the river. Michaux made his first attempt on 28 May, in company with a Dominican who had already been trying for a year to get to Ispahan; but the fleet of seventeen boats with which he travelled was attacked by thirty boatloads of Kaibs and obliged to retreat in disorder to Basra.

In spite of the fact that England and France were then at war, the English consul at Basra, William Diggin de la Touche, showed Michaux great kindness; and in September notified him that he was sending a boat down the river with a safe-conduct from the Arab chief Ali Khalfan, in which he might travel in complete security. In spite of the safe-conduct, however, they were attacked by eight Kaib boats on the second day, and pillaged 'to the bottom'. Michaux's luggage was searched and everything useful taken; only his books, medicines and some despised European shirts were returned to him. He was then taken to the camp of another sheikh, and again robbed – this time of two reams of botanical paper and some hardware that had been among the medicines. Then he was carried away into the interior.

The country was in a ferment, with thousands of savage tribesmen preparing for a major encounter with the Turks; but owing to the intervention of the powerful de la Touche, Michaux was taken a week later to the place of his capture and set free. His rifle and money were restored to him, but not his microscope and other instruments, clothing or cooking utensils. He re-embarked on the Consul's boat and escaped down the river, while Turks and Arabs fought a battle behind him. (He complained later that he was unable to botanize on the river banks, as the Arabs had taken his shoes, and the sands were too hot for bare feet.) They got safely to sea, and after a pause at the island of Kharek for repairs, reached Bouchir on 21 September.

It seems that Michaux's journals, which give full details of his Syrian adventures, became scrappy and incomplete when he got to Persia, and the most interesting part of his journey is that of which we know least. The worst of his troubles were over, for although travel in Persia involved many hardships, it was as safe as travel in Europe. The winter of 1783–4, however, was exceptionally severe, with a month of snow and a frost that scorched the orange-trees; and the lateness of the season and the sterility of the country made little botanizing possible before he reached Shiraz in March. Even so, he managed to send le Monnier a valuable consignment of seeds, and a box of seeds and specimens to Sir Joseph Banks.

On 16 March 1784, Michaux left Shiraz for Persepolis, Sivend, Yodekast and Ispahan, and thence to Hamadan, where he climbed Mt Elwend, but was unable to measure its height because of the loss of his instruments. The country improved and botanizing again became possible

– indeed, embarrassingly so; with his next consignment, sent from Kasvin about the end of June, he explained to le Monnier that he had been obliged to put up some of the less interesting seeds in mixed packets – assorted crucifers, assorted grasses, and the like – as he was always pressed for time, and anxious to gather as large a quantity as possible of 'the multitude of different plants which often inhabited a small space'. He found the wild originals of many cultivated plants, but few shrubs or trees, though it was these that were most in demand at home.

From Kasvin he went to Recht, and its small port of Enzeli on the Caspian Sea; and having thus crossed Persia from south to north, began to turn his thoughts homeward. He returned to Hamadan, and from there took a caravan to Baghdad on 3 August. On arrival he despatched his collections by the first caravan for Aleppo, but missed it himself through waiting to see Rousseau, whose return from a visit to Basra was daily expected. (It was probably through Rousseau that Michaux received his supplies, for he complained in a letter of the expensive delays incurred through being obliged 'to have recourse to M. Rousseau to continue my travels'.) The uncertainty and lack of funds prevented him from making a journey into Kurdistan; but he was already planning to return for an expedition across Persia to Kashmir and Tibet. Meantime, he went for some local excursions, and in the ruins of Ctesiphon made an important archaeological discovery – the *Caillou Michaux*, an inscribed stone in the form of a flattened cone, dating from about 1300–900 BC.

Michaux left Baghdad on 30 December, and reached Aleppo on 31 January 1785, having with him a small collection of Persian bulbs still in good condition, in spite of severe frosts experienced on the journey. He picked up some parcels he had deposited on the outward journey, added them to fresh collections, and, sailing from Larnaca on 13 April, reached Marseilles on 23 May and was back at Versailles by the end of June. He was well received at Court, given an official brevet as botanist to the King's nurseries and an honorarium of 2000 livres – and by 1 September was on his way to America.

Unfortunately some of the most interesting of Michaux's Persian plants have proved difficult of cultivation. They included the unique *Rosa persica*, so unlike any other rose that it is sometimes allotted a separate genus as *Hulthemia berberidifolia*. He also found *Dionysia michauxii*, truly an intoxicating plant for the few that can get or keep it; but even his lovely biennial Michauxia is the better for some protection.

Michaux took little interest in the plants of the ancients, but the theme recurs again with the travels of Dr **John Sibthorp** (1758–1796). In spite of the researches of Rauwolf and Tournefort, only about 400 of the 700 plants mentioned by Dioscorides had been identified; and his influence on medicine had been so great that it still seemed worth while to discover the remainder. Sibthorp was particularly interested in this subject; he had studied botany on the Continent, and had visited Vienna on purpose to see the famous illustrated copy of Dioscorides' *De Materia Medica*, made in AD 512; he hoped on his journey to find not only the plants but vernacular names and herbal traditions still lingering among the Greek peasants.

His mother's death in 1780 had left him in 'very affluent circumstances', and although he had succeeded his father as Professor of Botany at Oxford in 1783, he was able to arrange for his duties to be taken over by a deputy during his absence.

Sibthorp left England accordingly in February 1786, and went first to Vienna, where he collected the draughtsman Ferdinand Bauer, who was to accompany him on his journey. They proceeded, partly by land and partly by sea, through Italy to Naples, where they stayed nearly a month; then by way of Messina and Milos to that Mecca of botanists, Crete, arriving in June. After visiting Athens and several islands, they went to Smyrna, and thence overland to Mt Olympus and Brousa, and finally Constantinople, where they spent the ensuing winter. In the spring they made some local excursions including one to Belgrade, where they found *Hypericum calycinum* still growing where Wheler had discovered it, more than a hundred years before.

On 14 March 1787 they set out again, accompanied by a friend, Mr John Hawkins, and made their way by Mitylene, Scio, Cos, Rhodes and the coast of Asia Minor to Cyprus, where they stayed for five weeks. Then on 19 June to Athens and its neighbourhood – 'the snowy heights of . . . Parnassus, the steep precipices of Delphis, the empurpled mountain of Hymettus, the Pentele, the lower hills about the Piraeus, the olive grounds about Athens, and the fertile plains of Boetia'.[1] Sibthorp wished to return overland through Greece; but 'the disturbed state of this country, the eve of a Russian war, the rebellion of its bashaws and the plague at Larissa, rendered my project impracticable.'[2] Instead he went northwards early in August to Mt Athos and Salonica, then back to Corinth, and sailed on a Bristol vessel from Patras on 24 September, bringing with him Bauer, a large number of drawings, and specimens of some 2000 species of plants, about 300 of which were new.

Sibthorp's health had suffered severely from heat among the islands, and the confinement of a stormy ten-week passage home; but after the exertions of his travels his Oxford duties seemed 'more a recreation than a toil', and he quickly recovered. He was very conscious that his work in Greece was incomplete; but owing to pressure of affairs and also, perhaps, to the outbreak of the French Revolution in 1789, it was seven years before he was able to return to the field. Bauer remained in Oxford, still working up the drawings of their first trip, and Sibthorp took as assistant a young Italian, Francisco Borone, whom J. E. Smith had taken into his service some years before and trained as a botanical servant. He had already accompanied Afzelius to Sierra Leone, and had won golden opinions as an amiable, ingenuous youth and a promising botanist. By now England and France were at war, and the outward journey had to be made by sea; again the passage was stormy and slow, and it took from 20 March to 19 May 1794, to reach Constantinople. Sibthorp was very ill on arrival with a 'bilious fever and colic' and Borone too had an 'intermittent fever' during their stay.

[1] and [2] Smith, in Rees' *Cyclopaedia* (1819–20) under 'Sibthorp'

In August they were joined by Hawkins from Crete, and made an excursion to Bithynia, where they again climbed Mt Olympus. On 9 September they set out via Troy, Imbros and Lemnos for Mt Athos; they arrived on the 25th, but their departure was delayed by the presence of Barbary pirates in the offing. Borone was nervously anxious that Sibthorp should hide his money, 'that we might have something, he said, to return home with'.[1] When the coast seemed clear they sailed again, and after landing on the little island of Skiatho, reached Athens on 15 October.

A few days later, Borone was particularly gay. He had quite recovered from his fever, and had been singing in the evening to a tune played by Mr Hawkins' servant on the guitar. But during the night he walked in his sleep, climbed through a high and narrow window, and fell some eighteen feet to the street below, receiving injuries from which he died within an hour. He was only twenty-five years old.

Much shocked and saddened by this event, Sibthorp gave Borone the handsomest funeral that could be procured, and lingered in Athens till 16 November, when he went by Patras to Zante, where he spent the winter. In February 1795, he returned to Morea and spent the next two months in extensive explorations of the peninsula, including the ascent of Mt Taygetus. On the 29 April he took ship to Zante, and parting from his friend Hawkins, who returned to the Greek mainland, set out for home.

Sibthorp seems to have been particularly unfortunate in his sea voyages, and this last was no exception. The crossing from Zante to Otranto in Southern Italy, which should have been accomplished in five days, actually took twenty-four. The boat touched at Cephalonia and at Previsa, where Sibthorp went ashore to see the ruins of Nicopolis, near Actium, and where he caught a severe cold – 'The air of Previsa is even by the Greeks deemed infamous.'[2] Bad weather and violent north-west winds detained them for ten days at the little island of Fano off Corfu; six times they tried to put off and six times Sibthorp was obliged to return to the 'miserable hovel' where he lodged and where the north wind 'continued to nurse his cough and fever'.[3] At last they reached Otranto, on 24 May, and were detained for three weeks in quarantine. Having had enough of the sea, Sibthorp travelled home overland, by way of Germany and Holland, arriving in the autumn of 1795. But the harm had been done. He complained of 'a nasty low fever, with a cough that alarms me, from some affection of the lungs'.[4] In vain he underwent treatment in Brighton, Bristol and Bath; asses' milk and tepid sea-water baths failed to effect a cure, and he died at Bath in February 1796, aged thirty-eight.

Sibthorp was a botanical rather than a horticultural collector, but he sent seeds home to the Oxford Botanic Garden and elsewhere, and is credited[5] with about fourteen new introductions, including *Campanula versicolor*, *Linum arboreum* and one of Tournefort's discoveries, *Origanum tournefortii*, from the Isle of Amorgos. His most successful plant, in one

[1,2] and [4] Smith *Letters*
[3] Smith, in Rees' *Cyclopaedia*
[5] In Aiton's *Hortus Kewensis* (2nd ed.) 1810

sense, was *Saxifraga sibthorpii*, which has become naturalized as a garden-weed in Kent. He left a magnificent legacy to all plant-lovers, for he made a bequest to cover the cost of publication – some £30,000 – of his superb ten-volume *Flora Graeca*, with Bauer's plates; only twenty-five copies of the first edition were printed and the price of a complete set was 240 guineas. Sir J. E. Smith was chosen as editor, but he did not survive to complete the work, and the last three volumes were edited by Lindley. The editors had no light task, for Sibthorp's notes were often scanty and incomplete; 'he trusted to his memory, and dreamed not of dying'.[1]

Asiatic Turkey has a particularly rich flora, and this, combined with the difficulty of access, meant that plants of importance to gardeners were still being found at a relatively late date. In 1874[2] **Henry John Elwes** (1846–1922) visited Anatolia; he had previously been to Greece, Constantinople and the Crimea (1869) and to the Himalayas (1870), but he only became interested in gardening after his marriage in 1871, so this may be ranked as the first in his long series of collecting travels. He arrived at Smyrna early in March, and finding that the season was still not sufficiently advanced for travel in the interior, went by the islands of Samos and Cos to the more southerly port of Macri, which he made the base for two journeys through country known to archaeologists, but very little to botanists. The first took him northwards by Minara, Tortuca and Bayeerzan to Cassaba in the Gedis valley (east of Smyrna) and back by another route. It was an uncomfortable trip: he had no camping equipment, the villages were dirty, and the only food obtainable was bad bread, onions and sour milk. But the results were rewarding, for it was on this journey that he found *Galanthus elwesii*, and several other fine plants. Back at Macri he made a shorter excursion to Ahoory and Aisa; then, having missed his steamer up the coast, went overland by Mughla to Aidin, and thence by train to Smyrna. There he met a Greek who promised to collect for him, and who afterwards sent him *Galanthus elwesii* and other bulbs in very large numbers. This was fortunate, for nearly all his own collections were lost or stolen when changing steamers at Syra; and the country in which he had worked afterwards became unsafe through brigands, and was not visited again by naturalists for more than forty years.

Smyrna, as we have seen, had been an important centre ever since botanical exploration began, and had moreover been the residence from 1703 to 1715 of that impassioned botanist William Sherard; yet there were still new plants to be found, and it was near Smyrna in the spring of 1877 that **George Maw** made one of his most profitable journeys. The mountain of Nymph Dagh lies only about ten miles to the east; and it was here, near the village of Taktalie, that Maw's Greek and Turkish attendants 'became botanically excited' at the sight of a thick bank of *Chionodoxa luciliae*, growing with other flowers at 4300 ft and forming one of the most sumptuous displays of floral beauty Maw had ever seen.

[1] Smith *Letters*

[2] Several important but purely botanical journeys had been made in the interval, including those of Olivier, Grisebach, Boissier and Balansa

It had previously been found on Boz Dagh (Tmolus) much further east, but had never been introduced to gardens. Maw was able to bring home bulbs, which flowered the following year; he also collected two other species, *C. siehei* and *C. sardensis*. Later in the same year he made a second journey, to Italy, Corfu and the Greek island of Santa Maura, where he was set upon and robbed by brigands – to their detriment rather than his, for the shocked authorities combed the island for the culprits, and were not satisfied till more money was restored than had been taken from him.

The travels of Farrer, Forrest and Kingdon-Ward, and their dramatic adventures in distant China, have been so much publicized that those of plant-hunters nearer home in approximately the same period have been unfairly overlooked. **Edward K. Balls** (b. 1892) was a professional horticultural collector of the highest standards, who made eight journeys between 1932 and 1939, but his work is seldom recalled. He must not be confused with John Ball, who travelled in Morocco with Hooker, or with **Charles Frederick Ball** (1879–1915) a Kew-trained gardener who went to Bulgaria on behalf of the Glasnevin Botanic Gardens, Dublin, in 1911. With Herbert Cowley[1] as companion, C. F. Ball visited Sofia, Mts Vitosha and Moussala and the Shipka Pass. King Ferdinand was much interested in the expedition, and sent guides and his own gardener, Kellerer, to accompany it; but he also expressed the opinion that English nurserymen were doing their best to exterminate some of the rarer Bulgarian plants, such as *Lilium jankae*. A promising career was cut short when Ball was killed at Suvla Bay, Gallipoli, in September 1915.

At that time E. K. Balls was engaged on Quaker relief and reconstruction works which kept him abroad for eleven years; he came home in 1925 and in 1926 he joined Clarence Elliott's nursery at Six Hills near Stevenage, where he became particularly interested in alpine plants. In 1932 Dr T. L. Guiseppi was looking for a companion for a collecting-trip to Persia, and Elliott suggested that Balls should go, introducing him also to Major (later Sir) Frederick Stern, Sir William Lawrence and other potential subscribers. In April the two set off; Dr Guiseppi had only a month to spare, but Balls was away for four months and a half.

They travelled through Hamadan and Ispahan to Shiraz, and then west to Kerman as a centre for the mountains of the Kuh Banan. Here they parted for a week while Guiseppi went to see the volcano of Kuh-i-Taftan and Balls explored the ranges to the south-west. When Guiseppi returned they hired a car to take them to the village of Lalhezar, and from there went first on horseback and then on foot, to one of the main peaks, Kuh-i-Kha or the Kind Mountain. They were unable to reach the 15,000 foot summit of the neighbouring Kuh-i-Hazar, the Hill of a Thousand Flowers, owing to the onset of a thunderstorm, which forced them to take refuge in a nomad's tent. They went back to Ispahan by way of Yezd, and Guiseppi returned to England, while Balls went to Hamadan and spent three weeks in a thorough exploration of Mt Elwend and its neighbourhood, while waiting for seeds to ripen farther south. (Part of

[1] Subsequently Editor of *Gardening Illustrated*

the time he was the guest of some charming and hospitable Kurdish brigands.) Here he found *Tulipa pulcella violacea* and *Fritillaria verticillaris*; Michaux's *Hulthemia berberidifolia* (*Rosa persica*), 'so desired in England', was used in this locality for fuel. Then he went south again to Shiraz and to the mountain and village of Khalat about forty miles to the east, camping for three days farther north at Mt Kuh Ajub before crossing north-east to Yezd and exploring two more mountains, Schir Kuh (Lion Mountain) and Barf Khannich (Snow House). In September he returned to England by way of Baghdad – with seed of *Dionysia curviflora.*

This expedition launched Balls on a collecting career. The following year he was urged by Major Stern and others to go to Turkey, and Elliott introduced him to Dr William Balfour-Gourlay of Cambridge, who proved a most congenial travelling-companion. For the next three years Balls made an annual trip to Asia Minor, collecting plants for a number of subscribers of whom Stern was the chief, and also herbarium specimens for various museums. On the first and third journeys he was much hampered by restrictions imposed by the Turkish authorities; it was only on the second that he was able to accomplish all he had planned.

At the end of February 1933 Balls and Balfour-Gourlay started on their first trip together. They obtained permits to work in the Taurus mountains, and having sent their car by sea to Mersin drove off from there to a mountain village; but after two days they were recalled to Tarsus, and told that they were to be allowed to botanize only in and about the towns of Mersin and Adana. Dissatisfied, they returned to Ankara, and after a time got permits to visit the Pontine mountains south of Trebizond. Here the local authorities were more accommodating, and they stayed in the area until their return to England in September.

Next year, 1934, there was no trouble with permits; the two friends left England in March, and crossed from Istanbul to Mudanya on the Sea of Marmora, whence they travelled by car across the central plateau to the Taurus mountains and the famous pass of the Cilician Gates near Tarsus, where Balls found the desirable and distinct form of *Cyclamen cilicium* now designated var. *alpestris*. Here they were again watched with suspicion by the authorities and after one climb were summoned to the county headquarters for questioning, but were not otherwise hindered. From Tarsus they went east to Gaziantep (= Aintab) and from there started a thorough exploration northwards to Marash and up the valley of the River Jihan, between the Taurus and Anti-Taurus ranges, climbing Ahir Dagh and Beirut Dagh on the way. Near the deserted Armenian village of Zeitun, Balls was alarmed when photographing a plant by the rapid approach of 'two fierce-looking fellows brandishing guns', but they turned out only to be inquisitive Kurdish shepherds anxious to arrive before he stopped whatever he was doing. A little farther on, they were hospitably entertained by a Syrian sheep-dealer, whose black goat-hair tent did not keep out the rain that came on every evening, but only reduced it to a finer spray.

After penetrating up the Jihan as far as Albistan, Balls and Balfour-Gourlay returned to Marash, and then took a more westerly route by

Ergias Dagh to Sivas, east to Erzingan, and north again to Gümüsane, south of Trebizond. They then worked in the Lazistan mountains along the Black Sea coast as far east as Rize; and when the first snows of the season began to fall in mid-September, turned reluctantly homewards by way of Ankara, Mudanya and Istanbul. It had been a good season and resulted in the introduction of *Fritillaria crassifolia* 'Ball's Form', *Orphanidesia gaultherioides* and *Campanula betulifolia*, while the invaluable *Iris histrioides* var. *aintabensis* had been found near Gaziantep.

On the next expedition, in 1935, nothing went right. Balfour-Gourlay was unable to go, and it was arranged that Balls should be accompanied by an ornithologist appropriately called Charles Bird, who was collecting for the British Museum. They left at the end of February with all permits apparently in order, but in Istanbul passes granted on one day were rescinded the next, and it required a long visit to Ankara before permits which had been requested the previous September were eventually issued. Balls returned to Istanbul at the end of March, and proceeded by sea to Mersin, rejoining Bird, who had preceded him, at Adana. They then started eastwards; their permits allowed them to go right to Lake Van, but further restrictions were imposed at almost every step. They got as far as Gaziantep, with one hold-up on the way, but were not allowed to go south to the mountains or north to Marash; they were eventually permitted to go north-east to Malayta, but might only botanize by the roadside, and might not ascend above 5000 ft. At Malayta they were virtually prisoners, forbidden to go outside the town; and were presently recalled to Adana, where in late May or early June the expedition was abandoned. Balls could not understand the reason for this obstructiveness, but it later emerged that the Turks believed him to be a secret-service agent – perhaps even Lawrence of Arabia in disguise, in the report of whose death[1] they did not believe.

After his tour in Morocco in 1936[2], Balls made a four-month visit to Greece in 1937, again with Balfour-Gourlay. They seem to have worked chiefly in the Pindus mountains of Epirus, where on the same day (27 July) Balls found *Salvia haematodes* near Metsovo, and *Achillea clavenae* var. *integrifolia* on Mt Peristeria. Another good find was *Fritillaria graeca* subsp. *epirotica* on Mt Smolika. (The salvia is thought to have been grown in the Oxford Botanic Garden before 1699; but it had long been lost to cultivation.) The two friends planned to go next year to the Caucasus, but this trip never materialized, and they went to Mexico instead.[3]

The resources of Asia Minor, which has a particularly rich flora, are by no means exhausted, and there have been a number of post-war expeditions to the area, which must be left to some other chronicler.

The Mediterranean circuit

The countries south of the Mediterranean have contributed relatively little to our gardens. Egypt, for example, has never been a paradise for the

[1] On 19 May 1935 [2] See p. 36 [3] See p. 350

plant-hunter. A land divided between desert and intensive cultivation for some 5000 years cannot be expected to be very rich in wild flowers; and the principal plants of interest were observed by Herodotus when he travelled as far as the First Cataract, about 450 BC. The country was subsequently explored by a number of hopeful botanists, including two of Linnaeus' pupils, Hasselquist and Forsskål.

From childhood **Frederick Hasselquist** had a hard struggle with poverty and adversity, and it was only by the utmost resolution that he achieved the education and the means to realize his heart's desire, of travel in the East to obtain plants and information for his revered master, Linnaeus. After six months at Smyrna, nearly a year in Egypt, and about nine months in Palestine, he died in February 1752, aged thirty, from the tuberculosis that had threatened him before his departure. **Pehr Forsskål** had a more favourable start. A brilliant student, he was appointed to a Danish expedition, arriving in Alexandria in 1761. After a year in Egypt the party embarked to go down the Red Sea to Southern Arabia; but Forsskål died of malaria on the way in July 1763. **Alyre Raffeneau Delile** (1778–1850) was one of the team of scientists taken by Napoleon to Egypt in 1798; he survived, though his younger colleague de Montbret[1] died. But all these were botanists, and made no contribution to horticulture.

The Arabs were even less tolerant of Christians than the Turks, and the possibility of being sold as a slave to the infidel acted as a considerable deterrent to botanical pursuits in North Africa, until well on in the nineteenth century. Piracy was rife till about 1855, when the increasing use of steamships made it more difficult and less profitable. The conquest of Algeria by the French let loose a flood of botanists over the parched land; forty-eight are recorded as having worked in the area between 1830 and 1850. A new impetus was given by the appointment to the Algeria Commission in 1852 of **E. St Charles Cosson** (1819–1889), with nine 'collaborators' to assist him–a number that eventually rose to fourteen. For more than thirty years this indefatigable botanist ranged Algeria, Morocco, and (after it became a French protectorate in 1883) Northern Tunisia. His botanical work was continued by a series of distinguished Frenchmen; after a late start, hardly a leaf in North Africa remained unturned.

Morocco, never conquered by Turkey, retained its independent dynasty and remained dangerous for travel until it, too, became a French protectorate in 1912; in 1871 Hooker described it as one of the least-known countries of the world. Several bold botanists attempted its invasion, a pioneer among them being **Philip Barker Webb** (1793–1854). He was an eldest son, whose father, dying in 1815, left him a sufficient competency to enable him to follow his own inclinations. These ultimately led to the intensive study of botany; but his first voyage, in 1818, was in the nature of a classical pilgrimage. He went with an Italian friend to Corfu, Patras, Athens, Constantinople and the plains of Troy, and returned by Smyrna, Malta and Sicily, having, it is said, elucidated a number of points in

[1] Namesake of the montbretia

Homeric geography. (He seems to have travelled in Turkish costume, and rather fancied himself in it, judging by his portrait, which was probably painted in the Orient.) On this journey the collection of plants was only a secondary object, but during the next few years he is reported to have assembled many interesting species[1] in the garden of the family home at Milford near Godalming in Surrey, and in 1825 he set out on what proved to be a long series of botanical travels.

After wintering in southern France, Webb spent a year exploring the east and south coasts of Spain, collecting birds, fish and shells as well as plants; and in April 1827 crossed from Gibraltar to Tangiers. Foreigners were not permitted to travel in the interior of Morocco except in the suite of an ambassador or other plenipotentiary, and Webb had planned to accompany the British Consul, Mr Douglas, to the court of the Sultan; but for some reason the arrangement fell through and he was not able to go beyond the vicinity of Tangiers and Tetuan. By means of what Hooker called 'liberal expenditure' – starting with a present to the Governor of 40*s* – he obtained permission to climb the mountains of Beni Darsa and Beni Hosmar, and was the first European to reach the upper ridge of the latter. He was unable to investigate the Rif mountains eastward along the coast, as their 'savage and indomitable' tribesmen made the region too dangerous; but he was the first scientist to examine the barren Zafarines islands (owned by Spain) off the mouth of the Wadi Muluya.

After two months in North Africa, Webb returned to Spain. Even there, it was hardly safe to wander about alone, 'not so much on account of the professional robbers, as of what the Spanish call *raterillos*' – peasants armed with knives but not with firearms, who could, however, be outfaced by a show of boldness. Webb botanized 'throughout the Sierra Nevada, the Alpuxerras and a great part of the mountains of Andalusia and was never attacked. The tranquillity of Portugal, where the Botanist had no need of pistols in his girdle or gun at his shoulder was, after this, a paradise.'[2] Having toured Portugal on horseback, he sailed from Lisbon to Madeira; then came the long sojourn in the Canaries and the collaboration with Berthelot, on which his botanical reputation chiefly rests. Although ninety per cent a botanist, Webb was ten per cent a horticulturist and introduced many plants to the garden at Milford, either in person or through correspondents and friends; he also sent seeds to a local nurseryman.

George Maw (1832–1912) had a larger proportion of the gardener in his make-up. He was a tile-manufacturer whose interest in clays and in Roman tiles and pavements led to the study of geology and archaeology, and whose acquisition in 1852 of Bentham Hall near Broseley in Shropshire aroused a passion for plants and gardening. From 1869 onwards he visited the Continent – often two or three times a year – to collect plants in Italy, France, Switzerland, Portugal and Spain, and ventured as far afield as Asia Minor and North Africa. Latterly he specialized in the

[1] Including *Genista aethnensis,* sent to him by its discoverer, Baron Bivona-Bernardi
[2] Hooker *Companion to the Botanical Magazine* II 1836

genus Crocus, but he was an expert on all sorts of bulbs; 'Maw recognises the bulbs by leaf', wrote Hooker, 'however long the tall grass they grow amongst.'[1] In the first year of his travels he visited Morocco, but, like Webb, was unable to go much beyond Tangiers and Tetuan. Nevertheless, he found and introduced *Iris tingitana* and *Saxifraga maweana.*

Two years later Maw returned to Morocco, in the capacity of geologist to Hooker's expedition to the Atlas Mountains – (geologists, however, classified him as a botanist, and botanists as a gardener). Dr. **Joseph Dalton Hooker** (he was not yet knighted) was by then fifty-three years old, and had been for six years the Director of Kew; his travels in the Antarctic and the Himalayas lay far behind him, and this was to be his last collecting expedition. He undertook it at his own expense, to fill a gap in the knowledge of botanical geography; though it involved extreme discomfort and considerable danger. The only botanist to have visited the region was Benedict Balansa, an intrepid colleague of Cosson's, four years before; and his collections were limited and not generally available. Hooker, although backed by the permission of the Sultan of Morocco and the prestige of Queen Victoria, encountered many difficulties; the Sultan's protection was effective only in certain areas, and the local authorities responsible for the safety of the party, would not permit them to enter what might be dangerous regions, though the Englishmen were willing, and indeed eager, to take the risk.

Hooker and Maw had planned the expedition long before, and when **John Ball** (1818–1889), a good botanist and expert Alpinist, volunteered to join, he was gladly accepted. The party was completed by a Kew gardener called Crump, who also acted as Hooker's personal attendant. Their equipment included 'soups, tea, old watches, musical-boxes etc., no end of paper for drying plants, and so forth',[2] besides saddles, mattresses stuffed with cork shavings, and the 'Alpine Club' pattern of tent with a canvas floor – probably recommended by Ball.

Some preliminary time was spent in excursions, together and apart, in the neighbourhood of Gibraltar, Tangiers, Tetuan and Ceuta; the journey proper began on 20 April 1871 when they took ship for Mogador, with a short call at Casablanca on the way. In Mogador they reluctantly assumed Arab dress, and endeavoured to turn their solar topees into turbans by swathing them with muslin; but, like Tournefort, refused to abandon their practical European boots. Having thus, like characters in the Arabian Nights, transformed themselves by disguise, they proceeded by slow stages overland to Morocco (Marrakesh), which, used though they all were to the squalor of Oriental cities, they found 'utterly foul and repulsive'.

Hitherto they had experienced only the expected discomforts – hostility, dirt, delays, insects, extremes of climate, bad lodgings and bad food; but after Marrakesh more serious difficulties arose. The Kaib or Captain of the military escort with which they were now provided, turned out to be treacherous and obstructive, and distressed them by his impositions on

[1] and [2] Huxley *Life and Letters of Sir J.D. Hooker* 1918

the poverty-stricken villagers. Up in the hills, the natives, though desperately poor, seemed peaceable enough; but hostile tribes were supposed to lurk beyond the mountains, and the Europeans were never permitted to cross the passes. They reached the snow-line, with great difficulty, on only two occasions.

Their first setback came at Tasseremout on 9 May, when they were not allowed to go up the Ourika Valley. Instead, they went by Tassilunt, Reyara and Tassighirt to Asni, where they established a base camp under the charge of the gardener, Crump, while the botanists went up into the hills at Arround. Here the local Sheikh objected to the green-painted tin cases intended for the transport of plants; he said that the villagers would think they contained treasure, and kill the foreigners and himself in order to possess them. So the cases had to be abandoned, and with them all hope of making any large collection of living mountain plants. An attempt to cross the Tagherot pass was also prohibited.

After a few days they returned to their base at Asni, and then moved on to Sektana, where Maw left the party and returned to England, taking 'a considerable number of living plants, which, owing to his skill and experience in managing this difficult process, arrived in excellent condition, and have since thriven in his garden in Shropshire'.[1] Ball and Hooker continued westwards along the skirts of the mountains to Amizmiz and Seksaouna, but again were twice prevented from ascending the main range, though they climbed the detached peak of Djebel Tezah. An outbreak of local warfare forced them to abandon their proposed route at Mtouga and return direct to Mogador with only a slight pause to ascend Djebel Hadid on the way. On 7 June they embarked for Tangiers; Hooker had engagements in England and was in a hurry to get home, so he sailed again from Tangiers immediately after his arrival and was back by 21 June.

Throughout the trip they all worked exceedingly hard. 'It often happened that the solitary candle was in use through the entire night, Ball working till two o'clock or later, when Hooker would rise, more or less refreshed, and keep up work till daylight.'[2] Yet the arduous expedition covered only a relatively small area on the north side of the High Atlas, and the results were disappointing. The majority of the genera found were Mediterranean or European – with the more ornamental alpine ones omitted; they found no gentians, primulas, androsaces, rhododendrons, anemones or potentillas, and only lowland forms of saxifrage and ranunculus. Hooker was too busy after his return to prepare an account of the journey, and the *Journal of a Tour in Morocco*, eventually compiled by Ball from the journals of all three, did not appear till 1878; by this time interest had lapsed and it was published at a loss. As usual, it contains practically nothing about the activities of the gardener of the party, Crump. He probably returned to England in charge of a consignment of living plants, for the cost of their transport was the only expense that devolved on Kew from the whole expedition: and he was afterwards

[1] Hooker and Ball *Journal of a Tour in Morocco* 1878
[2] Huxley *Life and Letters*

employed in the Herbarium. Of the numerous plants introduced, the only one that has survived in general cultivation is the annual *Linaria maroccana*, which was found near Sektana. 'In some fields of corn not yet in ear, the spikes of numerous dark-crimson flowers all but concealed the green, and gave to the surface a tone of subdued splendour.'[1]

Beset as he was with difficulties, Hooker never really attained the heights; sixty-five years later **E. K. Balls** was able to reach a truly alpine flora, which included gentians, narcissi and fritillaries – though there were still areas in which it was unsafe to botanize without the protection of an armed escort. After its annexation by France in 1912, Morocco was extensively explored by French botanists, prudently following close in the wake of the army: roads were built, and by the time Balls arrived the distance between Marrakesh and Asni, which took Hooker and his party nearly a week to cover, could be traversed in a car before breakfast.

Balls' companion on this occasion was Dr Richard Seligman, an enthusiastic alpine gardener. They arrived at Marrakesh at the end of May 1936, and made short preliminary excursions to Amizmiz and Ourika while waiting for permits for the next part of the journey; but just when they were ready to start, Seligman fell ill, and Balls had to go to Ouarzazat without him. This was a military post situated between the main Atlas and Anti-Atlas ranges; to the south lay desert country, but the spring that year happened to have been unusually wet, and Balls found many attractive flowers. He made a five-day excursion from Ouarzazat to the oasis of Aguelmous and the volcanic mountains Djebel Siroua and Dj. Amezdour, and returned north on 11 June to Amizmiz (about forty miles south-west of Marrakesh) where Seligman, now recovered, was waiting for him. Next day they set out south-west to Dj. Erdouz, 'a tiresome mountain to climb', finding the nights in camp intensely cold, though it was hot by day; and returned after four days to Amizmiz again, and then to Marrakesh to pack and despatch collections. On 20 June they set off again, down a different spoke of the wheel of which Marrakesh is the hub – almost due south by a good motor-road to Asni on the Oued Reraia, then by a mountain trail to Arround, changing from car to donkey-transport when the track ran out. Arround, at about 6000 ft, lies at the base of Dj. Toubkal, one of the giants of the Atlas; and they lodged at an Alpine Club hut while they explored the neighbourhood, seeking, and finding, *Narcissus watieri*. (Balls climbed the mountain, but Seligman was not sufficiently recovered from his illness to reach the summit.) They returned to Marrakesh by a different route, crossing a pass to Tashdirt, in the next valley to the east, where there was another Alpine Club refuge, and then down to Asni by the Iminen river.

Seligman now had to return to England; he felt some anxiety at leaving Balls to carry on alone 'with his little Arab cook', but Balls met with no difficulties. He revisited both Amizmiz and Arround for seed-collecting, and also made a trip further east, with an ornithologist from Marrakesh for company. From the village of Demnat they went on mules for two

[1] Hooker and Ball *Journal*

days to Tirsal at the foot of Dj. Ghat, where Balls again found an attractive Alpine flora. He spent some days on the mountain, returning by the same route; and left Marrakesh for England at the end of July.

Balls found many plants that had been beyond Hooker's reach, and some that had not been discovered even by the intervening French botanists; but the permanent contribution to gardens from this expedition was small. *Colchicum triphyllum*, collected near Tashdirt, received an RHS Award of Merit in 1937, but seems since to have disappeared from cultivation. Some fine colour forms of *Chrysanthemum catananche* were collected and some welcome re-introductions made; but on the whole the flora of the Atlas mountains seems determined not to submit to the infidel European gardener.

The Iberian peninsula

Spain was always a country apart, isolated physically by the barrier of the Pyrenees and politically by the Moorish occupation, which lingered till the end of the fifteenth century. The last country in Europe to be botanically explored, it was the first to be exploited by a professional horticultural collector, at a time when gardens were still in a transitional stage, part physic-garden, part botanic-garden, and only incidentally assemblages of ornamental plants. The collector in question was a Dutchman, **Guillaume Boel** (sometimes called Dr Boel), 'in his time a very curious and cunning searcher of simples',[1] who worked for Coys and Clusius as well as for John Parkinson, by whom he was 'often before and hereinafter remembred'.[2] The references to Boel in Parkinson's first book (1629) are all complimentary, but in his second (1640) there are complaints that while travelling at Parkinson's expense Boel had sent seeds to a rival (William Coys), who had forestalled the author with descriptions of the new plants – 'while I beate the bush, another catcheth and eateth the bird'.[3] Nevertheless, Parkinson received from Boel some 200 packets of seeds, besides bulbs and 'divers other rare plants dried and laid between papers';[4] and the collector could hardly be blamed if they did not all succeed – '. . . manie of them came not to maturitie with me, and most of the other whereof I gathered ripe seed one year, by unkindly yeares that fell afterwards, have perished likewise'.[5] As this implies, many of Boel's introductions were annuals, and not all were new; but the plants he sent to England included *Scilla peruviana*, *Leucojum vernum*, *Armeria latifolia*, *Convolvulus tricolor*, and possibly *Linaria cymbalaria* and *Nigella hispanica*.

According to Pulteney,[6] Boel travelled and sent seeds from Germany as well as Spain, and resided at Lisbon and Tunis; but little is known of his movements. On his journey for Parkinson, which took place in 1607–8, he collected in 'Boetica', (Andalusia; from Boetis, the ancient

[1] and [2] Parkinson *Paradisi in Sole* 1629
[3], [4] and [5] Parkinson *Theatrum Botanicum* 1640
[6] *Historical and Biographical Sketches* 1790

name of the river Guadalquivir), the only place-names mentioned being the Island of Cales (Cadiz) and the near-by port of Santa Maria, where *Scilla peruviana* grew in such abundance that it seemed to 'cover the ground like unto a tapistry of divers colours'.[1]

Alicante on the south-east coast of Spain was visited by the elder **John Tradescant** in February 1621, when he accompanied Captain Argall's expedition against the Barbary pirates. Tradescant made a large contribution to seventeenth-century gardens, but it is doubtful whether he can be ranked as a collector in his own right. Between the conclusion of the siege of Algiers on 24 May and his return to England on 10 September his movements are unknown; his biographer[2] suggests that he may have been touring the Mediterranean, but of this there is no evidence. Parkinson's information that certain Mediterranean alliums were 'procured from thence' by Tradescant is capable of a dual interpretation, and though he 'observed' *Gladiolus byzantinus* growing in great quantity in Barbary, it does not necessarily follow that he introduced it. In the case of two cultivated plants there is no such ambiguity; he procured the famous 'Argiers' variety of apricot – the object for which he joined the expedition – and he brought back the Persian lilac, which had been introduced to Europe before 1614 through the Venetian ambassador to Turkey.

It was never very easy, nor very safe, for the foreigner (especially the Protestant foreigner) to botanize in Spain, and the Spaniards themselves were slow to take an interest in the subject. In 1751, after having dropped a broad hint about the backwardness of the country in this respect, Linnaeus was allowed to send one of his favourite pupils to investigate the Spanish flora; but **Pehr Loefling** soon found that though fêted and praised, his movements were much restricted. He was unable to travel anywhere in the country without the express permission of Joseph de Carvajel, the Secretary of State; and was relieved when he was eventually appointed botanist to an expedition sent by Spain to South America in 1754. Meantime another Linnaean pupil had arrived in Spain – **Claes Alstroemer.** The purpose of his travels was to study the breeding of sheep, but he also made natural history collections, and sent many plants to Linnaeus, including (in 1753) seeds of the Peruvian flower named after him *Alstroemeria pelegrina*. He found *Rhododendron ponticum* growing wild between Cadiz and Gibraltar; but unfortunately his journals were accidentally destroyed by fire.

Indeed, few travels in Spain or Portugal have been adequately recorded; we know the collectors were there, but little of what they did. During the War of American Independence, Sir Joseph Banks considered the situation 'almost to preclude the idea of Mr Masson being employed with success in any part of the world', and Francis Masson was given permission to accept a post offered by a merchant, Gerard Devisme, who had a botanic garden at Lisbon. Masson was in Portugal from 1783 to 1785, and from there made excursions to Spain, Tangiers and Madeira; but no

[1] Parkinson *Paradisi*
[2] Mea Allen *The Tradescants* 1964

records of these journeys remain. Of Webb's travels in Spain, already mentioned, few details survive; he wrote two books on Spanish botany, but they give no account of his adventures. The same is true of George Maw, who visited Spain and Portugal at least five times, finding *Draba dedeana* var. *mawei* (Pancorbo, 1870) and *Erica ciliaris* var. *maweana* (Portugal, 1872), and collecting material for his great work, *The Genus Crocus*, published in 1886.

A more modest monograph by another traveller in Spain had appeared the previous year – *Ye Narcissus*, by **Peter Barr** (1825–1909). Barr, the son of a Scottish mill-owner, established a nursery at Tooting in 1862 and specialized at first in hellebores and later in irises, paeonies and bulbous plants. But the ultimate love of his life was the Narcissus, and he set himself the task of rediscovering all the species mentioned by Parkinson in 1629, many of which had been lost to cultivation for more than 200 years. His subsequent travels took him wherever wild daffodils might be found, and from Spain, Portugal and the Maritime Alps he re-introduced among others *N.N. asturiensis*, *bulbocodium*, *cyclamineus*, *johnstonii*, *moschatus*, *nobilis*, *pallidiflorus* and *triandrus*. (Various stories exist of the means by which the latter acquired its name of Angel's Tears; they all relate to the distress of a guide called Angel or Angelo.) The only incident recorded of Barr's Spanish travels is that on one occasion the only room available at an inn had two beds in it, and was correspondingly expensive; to get his money's worth this very Scottish gardener spent half the night in one bed, and half in the other. At the age of seventy, Barr made a world horticultural tour which included Canada, the United States, Japan, China, Australia, New Zealand and South Africa. It lasted seven years, and on his return he started a new garden near Dunoon and began growing hellebores again; but even he could not repeat the rest of the cycle, and he died suddenly at the age of eighty-four.

Scandinavia and Russia

SCANDINAVIA

1732 Linnaeus

RUSSIA

1618	Tradescant	1800	Mussin-Puschkin
(1720	Messerschmidt)	(1826	Ledebour and von Bunge)
1733	Gmelin	1853	Maximowicz
1739	Steller	1861	Mlokosewitsch
1768	Pallas and S.G. Gmelin	1877	Regel
1796	Bieberstein		

'We could not avoid feeling extreme satisfaction at observing that the English style of gardening had penetrated even into these distant regions . . . Most of the Russian nobles have gardeners of our nation, and resign themselves implicitly to their direction.' WILLIAM COXE *Travels* 1784

'The Count smiled when we spoke of the facility with which he might obtain the Siberian plants. "I receive them all" said he "from England; nobody here will be at the trouble to collect either seed or plants, and I am compelled to send to your country for things that grow wild in my own." ' E. D. CLARKE *Travels* 1816

Scandinavia and Russia

Scandinavia

Scandinavia has not made any great contribution to our gardens, and travels in that part of the world must be regarded as of purely botanical interest. One journey, however, was so significant that it must be recorded: Linnaeus' renowned tour of Lapland in 1732. It was important to Linnaeus' own career, and a stimulus and example to many later naturalists.

In 1730 the young and impecunious **Carl Linnaeus** (1707–1778), a student at the University of Uppsala, was appointed assistant to Olaus Rudbeck, Professor of Botany, then nearly seventy years old; he was to live in Rudbeck's house and act as tutor to some of his twenty-four children. Thirty-five years earlier Rudbeck had been commissioned to make a tour in Lapland, and had 'wandered about, and made discoveries, observations and notes',[1] but almost all his materials and results perished in the fire at Uppsala in 1702, which also destroyed his father's great work, the *Campi Elysii*. It was probably Rudbeck's example which stimulated Linnaeus to make a similar journey. Capital in Sweden was then very short, after a series of wars, but at his second application the Scientific Society of Uppsala granted him the 400 dalar he thought he would require for the trip; this sum later turned out to be quite inadequate, but its disbursement left the Society with only about 7d in hand.

On 12 May, therefore, Linnaeus left Uppsala, naïf, self-confident and just twenty-five years of age. There is an atmosphere of youth and birdsong about his journals as he set out in the fine spring weather, deter-

[1] Quoted by Gourlie

Linnaeus in Lapland. Title-page *Flora Lapponica* 1737
By courtesy of the Royal Botanic Gardens, Kew *Photo by Kenneth Collier*

mined to miss nothing – bird, fish, insect or plant, mineral, agricultural method, local custom or antiquity – that came within range of his exceptionally keen sight. He followed the usual post-road from village to village northwards along the coast, his pleasure slightly damped by the fact that he was a poor horseman, and could only afford the most inferior nags. Gefle, only three days' journey from Uppsala, was the last town on his road that could boast an apothecary and a physician.

Linnaeus diverged from his main route to climb Mt Nyaeckersberg near Sundsvall, the descent from which was so steep that he frequently had to sit down and slide; fortunately he wore stout leather breeches, serviceable though second-hand. He made a second detour to explore a noted but rather inaccessible cave on Skulaberg, and from Hernösand sent back an already heavy collection of minerals. He reached Umea on 24 May.

It had already become apparent that he would have neither the time nor the money to go as far as Asila Lapmark, as he had hoped; but at Umea he obtained from the Governor a pass for the nearer parts of Lapland, and made his first attempt to strike inland to the Lapp settlements in the mountains of the north-west. But it was still too early in the season; the rivers were swollen by the thaw and by heavy rains, the marshes were flooded, and when the Lapp country was reached, the nomadic inhabitants were all absent at the pike-fishing. On horseback and by boat on the river Umea, Linnaeus penetrated to the tiny mission-station of Lyksele and for two days' journey beyond, but was then obliged to return to Umea, much the worse for fatigue, exposure and starvation.

After four days' rest he resumed his journey northwards along the coast through Skeleftea and Pitea to Lulea, and from there made his second and more successful attempt to cross the Lapp-inhabited mountains, which he was very anxious to reach in time to see the sun above the horizon at midnight, to the best advantage. Leaving after service on 25 June (he never travelled on a Sunday until after church) he went by boat up the river Lule for several days and nights, this time in comparative comfort, to Jokkmokk and up Lake Skalka to Tjamotis and Kvikkjokk. Here the pastor's wife supplied him with provisions for eight days, and two Lapp guides, one of whom served as servant and interpreter. Thus equipped, Linnaeus set off on 6 July to ascend Mt Wallaviri – his first Lapland alp. 'I scarcely knew whether I was in Asia or Africa,' he wrote, 'the soil, situation and every one of the plants being equally strange to me. . . . I sat down to collect and describe these vegetable rarities, while the time passed unperceived away, and my interpreter was obliged to remind me that we had still five or six miles to go to the nearest Laplander.' Swedish miles are long – approximately equal to six English ones – and it was not until the evening of the following day that they reached the Lapp encampment, having in the interval seen the midnight sun. The Lapps were hospitable, and Linnaeus shared a hut with sixteen naked inhabitants, but showed an unaccountable reluctance (which almost indicates that he was no true plant-hunter) to eat reindeer-whey from a communal bowl with a communal spoon. The next stage was also a long one, as they missed the Lapp hut at which they were aiming, which they found at last by following a fresh reindeer-track. Next day he sank to his waist through snow undermined by floods, and had to be hauled out by a rope.

Linnaeus is vague about the geography of this part of his wanderings, but he is believed to have followed the old road by which silver ore from the mines of Kiedevare was taken to Kvikkjokk, and to have come down to the SE angle of Lake Virihaure, having crossed what is now the Sareks National Park. On 11 July, in a biting east wind, they crossed the high mountains dividing Swedish from Norwegian Lapland, and descended to a valley farm – Linnaeus very tired, but his Lapp guides, aged fifty and seventy, as fresh as ever. This was at Norway's narrowest part, and he was very soon on the coast, staying at Torfiolme and Rorstad. He wished to see the famous Maelstrom off the Lofoten islands, but could find no boatman who would take him, so went fishing instead. When gathering wild strawberries on the hills behind Torfjorden, he hardly noticed a Lapp with a fowling-piece, apparently in pursuit of birds, until a bullet struck a stone near by; whereat 'the fellow ran away and I never saw him after, but I immediately returned home'. In later reminiscences, this became a dangerous attempt on his life.

The return journey was started on 16 July, but again his route is uncertain. By the 24th he was back at Kvikkjokk and resuming his travels by water, as he came. At Purkijaur, no boat being available, he and his guide contrived a raft – not very efficiently, for in fog and approaching night, and with a strong current, the raft began to disintegrate in midstream, and they had much ado to land on an island about a mile and a half from their starting-point, with the loss of some of Linnaeus' collections. However, they got safely back to Lulea by the end of the month.

Early in August our traveller proceeded round the coast by Kalix to Finnish Torneå, where he spent some weeks of apparently undecided comings and goings. It was getting late in the season for a sea-passage and many ships bound for Stockholm had already been waiting a long time at Torneå for a favourable wind. Linnaeus decided to return by way of Finland, and having learnt the names of a few essentials in the Finnish language, set out on the long coast road round the Gulf of Bothnia, seldom out of sight of the sea. This part of his journey seems to have been without particular incident; he passed through Jakobstad, Vaasa, and Kristinestad, by good roads and bad; reached Abö on 29 September, crossed to the Isle of Aland and was home again at Uppsala on 10 October.

Considering the shortness of the time he actually spent among them – no more than a fortnight altogether – Linnaeus gives a remarkably full account of the Laplanders, their food, clothing, huts, tools, games, diseases, and the natural history of the reindeer; all accompanied by crude but adequate sketches. The first published result was a paper in the Scientific Society's *Acta* in 1732; the complete *Flora Lapponica* appeared in 1737 and was the first book in which the plants were arranged according to the author's new sexual system. The actual journal, however, was not printed until Sir J. E. Smith's abridged English version appeared in 1811. 'The best botanical book', Richard Jefferies called it, 'written by the greatest of botanists, specially sent on a botanical expedition, and it contains nothing about botany. . . .'[1] To Linnaeus himself this tour was the greatest adventure of his life, often recalled – and much exaggerated – in retrospect.

Russia

Northern plants are on the whole more difficult to cultivate than southern ones; the conditions under which they grow are harder to imitate. Yet the northern flora is brilliant in its short season, and some of our most familiar and indispensable garden-plants are natives of Russia – the Iceland Poppy, the Scarlet Lychnis,[2] and *Scilla siberica*,[3] not to mention the ancestors of the Siberian Wallflower, and of the garden strains of delphinium and herbaceous paeony. South Russia and the Caucasus are reservoirs of good plants, still largely unexploited.

The first recorded investigation of the Russian flora was made by an Englishman. In 1618, negotiations were in progress between James I and Tsar Mikhael, who wanted financial and political help in his war against Poland, in return for trading concessions. Two Russian ambassadors visited England, loaded with costly presents; and the king sent an embassy under Sir Dudley Digges, of Chilham Castle, Kent, to accompany them on their return and continue the discussions. The gardener, **John Tradescant** (d. 1638), then working for Sir Edward Wotton at Canterbury, got leave from his employer to join the party.

Of all the travels of the elder Tradescant, this is the best entitled to be considered a true plant-hunting expedition, in the sense that the discovery and introduction of wild, not cultivated, plants was a main object. His opportunities on the journey were very limited. After six weeks in Russia, Digges, hearing that the Poles were at the gates of Moscow, prudently withdrew, his mission uncompleted. Tradescant's section of the party had left three weeks earlier, and his excursions were confined to the Dvina delta, in the neighbourhood of Archangel. His journal is largely concerned with the events of the voyage, and with observations on Russian manners and customs; but it is evident that he did his utmost to collect and introduce plants, in the short time available.

After a stormy six-weeks' passage, the *Diana* of Newcastle, with Tradescant on board, crossed the bar of the Dvina on 14 July 1618 with only a foot to spare, and moored on the 16th 'befor the Inglishe house' in the Bay of St Nicholas, where a trading-post had been established since 1591. Hardly had they anchored when Tradescant 'desired to have the boat to goe on shore whiche was hard by' where he found 'many sorts of beryes, on sort lik our strawberyes but of another fation of leaf; I have brought sume of them hom to show with suche variettie of moss and shrubs, all bearing frute, suche as I have never seene the like.' Later he refers again to his 'strabery' (*Rubus chamaemorus*) whose amber-coloured

[1] *Field and Hedgerow* ('Nature and Books') 1889
[2] *L. chalcedonica*
[3] Like the poppy and *Iris sibirica*, this species is not confined to Russia

fruits were locally used as a 'medsin against the skurvi' and describes it more fully: 'I dried some of the beryes to get seede, whearof I have sent par(t) to Robiens[1] of Paris.'

The following day the Ambassadorial party went on shore 'with all the showe we could make' and were allotted quarters, though 'havyng but four bedsteads' Tradescant was among those 'content to lay our bodi on the ground'. On Monday 20 July he had 'On of the Emperor's boats to cari me from iland to iland, to see what things grewe upon them, whear I found single roses, wondros sweet, with many other things whiche I meane to bring with me.' He describes this rose as being similar to 'our sinoment rose' (*Rosa cinnamomea*). '. . . I hop they will bothe growe and beare heere, for amongst many that I brought hom with the roses upon them, yet sume may grow.' This was on Rose Island, where he also found *Dianthus superbus*, already known to him – 'pinks growing natturall of the best sort we have heere in Ingland, withe the eges of the leaves deeplie cut or jaged very finely'.

On the return voyage, on 5 August, Tradescant again had an opportunity of landing on Rose Island, and 'gathered of all such things as I could find thear growing, which wear 4 sorts of berries, which I brought awaye withe me of every sortt'. They included *Cornus suecica*, of which he 'took up many roots, yet am afraid that non held'. This was because the ship grounded on the bar on a neap tide; to lighten her, some fresh-water casks were staved, and were afterwards refilled with salt water. Tradescant, unaware of the change, used the brine for watering his plants – with fatal results. On the following day it rained heavily, and the decks leaked – 'it rayned doune thourow all my clothes and beds to the spoyll of them all', but he does not say whether any of his collections or papers were damaged. They disembarked at St Katharine's near London on 22 September.

Some twenty wild plants are mentioned in the journals, the majority of them natives of Britain, though Tradescant, a Southerner, was obviously not familiar with such northern species as the Cornus or the Rubus. One or two were long-standing denizens of English gardens; for example *Veratrum album* – 'helebros albos enoug to load a shipe'. He seems to have been successful in introducing his Muscovy Rose, which is mentioned by subsequent authors, and he may also have introduced the larch, which was brought to Britain about this time; it has been identified as one of the '4 sorts of fir trees' that Tradescant observed.

After this precursor there was a long gap, and little more is heard of botanical exploration in Russia until the eighteenth century. The first traveller of note was Dr Daniel Gottleib Messerschmidt of Danzig (1685–1730), who in 1720 was sent by Peter the Great on an exploring expedition to Siberia. After seven years, during which he reached Chinese Mongolia and Lake Dalai Nor, he returned with large collections; but no results were ever published, and eventually the collections were lost.

[1] Jean Robin of the Jardin du Roi

One is inclined to admire the Russian empresses of the Romanoff dynasty for their enlightened patronage of the sciences, until one realizes that for seventy years there were virtually no emperors. Peter the Great, whose eldest son was unsatisfactory, abolished the law of primogeniture, with the result that the crown made an erratic progress from one to another of his male and female descendants. Neither Peter II, who succeeded at the age of twelve and died at fifteen, nor Ivan, who became Emperor aged twelve weeks, and was deposed at sixteen months, nor Peter III, Catharine the Great's unsatisfactory husband, who 'abdicated' after a year and died mysteriously seven days later, could have been expected to take much interest in botany. From 1725 to 1797, except for these short interregnums, Russia was governed by women. Naturally, then, it came about that the Imperial Academy of Sciences at St Petersburg, planned by Peter I, was actually instituted in 1725 by his widow, the Empress Catharine; was reinstated after a period of neglect by the Empress Anne and again by the Empress Elizabeth, and was finally brought to great fame and importance by the encouragement of Catharine the Great. It was through this institution that most of the great Russian exploring expeditions were launched.

The newly-civilized country was not yet able to produce its own scientists, and professors for the academy were recruited from all sources. Most of them were Germans, who found in Russia better chances of employment than were available in their own small and often impecunious States. One of these Germans, **Johann Georg Gmelin** of Tübingen (1709–1755), is said to have made himself so beloved that he was appointed Professor of Chemistry and Natural History in order to prevent his return to Germany; and indeed he had a most amiable character – easy-going, honest, generous and with an engaging sense of humour. In 1732 he volunteered to take part in the ambitious enterprise then being launched by the Imperial Academy and the Empress Anne.

It was all part of a fantastic exploration scheme – the Second Kamtschatka Expedition, under the management of Captain Vitus Bering. A survey was to be made of the coasts of Arctic Russia and America, the newly-discovered Bering Straits, and northern Japan. The naval contingent was to go by land across Siberia and build the boats to explore the Arctic section at Tobolsk on the Obi and Yakutsk on the Lena, and those for the American and Japanese explorations at the port of Okotsk. Meantime the Academy was to send out a small army of scientists, to travel to Kamtschatka overland, making astronomical, meteorological, geological, geographical, ethnological and natural history observations; it was to be independent of the navy, but the navy was to arrange for its supplies. This was to be done thousands of miles from civilization, in an undeveloped country where there were few roads, no agriculture and no industries, where most of the inhabitants were pagan nomads, and the local authorities indifferent or actively hostile. Yet the greater part of this impossible task, Bering actually managed to perform.

The scientific party was led by Gmelin as naturalist, G. F. Mueller, historian and ethnologist, and de le Croyère, astronomer – all Academy professors. They had six students as assistants, an interpreter, two artists, five surveyors, an instrument-maker, a huntsman, an escort of fourteen soldiers, and 'proper attendants'; together with a lavish supply of books and scientific equipment, provisions and even comforts. After the first part of the journey the astronomer and his party, with his nine cartloads of instruments, separated from the rest and went his independent way; but Gmelin and Mueller stayed together till the end. It was nearly ten years before they returned. The distance to be covered was vast,[1] and travel only possible during a portion of the year; long, long months had to be spent in winter quarters, waiting for ice to thaw on the rivers and snows to melt in the passes. Progress could not be anything but slow.

For the first two winters, however, they were still in civilized regions, and travel was little impeded. The party left St Petersburg on 3 August 1733, and proceeded by various means to Torjok on the Volga. After sailing down the great river for several months, they left it at Kazan on 12 December, and made their way by Kungur to Ekaterinburg (Sverdlovsk) in the Urals. Here there were mines, foundries and mineral deposits to be inspected, and Gmelin made a detour with the Governor of the district before rejoining the main party, which on 30 January 1734 had gone into winter quarters at Tobolsk.

Starting from Tobolsk on 24 May, the scientists made the first of two big loops to the south and east, ascending the river Irtich to Semipalatinsk and the Altai mountains, crossing to Kousnetsk on the Tom and down that river to Tomsk. It was a trying journey. As their excellent Tartar boatmen took them slowly upstream in the summer heat, the midges and mosquitoes were almost unbearable; they bit through stockings and shirts, and Gmelin had to wear two pairs of gloves while writing his journal. When the increasing shallowness of the stream made boating intolerably slow, they continued the journey on horseback, with an escort of ten light footmen as a defence against hostile Kalmuks; the steppe was swept by fires and the heat intense. They reached Semipalatinsk and continued up the valley to Oust-Kameno-Gorsk; inspected mineral deposits and foundries in the mountains, and crossing the river Toumich, considered themselves at last in Siberia proper. From Kousnetsk on the Tom they descended to Manicheva, where the Tartar women all fled on their approach, and to Tomsk, a miserable town, which they reached on 5 October. Two months later they crossed to Yenisiesk, where they went into winter quarters, while birds froze in the air and ice three lines thick formed on the inside of the windows.

In 1735 they made their second big sweep to the south-east – up the Yenisei to Krasnoiarsk, then across to Kansk on the Kan, Oudinsk on the Chuna, and Balachansk on the Angara. At Irkutsk they had trouble with the local authorities, who refused to supply horses; in the end, some had to be seized in the market-place. They boated across Lake Baikal, which, they were told, 'wanted' to be called the Holy Sea; if addressed by any less respectful name the waves would rise and overwhelm the blasphemer.

[1] Siberia covers nearly a tenth of the land-surface of the globe

Gmelin and his companions teased their boatmen by referring frequently to 'the lake', without evil results. When the Tchikoi was free of ice they left for Kiahkta on the very frontiers of China; then turned eastwards through the mountains to the river Ingoda – where their boatmen were astonished to see them eat that terrifying creature, the crayfish – and on to Nertschink and the salt lake of Sagan Nor. By mid-September they were back at Lake Baikal, which with high winds and intense cold nearly took its revenge, Gmelin said, for their 'puerile insults' in spring. For the first part of the winter they were at Irkutsk, removing at the end of January to Ilimsk.

In the spring of 1736 they crossed to Ust-Kut on the Lena – a great centre for the trade in sables – and when the ice thawed at the beginning of May, descended the river eastwards to Yakutsk, a journey which occupied the whole period until the river began to freeze again in mid-September. At Yakutsk they found Bering and the naval contingent already established and occupying all the best quarters. A good deal of social life was in progress, and on 8 November, when Gmelin and Mueller were at a party given by Bering, there was a cry of 'Fire!' and Gmelin emerged to see the cabin he had recently left, a mass of flames. Nothing could be done but watch it burn; it was three days before the ashes were cool enough to be examined. Gmelin lost everything except the clothes he wore – his recent collections, nearly all his MSS. and books, his personal possessions, most of his money and some of Mueller's that he was looking after. He particularly lamented the loss of Tournefort's *Institutiones Rei Herbariae* – yet here in the heart of Siberia he was able to borrow another copy, the property of an exiled nobleman.

Disaster followed disaster. The naval contingent, hard-pressed themselves, bluntly told them that they could not guarantee supplies and transport for the next part of the journey, to Okotsk, and referred them to the local authorities; these in turn replied that they could do nothing – the navy had taken all they had. It was the same with the shipping required to take their 'rather large party' from Okotsk to Kamtschatka – neither the naval nor the regional authorities were prepared, or indeed able, to supply it, more especially as Gmelin insisted on having 'our due comfort on board the ships', as befitted the dignity of travelling Academicians. Moreover, the road to Okotsk was hard; there was little fodder for horses, and the reindeer could not carry heavy loads.

Further progress seemed impossible. Mueller had been ill, and wished to spend the next winter in a less bitter climate, and Gmelin wanted to reduplicate his lost collections. They sent one of the students, Krasheninnikov, alone to Kamtschatka to make a beginning as best he might, and on 9 July 1737 turned their faces westward, retracing their route up the Lena and establishing their winter quarters in September at Kirenskoi Ostrog. In November, Mueller removed to Irkutsk for the sake of his health, and Gmelin joined him there in March 1738. Again they made great efforts to procure transport and supplies for a journey to Okotsk, until the sympathetic but harassed Vice-Governor gave them a written affidavit that for the next two years he would be unable to supply them. At the end of May, the weary and disheartened scientists wrote to the Academy explaining their difficulties and asking to be recalled. They worked in the Irkutsk neighbourhood during the summer, and at the end of August descended the river Angara to winter headquarters at Yenisiesk – the nearest to home that they dared venture until they received permission to return.

In January 1739 they were joined at Yenisiesk by yet another German physician-naturalist – **Georg Wilhelm Steller** (1709–1746), sent out by the Academy to assist them. Steller, much more than Gmelin, was of the true collector breed. Gmelin liked his comforts, and was inclined to stand on his dignity as an Academy professor; Steller, who ranked only as an 'adjunct',[1] was prepared to live hard and travel light. He was quick-tempered, capable, somewhat over-zealous and very impatient of authority. Gmelin saw in him the very man to fulfil his uncompleted mission, and Steller, secretly longing for independence, was only too anxious to comply. After much planning, he left for Kamtschatka on 5 March, Gmelin and Mueller depleting their own wardrobes to supply his deficiencies. After Steller's departure Gmelin decided to explore a more northerly region than any yet visited, and as soon as the ice left the Yenisei, he and Mueller went by boat down to Turukhansk, then the northernmost settlement in Siberia. It was still winter when they arrived on 10 June, but the spring arrived so rapidly that in less than a week the snow was gone and the meadows full of flowers. They returned to Yenisiesk, and spent the rest of the summer exploring a series of deserts between the Kan and the Tuba, where only Messerschmidt had been before. Even here, certain localities were very rich in flowers, *Lychnis chalcedonica* being particularly common. This time they wintered higher up the Yenisei at Krasnoiark.

1740 was spent in local excursions from Krasnoiark and in the country between there and their next winter-quarters at Tomsk; 1741 in the Baraba country between the Obi and the Irtich, and from there to Tara and Tobolsk; and 1742 in a further exploration of the mining districts of the Urals. After a short winter sojourn at the estate of Count Demidoff at Solikamsk they travelled home by a more northerly route than on the outward journey, reaching St Petersburg at last on 17 February 1743, after an absence of nine years and six months.

Meanwhile, Steller had fought his way through to headquarters at Bolsheretsk in Kamtschatka, and had succeeded in getting himself attached to Bering's naval expedition to America. Against Bering's wishes, he insisted on accompanying the first exploratory boat-party, and was probably the first white man to set foot in Alaska.[2] He was only ashore for half a day, but he collected over 140 plants, and sat down then and there to list them, with references (from memory) to botanical authors, including Tournefort and Parkinson. The plant that impressed him most was *Rubus spectabilis*; he vainly tried to take living plants of it back to Russia. As it turned out, this was his only opportunity to land, except in

[1] Gmelin, always regarded as Steller's much-to-be-revered senior, was actually born in the same year; neither had attained the age of thirty

[2] At Prince William Sound

the Schumagin Islands off the Alaskan Peninsula, for adverse weather and other circumstances obliged Bering to make for home.

With Bering's wrecked and scurvy-ridden officers and crew, Steller wintered on Bering Island, making important discoveries among marine animals, and reached Kamtschatka with the survivors in August 1742. During the next two years he made extensive explorations in the peninsula, living off the country and fraternizing with the natives. At last, in August 1744, he started for home – the last straggler of the great Kamtschatka Expedition.

During his stay at Kamtschatka the fiery, domineering German had quarrelled with two of the Russian naval authorities, and each side had written to St Petersburg denouncing the other. At Irkutsk on his homeward journey (autumn 1745) he was summoned before the Vice-Governor to answer the serious charges that had been made against him; he was completely exonerated, and a report to this effect was sent to the capital. At Tobolsk and again at Verkhoturie, Steller had trouble with the Customs, which Gmelin had also experienced; the officials made a practice of being obstructive, in the hope of being bribed. At last, in the spring of 1746, he reached the estate of Demidoff, just beyond the Urals – truly a sojourn in the Delectable Mountains for weary botanists. The wealthy Count Grigorij Demidoff was an enthusiast for the science, and famous for his greenhouse and gardens; and here Steller made a temporary plantation of over eighty species of plants, some of which he had brought with infinite pains from the other side of Siberia, while he assisted his host in making an exhaustive survey of the flora of the near-by Perm mountains.

On 16 August, the day after his return from a fatiguing local expedition, Steller was arrested, and told he must go back to Siberia. The report of exoneration from the Chancellery at Irkutsk had not yet been received at St Petersburg, whereas the reports from the Customs-houses had arrived, giving the impression that Steller had disobeyed the order to answer the charges against him and was trying to evade justice. The fact that he had so long passed the Siberian frontier and yet had not appeared in Moscow, made appearances doubly suspicious. Once more he embarked on the weary eastward journey, this time with no enticing goal beyond; he did not even understand the reason for his arrest. Some two months later, in Tara, Steller and his escort were overtaken by a fast courier from St Petersburg: his exoneration had at last arrived, and he was to be set free. In a week he was back at Tobolsk, where he had friends, and celebrating his liberation rather too well; he took fever, but insisted nevertheless on resuming his homeward journey. It was too much for even his hardy physique, and he died at Tyumen on 12 November 1746. Most of his botanical discoveries were published by Gmelin; in the foreword to his *Flora Sibirica* (1747–69) the generous author gave Steller unstinted praise.

Plants and seeds from this prolonged expedition were periodically sent to the Botanic Garden of the Academy, which was founded by Dr Johann Amman in 1736 and maintained by him till his death in 1741. Before coming to Russia in 1733, the Swiss-born Amman had been employed

Peter Simon Pallas – A silhouette by Dr E. D. Clarke, who visited Pallas in the Crimea. From *Travels in Various Countries* 1816 *Copyright photograph, British Museum*

by Sir Hans Sloane in London; he had many English correspondents, including Sloane, Collinson and Catesby, and through him a few Siberian plants arrived in Britain, though only *Cornus alba* and *Delphinium grandiflorum* were of horticultural value. There may have been others – *Lonicera tatarica*, *Cephalaria tatarica* and *Gypsophila paniculata*, for example, were grown by Philip Miller before 1758 'by seeds from St Petersburg', but their source is not given. Items sent by Amman were likely to be kept anonymous, as the exportation of the expedition's seeds and plant-specimens was strictly forbidden.

An equally ambitious project was launched by the Academy twenty-five years later, at the instigation of Catharine the Great. A number of scientists were sent out, and their meetings and partings, their travels singly or in groups over a period of several years, wove a net of exploration over the greater part of the Russian Empire. The principal botanists were two more Germans – Gmelin's nephew, **Samuel Gottlieb Gmelin** (1744–1774) and **Peter Simon Pallas** (1741–1811).

Pallas was the son of a prominent surgeon of Berlin, and after taking his medical degree at Leyden at the age of nineteen, and spending some months in England, he settled at The Hague. He was chiefly interested in zoology, and in 1766 published a book on tapeworms, madrepores, corals and other hitherto unclassified animalculae. Alarmed at his increasing neglect of medicine, his father summoned him home, just in time to prevent him from embarking for Cape Town and the Dutch East Indies; but his reputation by this time was such that he was invited by Catharine the Great to take the post of Professor of Natural Science at St Petersburg, and this offer he accepted, though against his father's wishes. He arrived at St Petersburg in August 1767, and soon afterwards volunteered to take part in the expeditions then being planned, to coincide with observations of the Transit of Venus.[1]

It is difficult to epitomize the journeys of Pallas, which occupied six years – not through lack of material but through its surplus. Two months after his return he began the publication of his *Travels*, which ran to five quarto volumes in German[2] each containing over 700 pages of purely factual report, very detailed and very dull; not a gleam of humour, not a glimpse of personality, is permitted to appear. The first year's travels were in any case of lesser interest, as they covered an area of European Russia which was well-developed both in agriculture and industry; it extended south-east from Moscow to Penza and north-east from there to Simbirsk, where he spent the first winter. From March to June 1769 he made his headquarters at Samara, farther down the Volga, where he was joined for a few days in May by two of his colleagues, Lepechin and Falk.

On 17 June he left Samara for a journey south-eastward over wilder country, inhabited only by military outposts and nomad tribes; it took him eventually to Orenburg, an important trading-centre where East met West, and produce of every sort was brought to market. After an excursion to Orsk at the foot of the Urals Pallas set off to follow the river Ural through saline deserts and marshes south to the Caspian Sea; it was desolate country but he was particularly interested in the plant-communities of the salt-flats, many of which afforded good pasturage for animals. At Gurief on the coast another rendezvous of Academicians had been arranged. They could hardly have chosen a worse spot; surrounded by marshes which were flooded most of the summer, the small settlement was notoriously unhealthy, but Pallas rejoiced to meet several of his colleagues, though they were together for little more than a week. He returned up the river to Uralsk, and from there, through heavy rain and floods already, in late September, turning to snow and ice, northwards to Ufa for the winter.

The open season of 1770 was spent in the investigation of the rich mining districts of the Urals, from Ufa up to Ekaterinburg (Sverdlovsk) and down again to Chelyabinsk for the following winter. There he was joined in a short time in March 1771 by Falk, Lepechin and Georgi, who among them had examined almost the whole province of Orenbourg. In the following two years he covered much of the country travelled by Gmelin twenty years before, but not, he is careful to explain, by quite the same routes; he claims that his own observations were both more comprehensive and more exact than those of his predecessor, who neglected natural history and 'only concerned himself with botany'. At Barnaul, later in the year, he again met Falk, and at Krasnoiark early in 1772 he met Georgi and several students; in the course of the summer he explored the Mongolian border from Kiahkta almost to the Amur, and

[1] The object also of Cook's first voyage to the Pacific

[2] Afterwards translated into French

returned again to winter at Krasnoiark. Next year he retreated westward by Tomsk and Tara to Kasan on the Volga – almost, as it were, within sight of home; but instead, turned abruptly southwards to Uralsk and thence right down to Astrakhan, before returning up the Volga to winter-quarters at Tsaritsyn (Volgograd). After some local excursions he at last started homeward, and reached St Petersburg on the last day of June 1774.

Some of his associates were less fortunate. J. P. Falk, a Swedish pupil of Linnaeus, was subject to hypochondria and melancholy, and committed suicide under particularly grisly circumstances at Kasan in 1773. Samuel Gottlieb Gmelin, who was in charge of another branch of the expedition, was led by his enthusiasm, in the third year of his travels, to cross the Caucasus to Azerbaijan and North Persia, where he spent nearly two years, and after returning to Astrakhan, made a second visit late in 1773. On his way home up the western shore of the Caspian, and within a few days' journey of the safety of the Russian fort of Kisliar on the Terek, he was captured by the Tartar chief Usmei Khan, and held as a hostage. He was sent from prison to prison, and died, chiefly from the fatigues and privations he had undergone, in a small Caucasian village on 27 July 1774. 'Though a man of genius and well-versed in natural history,' remarks Lemprière,[1] 'he was of a licentious turn of mind.'

For nearly twenty years after this journey Pallas lived in St Petersburg; he was in high favour with Catharine II and received many honours and official appointments. He continued to publish works on zoology, but his travels had stimulated his interest in botany, which he took up with enthusiasm. He corresponded with Sir Joseph Banks and sent him specimens and seeds, which were raised at Kew; they included *Pyrus salicifolia*, and the ancestor of the herbaceous 'Chinese' paeonies, *P. lactiflora*. Much of his time was occupied by the production of his magnificent *Flora Rossica*, the first volume of which appeared in 1784, and the second in 1788; unfortunately the public funds that paid for the publication were discontinued, and the work was never completed.

In 1793 Pallas asked and received permission to make another journey 'for the recovery of his health', though the unavoidable hardships of travel at this period were hardly likely to prove beneficial to a middle-aged man. This time he travelled at his own expense, with his wife and daughter, and an artist, C. G. H. Geissler, whom he had engaged for the trip. He left St Petersburg on 1 February and headed for the south, taking the shorter land-route from Vladimir to Saratov across the big bend of the Volga, and then by the 'usual winter road' of the frozen river, to Tsaritsyn. The ice had already been rendered difficult by an exceptionally early thaw, and they had to wait at Tsaritsyn till it broke up and they could continue their journey by boat. It was mid-April and flowers were beginning to appear; there are frequent references to the surrounding deserts being full of wild tulips – *biflora*, *sylvestris* and *gesneri*. At Astrakhan, where Pallas stayed till 5 May, he took Geissler on excursions to the neighbouring steppes in

[1] *Universal Biography* 1805

order to 'obtain exact drawings of the rare vernal plants'. Then he went by river and canal to Krasnoi-yar in the Volga delta, and leaving his family behind, made a perilous tour with a Cossack escort over the salty deserts to the north and east; fortunately there was only one night on which they had to camp without even the most brackish water. They crossed a gypsum plain where refraction added to the almost intolerable heat, but even here he found 'new and beautiful plants'. After a detour to visit Mt Bogdo he reached the Volga at Vladimirovka, and found his 'affectionate travelling companions' waiting for him at Tchernyiyar, farther up. Then they all went on to Sarepta (also on the Volga), where they stayed for the whole of June and July, partly because the daughter fell ill with smallpox, and partly because steppe-fires on both sides of the river made travel almost impossible.

On 4 August, the daughter having recovered, the party left Sarepta, and descending again to Astrakhan, began to make its way towards the Caucasus. They followed the coastline southwards and then struck inland up the valley of the river Kuma to Georgievsk, where they stayed for nearly three weeks while Pallas made excursions into the neighbouring mountains. It was now September, but in spite of the lateness of the season he found many plants, and collected seeds on Mt Beshtau of *Rhododendron luteum* (*Azalea pontica*) some of which he sent to two English nurserymen, Lee of Hammersmith and Bell of Brentford. Plants from two other sources followed shortly afterwards, but his was actually the first introduction of this invaluable shrub.

Anxious to reach a warmer climate before the winter, the party left Georgievsk on 23 September, and travelled without serious mishap by Stavropol and Tcherkask to Taganrog on the Sea of Azov, where they rested for a time, before undertaking another week of travel across the steppes to Perekop, at the entrance to the Crimea. A winter residence had been prepared for them at the seat of government, Akmetchet (Sympheropol) where they arrived at the beginning of November. Pallas' health had been much impaired by the fatigues of the journey, and he went little afield till the following March (1794) when he started a series of explorations of the mountainous south and south-east of the peninsula, on one occasion crossing the Straits of Kertch to the Isle of Taman, with its numerous volcanic springs. On 18 July they set out for home, by Perekop, Kherson and the Dnieper Valley, reaching St Petersburg on 14 September.

The Crimea had been 'annexed' by Russia only in 1783, and the territory was still comparatively little known. Catharine herself had made a tour of it in 1787, carefully stage-managed by Potemkin; and Pallas, anxious to please his sovereign, also presented the Crimea in the most favourable light, giving the impression that it was little short of an earthly paradise. The unforeseen result was that in 1795 Catharine bestowed on him an estate at Akmetchet, on which to pass the remainder of his days. She died in 1797, comfortably assured that she had done well for her favourite, and Pallas never dared to reveal that the estate, though beautiful, was exceedingly unhealthy, and that much of the countryside had been wantonly devastated by the Russian soldiers. Holding, as he did, strong views about absentee landlords, he was virtually the prisoner of his estates; he lived in palatial style but missed the intellectual life of the capital, and grew old before his time.

His exile was mitigated by numerous visitors, whom he received with great hospitality. Among them was Dr E. D. Clarke of Cambridge, a mineralogist and keen amateur botanist, on his way home in 1800 from a journey through Russia with his friend J. M. Cripps. Clarke tried hard to persuade Pallas – whose daughter had married – to come and live in England. He refused, as it would have meant the forfeiture of all his Russian property; but he sold Clarke his herbarium, for he said no-one in the Crimea would value it after his death. Other distinguished visitors were Baron **Friedrich August Marschall von Bieberstein** (1768–1826) and Count **Apollon Apollosevitch Mussin-Puschkin** (1760–1805).

'From a laudable zeal of exploring mineralogical objects,' Pallas tells us, Count Mussin-Puschkin had 'at his own expence undertaken a journey to the lofty regions of Caucasus'. His expedition lasted from 1800 to 1805, and he was the first Russian to take into the field a portable chemical laboratory, and to assay his minerals on the spot. He was not himself a botanist, but there were botanists – Germans, of course – in his party. For the first year he was accompanied by Dr J. M. F. Adams (b. 1780) – a member, like himself, of the St Petersburg Academy – and from 1802–5 by Marschall von Bieberstein.

From his military academy at Stuttgart, Bieberstein came to Russia as secretary to Count Kakhovsky. He then joined the Russian army, and for nearly three years (1792–5, the time of Pallas' first visit) was stationed in the Crimea. He left the army in 1795, and in the following year became a member of an expedition sent under Count Zubov to Persia, during which he made extensive explorations of the western end of the Caucasus mountains; and in 1798, after the death of Catharine, he made another journey to the more northern and eastern parts of the range. In the same year he published (in French) a *Tableau des Provinces . . . entre les fleuves Terek et Kur*; a German version of the same work appeared in 1800, to which were added descriptions of seventy-four new or rare plants – including *Crocus speciosus*. This was his first published botanical work, but he later wrote an important *Flora Taurico-caucasica* (2 vols. and suppl. 1808–1819) covering 2322 species of plants, many of them his own discoveries.

Mussin-Puschkin had been for a time Russian Minister in England, and at the outset of his expedition he wrote to Sir Joseph Banks and offered to send him seeds and natural-history specimens. He was as good as his word; consignments of seeds, dried plants and minerals were sent in 1800 and 1802, and in 1803 a further collection, including living plants of Rhododendrons *caucasicum* and *luteum*. Introductions credited[1] either directly to him or to Banks during this period include *Onosma tauricum* and *Galega orientalis*. In return, the Count asked Banks for seed of American trees, which he hoped to naturalize in the mountains. Writing from his camp near Acbala in November 1803, he mentioned that the plague had been

[1] In *Hortus Kewensis*

bad that season, especially among the members of the expedition. He died in the Caucasus in April 1805 – it is not said from what cause; and is commemorated in *Nepeta mussinii* (introduced before 1806) and in *Puschkinia scilloides*, named by Adams and introduced in 1819.

The importance to gardeners of this unofficial, unchronicled expedition has never been appreciated. In the first volume of the *Botanical Cabinet* (1818), George Loddiges mentions that his firm grew *Gentiana septemfida* from seed collected by Bieberstein, which they received in 1803 through a correspondent, 'the late Mr Stepan, of Moscow'. It seems fair to assume that other plants grown by Loddiges at this time 'by seeds from Mt Caucasus' – *Scabiosa caucasica* in 1803, *Chrysanthemum* (*Pyrethrum*) *roseum* and *Lilium monadelphum* in 1804 – came through the same channel.[1] Bieberstein subsequently became an inspector of silk-worm breeding for the whole of South Russia, and travelled annually from Merefa near Kharkov (where he made his home in 1807) to the Dnieper and the Volga, and over the mountains to Georgia; so he would still have opportunities for finding Caucasian plants, though he went no more on organized collecting expeditions. His discoveries included *Geranium ibericum* (introduced in 1802) and *Campanula lactiflora* (Loddiges, 1815).

Mussin-Puschkin's plant of *Rhododendron luteum* (if it survived) was actually the third introduction of this species, the second having been made by the Polish **Anton Hove** in 1798. After his plant collecting journeys for Kew in South Africa and India,[2] Hove had left England with the intention of setting up as a physician in Germany; but we hear of him next in St Petersburg in September 1795, writing to Banks that his plans in Poland had been upset by repeated revolutions, and that his 'institute of Botany' had been destroyed because of suspicions aroused by the fact that he had been presented the previous year to Her Imperial Majesty Catharine II. He had been obliged to take refuge in Russia, where he was given not only protection, but the offer of employment on the Turkish border, to investigate a part of the country not yet scientifically explored. His headquarters were to be in the small town of Uman in the Ukraine, and he expected to set out the following April. Under *Azalea pontica* (*Rhododendron luteum*) the *Botanical Magazine*[3] gives extracts from his journal, but the document itself does not seem to have survived, so we know only of those points on his itinerary where the azalea was found. On 9 June 1796, he was somewhere on the Dnieper; on the 20th at Mogilev on the Dneister; on 4 July at Ochakov on the Black Sea coast, and on the 15th at Trebizond. A month later he was back at Odessa, and writing to Banks about the medicinal properties locally attributed to the new shrub. He had himself found a decoction of its leaves effective in the treatment of a violent rheumatic pain in his arms and thighs, which he had 'contracted by sleeping on the swampy ground'. This is the last record we have; of Hove's subsequent fate nothing is known.

The expedition to the Altai mountains in 1826 of Professor Frederick

[1] The Caucasian *Achillea filipendulina* (1803) and *Anchusa azurea* (=*italica*) 1810, made their unexplained appearances at this time

[2] See p. 145 [3] VOL. 13 (1799), plate 433

von Ledebour and his pupils Alexander von Bunge and C. A. Meyer, must be regarded as of purely botanical importance; for though it resulted in the introduction to the botanic garden of the University of Dorpat in Livonia of more than 1300 species of living plants, none of them seems to have been of great horticultural merit or to have spread by this means into general cultivation. Von Bunge succeeded Ledebour as Professor of Botany at Dorpat in 1836, and in 1844 received as a pupil a brilliant young Russian of German extraction, **Carl Maximowicz** (1827–1891). After taking his degree, Maximowicz acted for a time as assistant to von Bunge, and in 1852 was appointed Conservator of the Herbarium of the St Petersburg Botanic Garden. In the following year he began a series of botanical travels which ultimately made him the greatest authority of the time on the flora of Manchuria and Japan.

Furnished with a commission to obtain living plants for the gardens, he embarked in 1853 on the frigate *Diana*, and touched at Rio de Janeiro, Valparaiso and Honolulu in the course of the voyage. The outbreak of the Crimean War in March 1854 caused the squadron with which he sailed to take refuge in Castries Bay, high on the Gulf of Tartary, where the unseaworthy flagship *Pallas* had to be destroyed, and the Admiral and his suite transferred to the *Diana*. Maximowicz in consequence was set on shore, and left to explore the flora of the Amur Province, recently annexed by Russia from China.

His first circuit was up the coast to the mouth of the river Amur and the vicinity of Nikolayevsk, then up the river to Marinsk and back to Castries Bay by land. In 1855 he made his base at Marinsk, and from there made two excursions up the Amur; the first only to its junction with the minor tributary, Dondon, but the second, in company with L. von Schrenk, to the great Ussuri, and up that river as far as the mouth of the Noro. After wintering at Marinsk, he set out for home in July 1856 – a three-month journey up the Amur to Ust Strielka, then by the ordinary post route across Siberia and European Russia to St Petersburg. During these years Maximowicz suffered many hardships, due chiefly to the shortage of supplies; but he brought back about 100 plants, new or nearly new, including *Paeonia obovata*, *Jeffersonia dubia* and *Actinidia kolomikta*.

Less than three years later Maximowicz returned to the same area. Leaving the capital in March 1859 he arrived at Irkutsk on 1 May, and left again on the 25th for Nerchinsk. From there he went down the river Shilka to the Amur, stopping at Ust Strielka and Blagovieshensk. On 13 July he reached the mouth of the Sungari, and explored this river southwards through Manchuria, until turned back by the Chinese, before reaching its junction with the Kulkha at San-sing Chen. Back, then, to the Amur and down to the Ussuri, which he explored to a much higher point than he had previously attained, before returning to the Amur and descending to Nikolayevsk near its mouth, which he reached on 2 October. He hoped to get a steamer from there to Japan, but the Amur was already beginning to freeze, and he was obliged to stay at Nikolayevsk for the winter.

Long before winter was over, however (in February 1860), Maximowicz went by dog-sleigh up the frozen river to Khabourovka, and up the Ussuri valley to the village of Busseva, where he made zoological collections while waiting for the snow to melt in the Sikhota Alin mountains. On 6 May he set out on horseback, followed the course of the Fudsi river eastwards, crossed the mountains, and descended by way of the Dadso Shui river to the bay of St Olga on the Sea of Japan (1 June). During July and August he made some excursions by sea – to Possiet Bay and elsewhere on the coasts of Korea, and then back to Peter the Great Bay where a new port called Vladivostock was just being built. In September he sailed from Possiet Bay to Japan;[1] it was 1864 before he got back to St Petersburg. Five years later he was appointed Chief Botanist and Director of the Botanical Museum.

Maximowicz introduced more than 200 new or little-known species to the St Petersburg gardens, some of them of great horticultural importance. He kept in touch with several European botanical institutions, including Kew, and many of his plants eventually reached the gardens of England. His best introductions were those from Japan, but his Manchurian finds included *Celastrus orbiculatus*, *Lonicera maximowiczii* and *Lilium hansonii*.[2]

Meantime, yet another German had been made Director of the St Petersburg Botanic Garden[3] – Dr Edward Regel, who came to Russia in 1855. He was an out-and-out horticulturist, to whom scientific botany was only of very minor interest; and had worked his apprenticeship in a practical way, in the botanic gardens of Goettingen, Bonn, Berlin and Zurich. He had a son, **Albert Regel**, who in 1875 was appointed district physician at Kuldja, on the borders of Russian and Chinese Turkestan. In this remote region he carried out a series of botanical travels, sending the plants he found to his father at St Petersburg. His wanderings during the years 1877–85 took him north-east to Lake Ebi-Nor and all about the Iren-Cha mountains; south-west to Lake Issyk Kul and the Alexandrovski mountains, and beyond to Aulieta, Tchimkent and Tashkent; south to Baldschuan on the R. Vaksh, near which he found the Russian Vine, *Polygonum baldschuanicum*; north-east again to Kaufmann Peak in the Trans-Alai range of the Pamirs, home of another of his introductions, *Tulipa kaufmanniana* – an amazing range of wild mountainous country, explored at a time when the local Kalmucks were said still to maintain the practice of human sacrifices.

On one occasion at least, Regel's travels extended into Chinese territory. In May 1879 he set out for Turfan, by the usual route, which led north-eastwards to Shikho in the Dsungarian desert. Here he was turned back by the Chinese authorities, who refused a pass to proceed further. Regel appeared to comply, and retreated westward; but knowing the Chinese ignorance of the mountain regions, he struck off to the south, and crossed the Iren-cha mountains to the narrow valley of the River Kash. This he followed eastwards to its source, but finding no pass, turned south-east and crossed the mountains to the headwaters of the Kungez river, which

[1] See p. 75 [2] Not introduced till later
[3] A horticultural post, which the botanist Maximowicz never held

runs parallel to the Kash. Southwards again he went, over another pass, to the Yulduz River, flowing in the opposite direction, to Lake Bagratch. From the lake another mountain pass took him into Turfan from the south-east – on 28 September; it had taken four months of travel to circumvent the political obstacle. Nowhere had he encountered any military posts, and it was only at Turfan itself that he again saw Chinese authorites; they were 'much surprised' but raised no objections to his sojourn in the town, which no European had visited since the seventeenth century. He stayed till 10 November, making local excursions, and then returned to Kuldja by the 'ordinary highway', arriving on 17 December.

From his travels in Turkestan, Regel sent his father great quantities of herbarium specimens, and seeds of a large number of species, which were successfully cultivated and distributed. Among his introductions were several eremuri; *Tulipa linifolia* and *T. praestans*, and that king of the bellflowers *Ostrowskia magnifica* (now rare in cultivation), which he found on Mt Favez in 1884.

Meanwhile the Caucasus was being further explored by **Ludwig Franzevich Mlokosewitsch**[1] (1831–1909), whose name is attached to one of the most beautiful of paeonies. He was born in Warsaw of a wealthy and aristocratic family and in conformity with a decree of Tsar Nicholas I, enlisted at the age of twenty-two in a Russian regiment which was stationed at Lagodekhi in the Signakhi district of the Caucasus. Here he laid out and planted a regimental park and orchard, including a fine collection of conifers and other trees, a water-garden and many exotic plants. Wearied by the jealousy, intrigues and frustrations he encountered, he resigned from the army in 1861, and went on a journey, to 'seek oblivion . . . in the deserts of Persia'. He collected seeds, roots and a fine herbarium, but on his return to Russia he was arrested on a trumped-up charge of having fomented a Polish intrigue in the Caucasus, and abetted a Lazgin uprising. His collections were confiscated (and subsequently lost) and he was sentenced to six years' enforced residence in Voronezh province.

On his release in 1876 Mlokosewitsch resumed his exploration of the Caucasus, chiefly in the Daghestan area, and in 1878 made another journey into Persia and Baluchistan. In 1879 he was appointed Inspector of Forests for the Signakhi district, and returned to Lagodekhi, where he remained for the rest of his life, except for frequent excursions to other parts of the range. He experimented with lemon, orange and other economic trees, and it is interesting to note that he tried to establish the cultivation of Chinese hemp or 'Ramie' (*Boehmeria nivea*), just as Roezl had done in Mexico some twenty years before. He sent zoological, entomological and botanical specimens to Russian museums, and eventually built up such a reputation that for a naturalist to go to the Caucasus without visiting Mlokosewitsch, was said to be like going to Rome and failing to see the Pope. His finds included *Paeonia mlokosewitschii* and *Gentiana lagodechiana*.

[1] or Mlokossjewicz

Mlokosewitsch had a large family and, after his retirement, a very small pension. He established a 'teaching garden' at his home, where his children might learn the elements of natural history 'as a guarantee of moral purity', and took them with him on botanical excursions when they were still so young as to require to be carried. They grew up to share his tastes, and it was his daughter Julia who discovered *Primula juliae*. He died in the course of one of his usual distant trips to the mountains of Daghestan – at the age of seventy-eight. His plants reached cultivation by various routes – the primula through the botanic garden of Dorpat, the paeony through that of Tiflis.

Japan

1690	Kaempfer	1861	Oldham
1775	Thunberg	1877	Maries
1823	Siebold	1892	Veitch, J. H.
1860	Veitch, J. G.	1892	Sargent
1860	Fortune	1914	Wilson
1860	Maximowicz	1926	Ingram

'Nature seems to have united in Japan the beautiful with the astonishing . . . so fond are they of flowers, that all their females are known by names taken from the most beautiful of them. . . .' SAMUEL CURTIS *Monograph on the Genus Camellia* 1819

' . . . Japan, which has not unaptly been described as the England of the East, so liberal and progressive are its people.' *The Garden* 1883

Japan

The world's richest sources of floral beauty, China and Japan, were kept by a beneficent providence to the last. Japan was closed to Europeans for more than two hundred years, because of the sixteenth-century encroachments of Spanish and Portuguese missionaries. Received at first with great tolerance, they repaid with intolerance and cruelty, till the rapid spread of the new religion alarmed priests and politicians alike, and an edict of banishment was put into force in 1614. It was not at once effective, but when it came to light in 1639 that the Christians were actively supporting a local revolutionary movement, the Emperor not unnaturally decided to be rid of them altogether. Those that did not flee were massacred; the ports were closed to foreigners, and no Japanese was allowed to leave the kingdom, or having left it, to return. Even the building of seagoing ships was prohibited. Only the Chinese and the Dutch, who had sent no missionaries, were allowed certain trading concessions, and these only at the port of Nagasaki. Here the Dutch were allotted a tiny artificial island called Deshima, connected to the mainland by a bridge, where they might build their warehouses and maintain a few officials, under the strict supervision of the Japanese: and they were allowed to send in three – later, only two – vessels a year. Japanese seas were stormy, and the Dutch reckoned to lose one ship out of every five; the trade must have been enormously profitable to induce them to face such hazards, and the restrictions and humiliations they were forced to undergo. Once a year they were expected to send an embassy – with gifts – to the court at Tokio, and it was in the course of these annual expeditions that Europeans had their only opportunities of seeing the interior of the country.

It must be admitted that the Dutch and Chinese merchants did as much as the missionaries to abuse their scanty privileges, and spared no efforts to introduce forbidden imports and obtain forbidden exports, and to suborn the native Japanese, who took only too readily to smuggling, in spite of the very heavy penalties imposed. Eighteen of them lost their lives through the activities of Andreas Cleyer – soldier, professor, physician and botanist – who was on Deshima as Governor for the Dutch East India Company in 1682–3 and again in 1685, and who organized a lively smuggling trade, for which he was eventually expelled.

All the notable physician-botanists employed by the Dutch Company in Japan were of foreign origin – Cleyer, Kaempfer and Siebold were Germans, and Thunberg a Swede. It is interesting to compare the experiences of **Engelbert Kaempfer** (1651–1715) and **Carl Peter Thunberg** (1775–1828) who held the same post[1] eighty-five years apart. In both cases the sojourn in Japan was the climax of a long series of travels. Kaempfer was more than seven years on the way (four being spent in Persia) and Thunberg tarried for four years in South Africa, reaching his ultimate objective five years after he started. They agreed in their estimate of the Japanese as 'a very reasonable and sensible People', naturally polite and curious, and hungry for European knowledge. Each ingratiated himself by medical practice and tuition, to which Kaempfer added instruction in astronomy and mathematics, and a 'cordial and plentiful supply of European liquors'.

The Dutch ships seldom stayed more than three months in harbour; they came with the south-west monsoon in August, and returned with the north-east monsoon in early November. A governor and his staff were left on Deshima, until relieved by a new governor the following year – not more than twenty unarmed Europeans, who were most sedulously guarded by hosts of Japanese, all bound not to fraternize with them by complicated oaths sealed in ink and blood. During this time was made the annual pilgrimage to Court, which was compulsory also for the native princes, lords and vassals. Kaempfer twice accompanied such an embassy (1690 and 1691) and Thunberg once (1776).

These journeys varied very little; they took place always at the same season (March to May) and followed the same route, even, except in cases of extreme emergency, lodging at the same inns. The party of three or four Europeans was provided with an escort of 150–200 persons – interpreters, guards, carriers, cooks and servants. Transport was by norimon, a large and comfortable palanquin carried by up to twelve bearers, by smaller carrying-chairs, or by horseback. Kaempfer gives full details of the route, but some of his place-names are hard to identify, especially as he spells the same name differently every time it recurs. The journey was divided into three parts; first, overland to Kokura on the north coast of Kiushiu and across the straits to Shimonoseki, where the heavy luggage and the presents, carefully censored by the Nagasaki authorities, had been sent in advance by sea; then by boat up the Inland Sea to Osaka, putting in at some convenient port every night; and then by land again – a short cut across by Otsu and Yokkaichi to Kuwana, and up the coast road from

[1] as physician to the Dutch Governor

there to Yedo (Tokio). The whole trip, with twenty or thirty days in the capital, took about twelve weeks. Thunberg was longer on the way, due to adverse winds during the sea-passage. He welcomed the delay, as the more advanced season would be better for botanizing; but he does not seem to have been allowed to go on shore.

Indeed, very little botanizing was possible for either visitor. Kaempfer carried before him on his horse 'a very large Javan box', which he filled with 'plants, flowers, and branches of trees, which I figured and described' – and with a compass concealed underneath. The Japanese escort highly approved of this occupation, and 'were extreamely forward to communicate to me, what uncommon plants they met with, together with their true names, characters and uses, which they diligently enquired into among the natives'; but he had little opportunity to gather flowers for himself. Thunberg, in his turn, was disappointed during the first part of the journey to find the Japanese fields so meticulously cultivated that not a weed was to be seen. His chance came in the Fakone mountains (also admired by Kaempfer for their vegetation and scenery) where the steepness of the way occasionally obliged the travellers to go on foot. He was not allowed far from the road, but by outdistancing his escort of interpreters and inferior officers, 'gained time to gather a great many of the most curious and scarcest plants, which had just begun to flower, and which I put into my handkerchief'. On the return journey he was allowed a little sightseeing, and was able to visit a Japanese nursery.

The principal difference between Kaempfer's experience and Thunberg's was that although in Thunberg's time the trading restrictions imposed on the Dutch were even more severe, the social restrictions were very much relaxed. Kaempfer had only been allowed to make excursions in the neighbourhood of Nagasaki once or twice a year, and was then accompanied by a large train of officials who expected to be treated to dinners and refreshments – making the outings expensive. Thunberg also was obliged to take an escort of 'head and sub-interpreters, head and sub-banjoses, purveyors and a number of servants'; but except for a short period when permission was withdrawn because the Japanese Governor could find no precedent for a *surgeon* to be allowed to botanize – only a surgeon's *mate* – he was free to make such excursions as often as he could afford, and in early spring, before the journey to Tokio, was going out about twice a week. He also examined carefully the hay-supply brought twice daily by the Japanese to feed the livestock[1] kept by the Dutch on Deshima, and by these means was able to add more than 300 species to the known Japanese flora.

On the journey to Tokio, Kaempfer was treated like a performing animal – closely guarded by day, locked in at night, and put through his tricks on arrival. He was obliged to dance, sing, jump and mime European manners and customs, for the edification of the Emperor and his hidden ladies, and similar exhausting performances had to be given in the houses of the nobility. Thunberg's reception eighty-five years later was very

[1] They kept their own stocks of calves, oxen, goats, sheep and deer – unprocurable in Japan

different; he was besieged by the local astrologers and physicians, who sought his advice about their patients (whom he was not, however, permitted to see); he was given a Japanese herbal and allowed to purchase several others, and even some prohibited maps. 'They loved me', he asserts, 'from the bottom of their hearts', and several of them afterwards kept up a correspondence and sent seeds and plants to him in Sweden.

Thunberg does not mention the 'ignivomous'[1] mountains in which Kaempfer took such an interest, nor the frequent conflagrations which swept the inflammable towns in which they stayed. (Kaempfer experienced four in Tokio, besides an earthquake shock.) He was extremely impressed, however, by the superiority of the Japanese roads, and their 'most excellent rule, that travellers should always keep on the left-hand side of the way'. This system, he thought, might well be adopted in Europe, where every year numerous people were injured or killed on the roads by being 'rode or driven over by the giddy sons of riot and dissipation'. During his sojourn, Thunberg spared no efforts to obtain Japanese seeds and plants. At the nursery he visited, he spent as much money as he could spare on 'the scarcest shrubs and plants, planted in pots', and he established on Deshima a collection of live shrubs and trees, which he looked after till the departure of the Dutch ships, when they were carefully packed and sent, via Batavia, to the Hortus Medicus at Amsterdam. He mentions in particular *Thujopsis dolabrata*, ornamental maples, and two plants of *Cycas revoluta* – the exportation of which, incidentally, was strictly prohibited.

Thunberg was almost exclusively a botanist; Kaempfer was a more general observer, but though botany only occupied a part of his attention, he was important because of the great influence of the book he wrote after he got home. Only the fifth section of his *Amoenitates Exoticae* (1712) was devoted to plants, but it included the first descriptions of many shrubs and flowers familiar to every gardener today – aucuba, skimmia, hydrangea, chimonanthus and ginkgo; Liliums *speciosum* and *tigrinum*, two magnolias, various prunus, azaleas and tree-paeonies, and nearly thirty varieties of camellia – all, at that time, quite unknown in Europe. Thunberg's much fuller *Flora Japonica* (1784) did not have anything like the same impact. The *Amoenitates* was intended by Kaempfer as a preliminary to a larger work; but by this time he was married, physician to the Count de Lippe in his native Westphalia, and involved in a busy practice, and he never had time to put his papers in order. After his death in 1716 they were purchased by Sir Hans Sloane, who arranged for the translation and publication of his *History of Japan* (1728).

Very different in character from the mild and scholarly Kaempfer and the naïf Thunberg, was Dr **Philipp Franz von Siebold**, (1791–1866), who held the same office of physician to the Governor fifty years later; and the conditions under which he worked had greatly changed. The Napoleonic wars had come and gone; Holland had temporarily lost her

[1] 'Fire-belching' – *ie* volcanic

possessions in the East, and for a few years tiny Deshima was the only place that still flew the Netherlands flag. The Dutch, anxious to restore their almost vanished trade, planned a special embassy to Tokio to ask for better conditions. For this, a skilled physician was an important asset, and their choice fell on von Siebold, a German doctor who had been employed since 1822 by the Dutch East India Company in Batavia. Skilled in several sciences, he was particularly well-qualified for the post, as he had graduated as an eye-specialist, and could perform the operation for cataract; and his ability to make the blind to see gave him tremendous prestige among the ophthalmic Japanese. Siebold was arrogant, power-loving and unscrupulous; his main interests were politics and ethnology, but plant-collecting (for profit) and natural history ran them very close. Medicine seems to have been only a means to an end, for he never resumed practice after his return to Europe.

He landed on Deshima on 11 August 1826, and found all the old restrictions still in force. He soon built up a following of pupils and grateful patients, and with these to escort him, was able gradually to extend his excursions farther and farther afield (on the pretext of visiting the sick) and to acquire a good deal of forbidden knowledge about Japanese politics and economics. The embassy to Tokio probably took place, according to precedent, in the spring of 1827; Siebold may have stayed on in the capital, or more probably, accompanied the next embassy the following year – accounts of his movements at this time are contradictory and vague. It was in the autumn of 1828[1] that the events took place that led to his expulsion from Japan.

In Tokio he had made contact with the Court Astronomer, through whom he had obtained maps not only of Japan itself, but of adjacent regions such as the Amur province, the island of Saghalien, and the Liu-Kiu islands between Japan and Formosa. Maps were the most strictly prohibited of all Japan's exports, and their possession ranked as treason. The story goes that Siebold's collections were already packed on the vessel that was to take them from Tokio to Deshima, but she was beached in a storm, and before she could be refloated, the transaction was discovered. The astronomer and several of Siebold's pupils and friends were imprisoned; some were tortured, or committed hara-kari. Siebold fought hard to retain his prizes, but was obliged to give up the maps – having first made hasty copies by night, which he hid among his zoological collections. He was imprisoned from 18 December 1828 to 28 December 1829; the fact that, although a foreigner, he was not a Hollander made him doubly suspect, and the Dutch, for whom all this came at a very inopportune moment, dared give him no support. He was released at last, but sentenced to permanent banishment from Japan.

Nevertheless Siebold contrived to send a consignment of plants in January 1829, and when he sailed from Nagasaki on 2 January 1830, took home a further collection of 485 plants, which had presumably been established in the Deshima garden before all the trouble began. After a

[1] according to the account in *Allgemeine Deutsche Biographie*

short stay in Batavia he was given permission to return to Holland – where further trouble awaited him.

Holland and Belgium had been united at the peace of 1815, but Belgium wished to regain her independence, and war broke out between them in the summer of 1830. It is said that Siebold arrived during the siege of Antwerp; that the place where his collections were housed was used to stable cavalry horses, and many of his plants trampled and destroyed; and that to save the remainder he presented them to the Botanic Garden at Ghent. But this story, told nineteen years later,[1] does not ring true, for the siege of Antwerp did not begin till 23 December, and Siebold, leaving Japan in January, must surely have reached Europe long before then. It seems likely that the other version is correct – that he arrived at Antwerp in July, a part of his collections being assigned to a patron, the Duke of Ursal, who had an estate near Brussels; and having quarrelled with the Antwerp authorities, removed to Ghent, and was there when war broke out in August. His collections were confiscated, and were eagerly seized by the local horticulturists, to whom the war gave an ideal pretext. Siebold had to fly to Holland, but on the cessation of hostilities claimed the return of his property; and some attempt was made to give him back at least one plant of every different kind he had lost. He was only able, however, to reclaim about eighty out of some 260 species, and with these set about the establishment of a 'Jardin d'Acclimatation' at Leyden.

Siebold received many honours from William II of Holland, including an official post in the Dutch East India Company, and a patent of nobility, with the title of 'Jongheer'. Among other favours was a Royal Decree in 1842, which entitled him to form a 'Royal Society for the Encouragement of Horticulture'. The Government was to pay and send out collectors; the ships of the Dutch East India Company were to carry the plants at reduced rates; the Botanic Gardens in Java were to refrain from distributing any plants in which Dr Siebold took an interest; 200 or more subscribers paid five florins a year; and all the plants went to Siebold and Co. at Leyden, from whom alone they could be obtained. The subscribers eventually rebelled at this monopoly, and connection between the Society and the nursery was severed in 1847, though Siebold retained his office as President. Meanwhile, he was producing a flow of publications on the geography, language, economics, politics, ethnography, bibliography and natural history of Japan – including (in collaboration with J. G. Zuccarini) a two-volume *Flora Japonica* (1835–42) magnificently illustrated with coloured plates after drawings by anonymous Japanese artists.

The Leyden nursery continued to flourish until after Siebold's death, and the flow of Japanese plants thereby introduced to Europe – bamboos, azaleas, lilies, camellias and hydrangeas – was his first great service to horticulture. The second was the part he played in the reopening of Japan – though this was perhaps less influential than he would like us to believe. He claims that at his instigation William II wrote an 'epoch-making' letter to the Shogun (about 1853), advising the opening of the country to

[1] in *The Cottage Gardener*, VOL II (1849) p. 59

European trade, and that it was to Holland and Russia rather than to America that credit should be given for the eventual result. It was more probably the American warships, however, that were the deciding factor. In 1853, Commodore Matthew Perry in the flagship *Vincennes*, sailed into Tokio Bay, and with hardly an impolite word, put on such an imposing show of force, that the Japanese Government, already alarmed by some previous manifestations of Western power, realized that the policy of isolation could no longer be maintained. In March of the following year a treaty was signed, providing for the opening of two ports and the establishment of an American consul at Shimoda. The first consul, Townsend Harris, was a man of tact and diplomacy, who procured the opening of six more ports between 1858 and 1863. Consulates of other nations were soon established, and although travel inland was still restricted and anti-foreign demonstrations occasionally broke out, the emancipation of Japan had begun.

Siebold had always wanted to return, to complete his unfinished ethnographical studies; but although his sentence of banishment was lifted when the treaties were signed, the Dutch government felt it would not be fitting to send him in any official capacity. Eventually he obtained a post as Advisory Councillor to the Netherlands Trading Company, and in April 1859 he sailed again for Japan, at the age of sixty-three. He arrived at Nagasaki in August, and settled in a house on the outskirts, remote from the rest of the European community, with whom he appears to have been unpopular, though he was welcomed by many old friends among the native Japanese. In 1860 he was invited to Tokio to act as Confidential Adviser to the Japanese Privy Councillor. He was to smooth foreign relations, introduce European sciences, and also advise on internal affairs – and Siebold could never resist meddling with politics. Exactly what he did is not quite clear, but his behaviour so alarmed the Dutch authorities (who may also have been concerned for his safety) that they ordered his recall; and when he refused to comply, requested the Japanese Government to dismiss him. Eventually he was beguiled to Batavia, to discuss a possible diplomatic appointment; only to be told on arrival that such posts were no longer at the disposal of the Governor of Java, since the control of Japanese affairs had passed from the Dutch East India Company (where Siebold had influence) to the Foreign Office at the Hague (where he had none). Disgusted, he returned to Europe in 1862, severed his connections with Holland and retired to his native Bavaria, to work at his books and collections. Even at this crisis of his affairs, he brought home plants: *Hydrangea paniculata*, *Malus floribunda*, *Spiraea thunbergii*, and *Prunus sieboldii* seem to have been among these later introductions. He still hoped to return to Japan, and was planning another voyage at the time of his death in 1866.

No sooner had the ports opened than eager botanists came flocking in: Veitch, Fortune, Wichura[1] and Maximowicz all visited Japan in 1860 and Oldham in 1861. The first to arrive was **John Gould Veitch** (1839–

[1] The German botanist Max Ernst Wichura stayed only for four winter months

1870) great-grandson of the founder of the famous firm of Veitch of Exeter and son of the James Veitch who had recently started the Chelsea branch. When he heard Japan was to be opened, young Veitch 'eagerly sought the means of proceeding thither',[1] and pulled every string in his power to procure recommendations to the English representatives on the spot. He sailed on the *Malabar* in April, was shipwrecked at Galle in Ceylon, losing everything on board he possessed; immediately took passage in another vessel and arrived at Nagasaki via Hong Kong, Canton and Shanghai, on 20 July.

At Nagasaki Veitch was lodged, along with three other Englishmen, in a temple surrounded by a garden, in which he was given a place to put his plants. His excursions were confined to a ten-mile radius, but within this limit he was able to move freely, accompanied only by an interpreter who carried his boxes and basket. The Japanese officials were tiresome but the people were friendly, and willingly gave him any plant he admired in their gardens, and the priests in the temple frequently brought him plants, which he accepted gratefully, though many were quietly discarded afterwards. It was between seasons, and the nearest nursery, fifteen miles away, was beyond his reach; nevertheless, within a fortnight he had assembled between forty and fifty plants in pots in his temple garden – he could almost fancy himself at Chelsea when he was watering and tending them – and a box of seventy kinds of seeds. He was learning Japanese, and remarks incidentally that Japanese ladies were friendly and not shy, but that they did not like European whiskers. One can hardly blame them; Veitch's portrait shows his attractive sensitive face framed in long, black 'weepers'.

At the end of August, he left his Nagasaki plants in the charge of one of his compatriots, and took a passage in the man-of-war steamer *Berenice* up the Inland Sea to Yokohama. At that time, the only Englishmen permitted to reside in Tokio were the British consul, Mr Rutherford Alcock, and his staff, but Veitch hoped by means of his letters of introduction to get a footing in the household. Two days after he disembarked at Yokohama, Alcock arrived there en route for an excursion to Fujiyama; the introduction was made, and Veitch was invited to join the party. In order to qualify him for inclusion, the Consul appointed him 'Botanist to Her Britannic Majesty's Legation at Yedo' – whereupon, Veitch reports, 'I at once grew six inches taller'.

This was the first time that Europeans had been allowed to visit the sacred mountain, and respectful, orderly crowds assembled to see them in the villages along the way. The trip, by the ordinary pilgrim route, took about a fortnight, their last stop being a rest-house at 8500 ft, where eight English, their servants, and the Japanese host, hostess and family slept in a room 25 × 11 ft without window or chimney and infested by hosts of fleas. In the temple on the summit the party drank the healths of the Queen and the Consul in champagne; fired a salute of twenty-one guns (revolvers) hoisted the British flag and sang 'God Save the Queen'.

[1] Lindley in *The Gardener's Chronicle*, 15 December 1860

(One wonders at the tolerance of their Japanese escort.) Throughout the journey there was a great deal of rain, and the Europeans were not allowed to leave the main road; but Veitch managed to collect seeds and specimens of some twenty-five species of conifer, on which he afterwards wrote a descriptive paper.

On the way home (about 20 September) the Consul stopped at Atamo to sample the famous sulphur baths; and he invited Veitch to stay with him after his return to Tokio in mid-October. In the interval our collector returned to Yokohama, and having sent out four men to obtain seeds and plants, embarked on a quick trip to Hakodate, the most northerly port yet open. Opportunities for transport were few, and in order not to be indefinitely delayed, he was obliged to go and return by the same steamer, which afforded him only three or four days ashore.

The rest of his stay was spent with Alcock at the Tokio consulate, but here his movements were very restricted, as no-one was allowed to leave the gates without a large escort of Japanese officials; he found it tantalizing to ride past trees and shrubs loaded with seeds and be unable to stop and gather them. As before, he sent out native collectors; communicating with them first by words, then by signs, and when these failed, by drawings; and Alcock organized a special tour of the local nurseries, where he saw chrysanthemums which 'would not disgrace even a London exhibition'. Before he left the Consulate, Robert Fortune had arrived there. Middle-aged and with three celebrated journeys in China behind him, Fortune must have seemed a worshipful figure to the twenty-one-year-old Veitch; but he remarks only 'Mr. Fortune arrived here on the 12th inst. He is quite well.'

Veitch was then busy packing up for his departure; he left about the end of November, called at Yokohama and at Nagasaki for the plants deposited there, and sailed for Hong Kong, where he shipped his collections to Europe but went on himself to the Philippines in quest of certain orchids. His Japanese introductions included *Lilium auratum*, *Magnolia stellata*, *M. soulangiana nigra*, *Primula amoena*, *P. cortusoides* and *P. japonica*, the vine called after him *Ampelopsis veitchii* (now *Parthenocissus tricuspidata*) and seventeen new conifers – among them the commercially-valuable Japanese larch, the beautiful *Cryptomeria japonica* var. *elegans*, and *Juniperus rigida*.

This gentle, humorous, friendly yet determined young man made a second voyage, to Australia and the South Sea Islands in 1864–6,[1] bringing back with him a valuable collection of the stove-plants then so much in vogue. Soon afterwards, he developed tuberculosis, and died in 1870 at the age of thirty-one, leaving a widow and two sons, the elder only two years old.

When Siebold had landed on Deshima in 1826, he thought himself back in the seventeenth century; all the old restrictions were in force, and on ceremonial occasions the Dutch still wore embroidered velvet and swords. By the time **Robert Fortune** arrived on 12 October 1860, all

[1] See p. 229

had changed; flags flew from half a dozen European consulates, and Deshima was almost derelict, its guard-house empty and its walls pulled down. It is surprising, therefore, to realize that Fortune – almost a modern collector, accustomed to Wardian cases and travel by steamboat – was able to visit Siebold in his house on the outskirts of Nagasaki; ancient had changed to modern in so short a time. Although Siebold had only been established for fourteen months, his house was already surrounded by 'small nurseries for the reception and propagation of new plants, and for preparing them for transportation to Europe',[1] and the veteran was clearing brushwood from an adjacent hillside to provide suitable sites for other species. He was locally popular, and boasted that he did not have to carry a revolver in his belt, 'like the good people in Deshima and Nagasaki'.

Fortune stayed only a week at Nagasaki, and then went on in the same vessel that had brought him from China, to Yokohama and Kanagaura. With a native guide, he visited a number of temples and monasteries in the vicinity – it was only in the gardens of such places that fine specimens of the indigenous trees were preserved, and good seed was to be had. In return, he presented the priests with pictures from *Punch* and *The Illustrated London News*, 'with which they were highly pleased'.

On 13 November,[2] at Alcock's invitation, Fortune removed to the British Legation at Tokio, where Veitch was already installed. The grounds were surrounded by a high fence, and a contingent of Japanese troops was encamped at the gate, ready to provide an escort of 'yakoneens' for anyone who left the Legation, either on foot or on horseback. Fortune did not find his bodyguard too great an encumbrance; they were amiable, often useful as guides, and when they attempted (chiefly through boredom) to prevent him from going further, they could be circumvented by firmness and tact. He soon found his way to a very large district of nursery-gardens which lay on the opposite side of the city; and visited one after another, always accompanied by a curious crowd that waited patiently behind the shut gates of every nursery he patronized, until he came out again and went on to the next. One area was particularly noted for its chrysanthemums, and here Fortune found some particularly fine varieties, distinct from the Chinese types, which he thought would revolutionize chrysanthemum growing in England. Taught by experience, he insisted on the plants of his choice being dug up and the 'suckers' detached under his own eye and given to him in person, and not, as in the case of other plants, delivered at the legation next day. Towards the end of November he packed his collections and sent them by sea to Kanagawa, following by land on 28 November, then transported them across the bay to Yokohama, where the plants were put in Dr Hall's garden until the Wardian cases he had ordered were ready for their reception.

Fortune had as little to say about young John Veitch, as Veitch had about him; perhaps he regarded him as a rival, for his own plants on this journey were consigned not to the Horticultural Society, but to the firm of Standish and Noble of Bagshot. But when he came to put his collec-

[1] Fortune *Yedo and Peking* 1863
[2] Veitch says the 12th

tions on board the steamer *England* for Shanghai, he found Veitch's plants (and presumably their owner) already installed on the same vessel, so that between them 'the whole of the poop was lined with glass cases crammed full of the natural products of Japan'. The *England* was carrying presents from the Shogun to Queen Victoria (in return for handsome presents bestowed) and for reasons of prestige and safety was allowed to take the less stormy route by the Inland Sea; but her Japanese pilots twice ran her aground, apparently thinking that the wonderful ship that moved without sails or oars, could pass as easily over land as water. (The scenery reminded the Scottish Fortune of Loch Lomond and the Kyles of Bute.) They stopped for three days at Nagasaki, where he obtained still more plants; his Wardian cases being full, an unusually complaisant captain allowed him to stow his new acquisitions in the starboard lifeboat. Fortune at first regretted that he had not put them in the more sheltered port lifeboat, but this was later swept overboard in a storm. Perhaps fortunately, neither boat was required for its original purpose; all arrived safely at Shanghai on 2 January 1861, and soon afterwards Fortune's collections were on their way to England.

Three months later he returned to Japan, in order to see the spring flowers, and also to collect other natural history objects such as insects and shells; and with only a two-day stop at Nagasaki, reached Yokohama in the middle of April. Kanagawa, on the other side of the bay and on the main road to Tokio, was the actual treaty-port; but the anchorage there was shallow, and although the consulates remained there, the merchants found it more convenient to remove to the better harbour of Yokohama. Fortune was therefore able to take over a whole temple in Kanagawa, recently vacated by the officials of a trading company, which gave him ample room for himself, his Chinese assistant and his collections; he moved in on 7 May.

He was anxious to revisit Tokio before the season became too advanced, but unfortunately the British Consul, Rutherford Alcock, was absent on a visit to China. Fortune did not wish to embarrass his deputy by asking for a permit that he might not be empowered to grant, so he wrote instead to the American Consul, Townsend Harris, and received from him a very kind invitation to stay at the American legation. He arrived there on 20 May – and thereby mortally offended the officer left in charge of the British legation, who insisted that only he (in Alcock's absence) could authorize a British subject to reside in Tokio. No explanation or apology would placate him; he accused Fortune of discourtesy and of acting 'in an improper manner', by coming to Tokio 'without even my knowledge', and ordered him to leave at once. And after two days, he had to go, though even in that short time he had managed to get together a collection from the local nurseries 'of great interest and mostly new to science'. This was the only time in all his travels that he met with obstruction from petty officials of his own nationality. He returned to Kanagawa, where for a time the rainy season limited his excursions; when the weather improved at the beginning of July, he went with three other Englishmen to visit the famous temples of Kanasawa and Kamakura.

Compared with his adventures elsewhere, Fortune's exploits in Japan sound remarkably tame. Nearly all the plants of which he was so proud were purchased from nurserymen; he exults in a new 'find', *Primula japonica* – but an enterprising Japanese brought it in a basket to his door. On the way to Kamakura 'a very beautiful new lily [*Lilium auratum*] was met with on the hillsides in full bloom, and its roots were dug up and added to my collections', and this is almost the only reference he makes to anything collected in the wild. But the dangers by which he was surrounded were very real. The Daimyos, or ancient feudal princes of Japan, with their swarms of two-sworded retainers, had a fanatical hatred of foreigners. Their higher ranks owed allegiance only to the spiritual ruler, the Mikado at Miaco; and the temporal ruler, the Shogun or Tycoon at Tokio, had little power over them. (Japan had had this dual-control system ever since the twelfth century.) It was the practical Tycoon[1] who had made the treaties with the foreign powers, and these had not yet been endorsed by that sacred and mysterious recluse, the Mikado. The Daimyos therefore felt perfectly at liberty to murder any foreigner they met, and regarded it as a patriotic duty to do so. Tokio was full of these princes and their retinues, and was therefore particularly dangerous to Europeans;[2] the military escorts forced on them by the Japanese government were necessary for their protection. On 6 July, a few weeks after Fortune's visit, the British legation suffered a night attack in which several of the inhabitants were wounded, before their guards, who had been taken by surprise, rallied to their defence. Alone at night in his temple at Kanagawa, Fortune felt considerable apprehensions, and he was relieved when the packing of his collections was completed and he was able to rejoin the larger European community at Yokohama. On 29 July he sailed for China.

On arrival at Shanghai, Fortune again put his plants in the garden of a friend, while he went north on a rather unproductive visit to Pekin, recently opened to Europeans by the Treaty of Tientsin. By 20 October he was back at Shanghai, packing his collections for shipment via the Cape of Good Hope. Some special favourites, however, including 'a charming little saxifrage, having its green leaves beautifully mottled and tinted with various colours of white, pink and rose' – (*Saxifraga sarmentosa* var. *tricolor?*) – he transported personally by way of Suez,[3] cherishing them during the voyage with solicitous care and taking them ashore for an airing at the principal ports of call. 'More than one of my fellow-passengers,' he remarks, probably with truth, '. . . will remember my movements with those two little hand greenhouses.' He reached Southampton on 2 January 1862.

Fortune himself was delighted with his Japanese discoveries. 'Never at any one time,' he wrote, 'had I met with so many really fine plants' – and all likely to be hardy in Britain. His first consignment arrived in such good condition that Standish was able to exhibit some of the plants at the

[1] Tai-Kun, Great Prince

[2] Nagasaki, less aristocratic and more accustomed to foreigners, was comparatively safe

[3] This was before the opening – in 1869 – of the Suez Canal

Horticultural Society only three days later, looking 'as if they had been luxuriating in the pure air of Bagshot all their lives'. Many were conifers, or variegated forms of plants already in cultivation. Besides the species already mentioned his novelties included *Arundinaria fortunei*, *Deutzia scabra* (*crenata*) *fl. pl. Euonymus radicans* (= *fortunei*) *variegatus*, *Lonicera japonica aureo-reticulata* and *Saxifraga fortunei*. He considered his greatest prize to be the male form of *Aucuba japonica*, hitherto known in England in its female form only, and therefore incapable of bearing fruit. 'Only fancy all the Aucubas which decorate the windows and squares of our smoky towns covered during winter and spring with a profusion of red berries!' he wrote; 'Such a result . . . would be worth a journey all the way from England to Japan.'

The introductions of Veitch and Fortune were valuable to British horticulture, but far greater both in magnitude and scientific importance were the collections sent to St Petersburg about the same time, by **Carl Maximowicz.** (This was understandable, for he stayed in Japan almost as many years as the others did months.)[1] After a long spell in the Amur region and Korea,[2] Maximowicz arrived on 18 September at Hakodate on the northern island of Yezo, where no botanist had been before.[3] He was allowed to travel within twenty miles of the port, and stayed there for fourteen months, sending home in due course 800 plant specimens, 250 sorts of seeds and many bulbs. Among his finds was *Ligularia clivorum*; but many of his discoveries were not introduced until Sargent visited the same area in 1892. He left on 28 November 1861, and after a short stay at Yokohama, returned to Nagasaki for the winter.

The next year, from April to December, Maximowicz spent in Yokohama and its neighbourhood, where he seems to have been able to move about fairly freely, making a number of local excursions besides visiting Fujiyama and the Hakone mountains. In the autumn he sent a native collector to the mountains – probably Tchonoski, whom he trained to collect for him, and who continued to send him plants till he died in 1887. On 21 December he returned to Nagasaki, which he made his base for the remainder of his stay; he sent his collector the following spring to the northern parts of Kiushiu[4] and other places, besides himself visiting such regions as were accessible to him. Bretschneider gives only the briefest account of his movements, and no personal details except that 'he had formed a great collection of bulbs of Japanese liliaceous plants. Unfortunately these were eaten by some pigs'.

Maximowicz returned to Yokohama early in January 1864, and after a stay of about a month, embarked for Europe with seventy-two chests of specimens, 300 sorts of seeds and about 400 living plants. He travelled via the Cape – and England; and was then reported as 'suffering from the remnants of a fever taken in Japan, from which he was never afterwards entirely free'. He reached St Petersburg on 10 July. Of the plants he

[1] Three years and a half. Veitch stayed four months and Fortune three on each of his two visits
[2] See p. 58 [3] Veitch's three-day visit hardly counts
[4] The south part, under the Prince of Satsuma, was prohibited even to the Japanese

brought home some, such as *Rosa rugosa* and the common Japanese privet, were already well-known in Britain; others, like *Enkianthus campanulatus, Rodgersia podophylla* and the ligularia, did not reach us till much later, through Maries or Wilson. The list of his species is too long to quote – oaks, maples, conifers, hydrangeas and lilies, *Rubus phoenicolasius, Spiraea bullata* and *Elaeagnus pungens variegata*; many were afterwards distributed to the gardens of Europe. For the rest of his life Maximowicz specialized in the study of the flora of eastern Asia; he acquired the herbaria of a number of other collectors, and became so great an authority, that it was said that the best place in which to study the flora of Japan was St Petersburg.

When Veitch set out for Japan, Dr Lindley wrote in *The Gardener's Chronicle*: 'Thus we shall again see the value of private enterprise in English hands, and how much more efficient it proves than missions entrusted to mere Government agents.' The despised Government agent followed in less than a year in the shape of **Richard Oldham** (1838–1864), the last collector to be officially sent out from Kew. He was one of those to whom fate seems to have been consistently unkind. His best introductions were anticipated by collectors a few months earlier in the field; his best botanical discoveries were overlooked and accredited to later travellers. He had the misfortune at the outset to incur the displeasure of the Director of Kew, Sir W. J. Hooker, who seems to have treated him with quite unmerited severity. Inexperienced, underpaid, and, as it turned out, not constitutionally robust, he was pitchforked into the Orient at the age of twenty-three, and died there three years later.

Oldham came to Kew from Macclesfield in 1859, and in the following year was promoted to the Succulent House,[1] with a salary of 14s a week. Not surprisingly, he gave notice in the spring of 1861 that he wished to leave; this seems to have brought about some sort of crisis, for on 1 April (an inauspicious day!) he was appointed Botanical Collector for a period of three years, at a salary of £100 a year. This seems to have been the standard; but what was enough for Masson in 1772 had already proved inadequate for Kerr in 1804,[2] and was not likely to be sufficient to cover all expenses of food, lodging and clothing in 1861 – especially, as Oldham had later to point out, in a country where extremes of climate made a fairly large wardrobe necessary. He was given an expense allowance of £80 a year (later increased to £150) a very meagre amount of equipment, some letters of introduction and a few books, and was launched.

The dates of his departure from England and his arrival in Japan have not been recorded, but in a later letter he refers to having been from 12 August to 17 November 1861, on H.M. Survey Ship *Acteon*, then employed, with her tender the *Dove*, in 'surveying several places in the Gulf of Yedo'. Besides the approaches to Tokio, the *Acteon* charted Simoda Harbour and Ajiro Bay, Tomo roadstead and the Inland Sea, and Oldham would have good opportunities of exploring at least the

[1] Appointment to this department was believed to indicate Hooker's displeasure, and was regarded as a penance

[2] See p. 100

coastal areas. By the spring of 1862 he seems to have been established at Nagasaki, and it was from here on 2 October that his first surviving letter was despatched. It was in answer to one from John Smith, the head-gardener at Kew, acknowledging the receipt of his first collections. Smith seems to have written in kind, even fatherly terms, but by the same mail Oldham had received a severe reprimand from Hooker. 'I deeply regret that the tone of Sir William's letter is so harsh and unfavourable towards me,' Oldham wrote, 'and that he makes no allowances for the troubles I have all along had to encounter. I did expect that my letter of 5th May to Sir William, so supplicatory and penitent, would at least have had a more conciliatory effect, and would have been sufficient atonement for the omission of which I had, through inexperience and ignorance, been guilty.' This refers to his failure to advise Hooker by the first mail of 'certain Bills which I drew'. Oldham had obviously no head for money matters, and his accounts caused him continual worry and anxiety. He was also hampered by the inadequacy of the out-of-date books with which he had been provided, and complained that he was 'perfectly unable to find out, with that accuracy which is necessary in the description of vegetables, the names of the plants I am constantly collecting'.

It had been arranged that Oldham was to wait at Nagasaki for the arrival of the *Swallow*, successor to the *Acteon*, which he was to join for her next voyage; but she was delayed, and by November he was heartily tired of the place and finding the waiting tedious. He sent three more lots of seeds – one of eighty species 'the majority of which are trees and handsome shrubs', and one of some seeds purchased at considerable expense from the interior. By a ship leaving in December he sent four boxes of seeds and bulbs, one containing 'what I think will prove a very good collection of lilies' packed in soil; and finally, at the beginning of January 1863, he despatched by the *Excelsior* ten cases of dried plants and some Wardian cases of living ones, which the Captain had promised to look after on the voyage.

In January the *Swallow* at last arrived, and Oldham was installed[1] with a cabin to himself and one for his plants, but room could not be spared for any Wardian cases. She sailed on the 28th for Yokohama and stayed there for nearly three weeks; then returned to Nagasaki and only on 21 March left Japan for Port Hamilton, a small island on the south-east margin of the Korean Archipelago. Except for short visits to Shanghai and Yokohama, the following nine months were spent among these islands; Oldham went ashore on at least nine of them, but seems only once to have entered a port on the Korean mainland.[2] His finds included a particularly beautiful rhododendron, but the dried specimens he sent home were overlooked until after the shrub had been rediscovered in Korea and named *R. schlippenbachii* after its second finder, Baron von Schlippenbach.

On 11 December the *Swallow* entered the South China port of Amoy. Oldham was feeling unwell, with neuralgia and rheumatic pains; he

[1] Unlucky Oldham was thirteenth in the officers' mess
[2] Kuper Harbour, opposite Selby Island

sailed again with the ship on 24 January 1864, but immediately fell so extremely ill that he had to be put back on shore next day. He had difficulty in finding a room, as his illness was suspected to be smallpox; and for more than a week he was confined to bed in an empty house, suffering from fever and dysentery and visited only by the doctor. A kind-hearted member of the European colony then offered him hospitality for his convalescence. During this stay at Amoy, Oldham met Robert Swinhoe, the British consul on Formosa and himself a keen naturalist; and at Swinhoe's invitation he arranged to leave the *Swallow* (now once more in port) and accompany the consul to Formosa, which was almost untrodden ground for botanists. He was now becoming anxious about his future; his three-year contract was almost at an end, and, although before his departure he had been promised employment on his return, he could obtain no confirmation of this, and no post of any sort was offered. He decided to remain for a time in the East, and either to work as an independent collector or to make a fresh start at some altogether different occupation. On 1 April, the day his contract terminated, he wrote from Formosa a formal and rather dignified letter of resignation, expressing his regret, and pointing out that in spite of 'the enormous expense to which the traveller in Japan and China is usually put', he had kept well within the three-year salary and expense allowance of £750, having spent only £603 9s 7½d. (How he must have toiled to arrive at this sum!) This letter crossed another censorious missive from Hooker, which arrived in June, chiding him for staying on after being recalled – a letter he had never received – and for leaving the *Swallow* without permission; although he had in fact obtained the consent of her officers, as stipulated in his agreement.

Oldham continued his exploration of Formosa during the spring and summer of 1864, working from Tam-sui and Ke-lung, and discovering, though not introducing, *Rhododendron oldhamii* and *Deutzia taiwanensis.* In the late autumn he fell ill again; he was taken to Amoy for better medical care, but died there on 13 November, aged twenty-six. In addition to seeds and plants, he had sent to Kew 13,700 excellently-prepared plant-specimens, including about ninety new species, and had introduced *Elaeagnus multiflora*, *Rhodotypos kerrioides*, and *Styrax japonica.* He got on well with his fellow-officers on board the *Swallow*, and evidently won Swinhoe's regard, for the consul afterwards organized a subscription among friends and erected a handsome tomb to his memory. There seems nothing to account for Hooker's animosity – except, perhaps, the prejudice of age; Hooker was seventy-six when Oldham left England, and died a year after him, in 1865. It seems all the more difficult to understand, since a predecessor, Charles Wilford, also from Kew, had spent three years in the East, visiting the coast of China, Formosa, Korea, Manchuria and Japan, and had achieved very much less.

Meanwhile, Japanese plants were flowing to America, as well as to England and Russia. Commodore Perry's expedition in 1853 was accompanied by a botanist, Charles Wright, who had already collected in Texas and Mexico. He made discoveries but no introductions; these were to follow shortly afterwards from enthusiastic amateurs. Chief among them was Dr **George Rogers Hall** (1820–1899), who soon after his graduation from Harvard in 1846 went into practice in Shanghai, but who later exchanged medicine for commerce. His first visit to Japan was made in 1855, and by 1860 he had a well-established garden in Yokohama, full of interesting Japanese plants; it was here that Fortune found his coveted male aucuba. Hall sent a consignment of plants to the USA early in 1861, and took home a larger one on his own return in 1861–2, about eighteen of which were new to American horticulture. They included *Wisteria floribunda, Hydrangea paniculata grandiflora,* the beautiful *Lycoris squamigera* (= *Amaryllis hallii*) and magnolias *kobus* and *stellata,* the latter known for a time as *Magnolia halliana.* Although some names have now been superseded, Hall is still commemorated in *Malus halliana* and *Lonicera japonica* var. *halliana.*

Another exporter of Japanese plants to America was **Thomas Hogg.** His father, a celebrated florist of Paddington, emigrated in 1820 and established a nursery near New York, where he befriended David Douglas on the occasion of his first visit. He died there in 1855, leaving the nursery in charge of his two sons, Thomas and James. In 1862, Thomas was appointed a US Marshal, and sent by Lincoln to Japan;[1] he was there till 1870 and again from 1873–5, and collected and sent many Japanese plants to the nursery of his brother. He made a special study of lilies and contributed information to Elwes' book on the genus.

With so many collectors working a restricted area within the same short period of time, it is natural that introductions should overlap; *Magnolia stellata,* for example, was sent almost simultaneously by Maximowicz to Russia, Hall to America and Oldham and Veitch to England. Four out of six plant-hunters sent home *Lilium auratum* and *Sciadopitys verticillata,* and several other conspicuous plants were introduced about the same time by two collectors or more.

After this botanical invasion in 1860–2, the list of plantsmen in Japan is surprisingly short. Perhaps the activities of native nurserymen and collectors made further incursions of foreigners unnecessary. The Japanese were quick to realize the commercial value of their plants. Oldham reported from Yokohama, early in 1863, that since his first visit there eighteen months before, 'several of the natives have commenced dealing in living plants for Wardian cases, as they now find it pays very well to procure new and rare plants from Yedo [Tokio] and to bring them down to Yokohama for sale'; and by the 1870s a flourishing export-trade in lily-bulbs had been built up. Perhaps, too, attention was distracted from Japan by the floral riches that were being discovered in China. Fortune and Wilson were primarily China collectors, whose visits to Japan were only, as it were, an afterthought; Maries was the only one to reverse the order, visiting China but accomplishing his most important work in Japan.

Charles Maries (1851–1902) was born at Stratford-on-Avon and

[1] Perhaps to succeed Townsend Harris, who resigned in 1861

educated at Hampton Lucy. He served his gardening apprenticeship with a nurseryman-brother at Lytham, Lancashire, and then obtained employment with the firm of Veitch, where he rose to be foreman. According to his own account, 'I was asked one day by Messrs. Veitch if I would like to collect plants for them in Japan, and without a second thought I said yes, my greatest ambition being to go abroad'. He left for the East on 1 February 1877, and was away for three years; most of the detailed information we possess refers to the first year. A few months of each year were spent in China, and the rest in Japan.

After touching at Hong Kong, Maries went to Ning-Po, his object being to obtain a 'lilac' said to grow in a garden in that town. He found the garden, but not the plant; so he embarked on an expedition about sixty miles up the Ning-po river to the Snowy Mountains. At his first anchorage he found the natives 'rather troublesome', so he moved on to Ning-cum-Jow and searched the vicinity, finding many good plants but no novelties and no 'lilac'. He returned therefore to Ning-Po, and thence to Shanghai, to make arrangements for his voyage to Japan. As he came out of the steamer ticket-office, he saw a native with a bunch of lilac-like flowers in his hand. To Maries' 'inexpressible delight' it proved to be the very flower he had been searching for – not a lilac, but *Daphne genkwa*. The man took him to his small nursery-garden, where the shrub grew in a profusion of bloom, and Maries returned to his hotel 'perfectly happy now I had found the very plant I almost came from England to discover'.

He sailed on 18 April and two days later was in Japan – but a Japan so changed as to be unrecognizable. An industrial exhibition was being held in the former palace of the Mikado at Kioto, which twenty years earlier it would have been death to enter; the Daimyos' gardens in Tokio were abandoned and derelict, and Maries seems neither to have expected nor encountered any difficulty in traversing (on foot or on horseback) half the length and breadth of the island of Nippon.

His ship stopped only for a few hours at Nagasaki, and then went on to Shimonoseki and up the Inland Sea to Kobi. Here he stayed for several days, and made his first trip into the mountains, finding almost at once a new white-flowered azalea and *Aronia asiatica*. He made excursions to Kioto and elsewhere, before taking ship for Yokohama and thence to Tokio. Here he visited the gardens of the Shogun, and two former Daimyos' gardens, in one of which he found a fine collection of aquatic irises, many of which he was able to obtain, and also the 'curious and ornamental' square bamboo (*Chimonobambusa quadrangularis*), a native of China, but long cultivated in Japan. Gardening, Maries thought, was 'fast dying out in Japan, and can only be spoken of as a thing of the past'. He paid only flying visits to the Tokio nurseries, partly because most of their plants were already well-known in Europe, and partly because he was very busy, preparing for the next part of his journey.

Owing to civil war in the south, no steamers were running between Yokohama and Hakodate on Yezo. Maries therefore set out to travel overland, a matter of some 500 miles. He left Tokio in the last week of May, with his luggage carried by horses and bullocks. As far as Nikko the road was good; from there to Sendai it was less good, and beyond Sendai so rough that he had to go mostly on foot. Beyond Morioka he found the country very thinly inhabited. Fourteen days' hard travel brought him to Aomori, the most northerly port in Nippon, where he had to wait some days for a steamer to cross the straits to Hakodate. In the interval he went with a coolie and a guide to explore Mt Hakkoda, and spent a most exhausting day trying to get to a new species of abies, which he could see, loaded with cones, but could not reach owing to the density of the bamboo scrub with which the trees were surrounded. He got home, very late, in a violent thunderstorm, having walked some thirty-four miles; but undeterred, set out again next day, this time on horseback and by a different route. It was wild country, seldom visited even by the Japanese; Maries disturbed two black bears and a number of harmless snakes – but he got his fir (now deservedly named *Abies mariesii*) and also seed of *A. sachaliensis* (= *veitchii*) which Veitch had found but had been unable to introduce. When he returned, he found that there had been a conflagration during his absence, and half of the smiling town he had left in the morning was now a mass of smouldering ruins. (A thousand houses are said to have been destroyed.) Fortunately his baggage had been rescued, and transferred to a house on the outskirts, where an amiable host was delighted to show him his cherished garden. Next morning he took ship for Hakodate.

He made his headquarters at Horidzuma and from there explored the coastal areas of the island of Yezo, northwards to Sapporo, westwards to Tokachi. Sapporo seems to have been a particularly good locality, for it was there that he found *Platycodon grandiflorus* var. *mariesii*, not to mention *Actinidia kolomikta*, *Schizophragma hydrangeoides*, and a number of maples. He also made an extensive collection of insects. Towards the end of the season he loaded his collections on a Japanese vessel which was carrying a cargo of seaweed from Horidzuma to Hakodate. She was not a well-found ship; the damp leaking in caused the seaweed to swell and to burst the timbers apart, and the captain had to run her ashore. The box containing Maries' seeds[1] was rescued and placed on a boat, which immediately capsized and sank. (Fortunately, he was able to replace his losses by fresh collections.) He left Hakodate towards the end of the year in HMS *Modest*, which took him to Niigata on the west coast of Nippon; from there he crossed overland (in snow) to Yokohama, and having packed and despatched his collections, sailed on Christmas Day for Hong Kong.

A few days after his arrival in January 1878 Maries left Hong Kong for Tai-wan-fu on Formosa, but was unable to penetrate into the savage country of the mountainous interior. Nevertheless, during his short visit he managed to obtain seeds of the rhododendron Oldham had found but failed to introduce, and bulbs of *Lilium formosanum*. After an interval in Shanghai and its neighbourhood, he went north-west to Chin-Kiang on the Yang-tse, and visited Kiu-Kiang and the Dragon's Pool in the Lu-Shan mountains. He now found *Daphne genkwa* abundant in the wild

[1] or, according to Bretschneider, his entomological collections

(including a white form) and many other flowering shrubs, among them *Hamamelis mollis*; but he also sustained a severe sunstroke which laid him up in Kiu-Kiang for two months. In the summer he returned to Japan, but of this visit nothing is recorded, except that he was principally engaged in collecting conifer seed in northern Nippon.

In December Maries again went to China, this time to Hankow, and in the spring of 1879 ascended the Yang-tse as far as Ichang, and beyond to the famous gorges. Here something happened – though exactly what, is not clear. According to *Hortus Veitchii*, Maries did not get on with the Chinese, 'with whom he was not sufficiently gentle, and was often threatened and sometimes robbed of his baggage', and on this occasion the natives, resenting his attitude, 'destroyed his collection, and he returned to the coast, reporting the people hostile'. Yet from other sources we hear that he 'had a glorious time' in the Yang-tse Gorges, where he stayed for a week and obtained large quantities of seed of *Primula obconica*. In the summer he was back again in Japan, looking especially for evergreen oaks and bamboos; but again we have no details of his travels. He returned to England in February 1880.

Maries' work as a collector has been decried by several authors, but all the criticisms seem to be derived from Sir Harry Veitch's verdict that he 'had enthusiasm, but lacked staying power', and this in turn seems to be based chiefly on his unsatisfactory visit to Ichang. E. H. Wilson, who succeeded where Maries failed, was among those who criticized Maries' achievements; he thought he had been too easily discouraged by the intensive cultivation and limited flora about Ichang, whereas if he had pressed on for another three days' journey north, south or west he would have discovered undreamed-of botanical riches. But Wilson, twenty years later, had Augustine Henry's experience to guide him; moreover he got on exceptionally well with the Chinese, whom other collectors besides Maries found difficult to handle.[1] Maries' first season's work in Japan shows no lack of endurance and enterprise; it is a pity so little is known of the other two seasons. He sent home about 500 living plants, including many new species of great merit, and quantities of seed, especially of trees. Besides the plants already mentioned he introduced *Enkianthus campanulatus*, *Hydrangea macrophylla* var. *mariesii* and *Viburnum tomentosum mariesii* – a trio of shrubs of which any gardener might be proud. He did no more collecting; in 1882, on Sir Joseph Hooker's recommendation, he received the post of Superintendent of Gardens to the Maharajah of Durbhungah; from there he transferred to the service of the Maharajah of Gwalior, and remained in India till his death in 1902.

Surprisingly little had yet been heard in Europe of the flowering cherries that were so prominent a feature of Japanese horticulture.[2] The first double cherry to reach Europe, *Prunus serrulata*, arrived by way of China in 1822; Siebold introduced the species that bears his name (*P. sieboldii* or 'Takasago', still in cultivation); Fortune sent home one or

[1] Maries had musical talents, which are said to have endeared him to the Japanese

[2] Fortune, Veitch and Maries were not in Japan at cherry-blossom time; but the other collectors must have seen them

two, but of nearly 150 species and varieties cultivated in Japan only a handful had reached European gardens. This omission was soon to be remedied.

In 1891–3, **James Harry Veitch** (1868–1907), elder son of the John Gould Veitch who had died so tragically young, made a world tour, starting in November with Ceylon and India. The first part consisted of a stately progress from one botanic or public garden to another, each of which he mercilessly describes[1] down to the very bandstand; he does not seem to have taken one wild flower in his hand for the first six months of his travels. He stayed with the firm's former collectors – Maries, now eight years at Gwalior, and Curtis, for seven years director of the botanic gardens at Penang. Veitch was only nine or ten years old at the time these comparative veterans left England. After visiting Java and Singapore, he sailed in March 1892 from Hong Kong to Yokohama. In Japan he became much more enterprising and emancipated, and had his first experience of collecting in the wild.

After looking about him in Yokohama and visiting some local gardens, James Veitch went to Tokio, and was somewhat dismayed by the quantity and extent of the local nurseries, more numerous even than those of Flanders or Holland. The cherries were in bloom, and he was greatly impressed by them, especially the famous mile of trees at Mukojima; though he makes the surprising statement that 'the species is known scientifically as *Prunus mume*; it is really an Apricot'. (He promised to send home tinted photographs – the only way to convey an idea of their fugitive beauty.) In April he returned to Yokohama, and towards the end of the month set out for Nagoya, partly by road and partly by rail, but was prevented by atrocious weather from climbing Mt Ozama. From Nagoya he went on to Otsu and Kioto, and then back to Tokio again.

On 21 June, Veitch sailed from Kobe for Chemulpo on the western coast of Korea, and from there made a remarkable journey northwards through Seoul to Wensan on the east coast, then west to the ancient capital of Phyen Yang, closing the triangle (having twice crossed the peninsula) by returning to Seoul by another route. Korea had been even more rigorously closed to foreigners than China or Japan, and in most of the places visited, Veitch was the first European the natives had ever seen. It was an interesting trip, but botanically unproductive – too late for flowers, too early for seeds, and difficulties of transport precluded the formation of large herbarium collections. On 5 August he sailed again for Japan.

Shortly after his return he made an excursion to climb Fuji-yama – only possible in the summer, when the summit was free of snow. He started the ascent from Gotemba, and came down by another route, skirting through the villages of Yashida and Hikoana to the railway at Suzu-kawa. He found the vegetation on one side of the mountain very different from that on the other, though at the lower levels *Platycodon grandiflorus* was everywhere as common as buttercups. His experience

[1] In *A Traveller's Notes*

of a night at a Japanese rest-hut was much the same as that of his father thirty years before; but this time there was no flag-waving and saluting at the summit.

Veitch tells us nothing about the circumstances of his meeting with Professor **Charles Sprague Sargent** (1841–1927) of the Arnold Arboretum, who was then travelling in Japan. All he reveals is that his September journey to Mt Chiokaisan was made on Sargent's advice, and a little later he refers to him as his travelling-companion. They went by train north to Sendai, then by 'jinrickshaw' across the central range to the valley of the river Mogami and the port of Sakota at its mouth. From near-by Fukura, a miserable village, they explored Mt Chiokaisan, and got many seeds, but nothing particularly new. After skirting the coast north to Akito, they again crossed the central range to Kurosawajiri, and from there went by train to Aomori. Both were anxious to see *Abies mariesii* growing in its restricted station on Mt Hakkoda; less tough than Maries, they spent three or four days on the excursion, passing two nights in some straw huts near the summit in which neither the tall Veitch nor the burly Sargent could stand upright, but which afforded some needed shelter from the intense cold. From Aomori they took the nightly steamer across to Hakodate on Yezo.

Although separated only by a narrow strait, the flora of Yezo differs greatly from that of Nippon and Kiushiu, and it still possessed areas of primeval forest, inhabited by bears and aboriginal Ainus. Sargent was particularly interested in trees (he afterwards published a *Forest Flora of Japan*), and on Yezo he found many new species, including *Malus sargentii* (near Mororan) and *Prunus sargentii* (near Sapporo); while Veitch collected seed of *Vitis coignetiae*, previously known but rare in cultivation. From Sapporo, Veitch hastened back to Tokio and thence to Nikko, another region with a very varied flora, to collect such seed as the unusually wet season could supply. He visited Lake Chujenji and Yumoto, and then returned to Tokio until his final departure in January 1893.

Except for seven weeks in Korea, Veitch spent nine months in Japan – the longest stay of his tour; five months had sufficed for India and the East Indies, and he subsequently devoted another five months to Australia. His Japanese introductions included *Rhododendron schlippenbachii* (as a cultivated plant) and *Physalis franchetii*; but perhaps the most important result of his visit came afterwards. In Tokio he heard of a nursery that specialized in cherries, which he was unable to visit, and after his return he asked this nursery to send him a representative collection. Of those that arrived, many were worthless; but one, afterwards named *Prunus pseudocerasus* var. 'James H. Veitch'[1] was a beauty; and this and Sargent's cherry, shortly afterwards introduced to England from America, at last created a demand for the cherries of Japan. Many species and varieties were introduced in the first decade of the present century, and when Sargent sent E. H. Wilson to Japan in 1914, it was with cherries as his main object.[2] Compared with Wilson's previous travels in China, this

[1] 'Fugenzo', still in cultivation
[2] Wilson later published a monograph on *The Cherries of Japan*

was a civilized journey; he stayed in a hotel, and was accompanied by his wife and eight-year-old daughter.[1] Almost you might call him a tourist.

The nurserymen of Tokio and Yokohama specialized in a rather limited range of plants, and had very little idea of what was going on elsewhere in Japan. In a nursery at Hatagaya (north of Tokio) however, Wilson saw some floriferous, low-growing, evergreen azaleas which had come from Kurume on the island of Kiushiu – a place little known even to the local Japanese. He arranged for a collection to be sent to one of his subscribers, John S. Ames of Massachusetts, and these caused such a sensation in America that Wilson made a second visit to Japan in 1918 in order to obtain a further supply. The species had been introduced to cultivation by Motozo Sakamoto of Kurume about a hundred years before, from a plant found on the sacred mountain Kiri-shima. His stock had been inherited and his work continued by the nurseryman Kojno Akashi, who now had more than 250 varieties. Wilson visited him and carefully selected fifty of the best and most distinct – the famous 'Wilson 50', collections of which are still preserved (with some difficulty, for not all of them are hardy) in England and the USA. On this trip he also visited Korea and Formosa.

It was left for Captain **Collingwood Ingram** (b. 1880) to complete the work on cherries that Sargent and Wilson had begun. A tireless traveller and ardent amateur plantsman, Ingram made three visits to Japan, the first as a tourist, the second chiefly as an ornithologist and the third, in 1927, as a plant-hunter. As befitted a man of Kent, he took a particular interest in cherries, and had long made a study of the genus, although his book on the subject did not appear till 1948. He travelled extensively both in Nippon and Kiushiu, and found a number of new varieties, some of which appeared to be unknown even to the best Japanese authority on the subject. It was in spring, and the trees were in bloom; seed at that season being unprocurable (and probably unreliable in any case) he adopted the practice of seeking out the nearest local horticulturist – there was always one to be found – and asking him through an interpreter to send him scions of the desired tree in the winter, part-paying for them in advance. In due course he received scions of every variety he had asked for, which he grafted on stocks of bird-cherry (*Prunus avium*); by 1929 fifty-nine species and varieties had flowered in his Kentish garden, and many more were still to bloom. The best ones were propagated and generously distributed. Some of the species now cultivated in Europe have apparently been lost in Japan.

Ingram did not concentrate exclusively on cherries; he introduced other plants too, one of them being a fine maple which was at first classed as a variety of *Acer diabolicum* but has since been given specific rank as *Acer purpurascens*. In all his travels he only once saw a specimen of this tree, in the province of Shimochuke, but was fortunate in finding a seedling beneath it which he was able to bring home.

[1] The namesakes of *Rosa helenae* and *Sinarundinaria murielae*

China

1698	Cuninghame	1862	David
(1735	Ternstroem)	1867	Delavay
1740	d'Incarville	1881	Henry
(1750	Osbeck)	1884	Potanin
(1753	Torin)	1899	Wilson
1792	First Embassy, with Haxton	1904	Forrest
,,	James Main	1905	Meyer
1804	Kerr	1909	Purdom
1816	Second Embassy, with Abel	1911	Kingdon-Ward
(1821	Potts)	1914	Farrer
(1823	Parks)	1922	Rock
1843	Fortune		

'The plant-collector's job is to uncover the hidden beauties of the world, so that others may share his joy. . . . It is no unworthy aim, to reveal what God has planted in the lost mountains, since thereby may also be revealed what he has hidden in the hearts of men.' F. KINGDON-WARD *From China to Hkamti-Long* 1924

China

No country, with the possible exception of North America, has made so great a contribution to our gardens as China; and by a freak of the great forces that shaped our globe, the two floras have much in common, certain genera such as magnolia, hamamelis and wisteria occurring only in the eastern USA and in China and Japan. The flora of China might have been even more rich, had not its ancient civilization and large population entailed the destruction of all wild plants by intensive cultivation over vast areas, long before the first plant-collectors arrived. Probably, however, the cultivable plains were never so rich in species as the wild mountain areas, which remained virtually untouched until the late nineteenth century.

China is split horizontally by the Yang-tse-kiang, as America is split vertically by the Mississippi, though less neatly – a tear rather than a cut. North of the river are the hush-and-shush provinces – Hupeh, Hopeh and Honan, Chi-li, Szechuan, Shansi, Shensi and Shantung; while to the south are the knife-edged Kwei-chau, Kiang-si, Kwang-si, Kwang-tung, Kokien and Che-kiang. This useful generalization is unfortunately spoilt by the presence of Kansu and Kiang-su in the north and Hunan and Yunnan in the south; if these could change places, Chinese geography would be much simplified. The difficulty of tracing the unfamiliar place-names is increased by the fact that no two travellers and no two cartographers spell them alike; phonetic transcription from the Chinese characters giving a wide latitude. It is perhaps a help to know that the suffix Pu indicates a small village, Ch'en a larger one, Hsien a walled town with a magistrate, Chow or Chou and Kau progressively larger towns, and Fu a provincial capital, the seat of a viceroy. Shan is a mountain or mountain-range, Ho a river, and La in Tibetan means a pass.

China

Plants from China began to filter through to Europe at a very early date, and not economic plants only; the 'Persian' lilac and the 'Syrian' hibiscus came overland with the trading-caravans in the dark of history. Recorded introductions by sea, however, came late and slowly. Europeans were never welcome in China, though the country was not actually closed to foreigners till 1755; and would-be British merchants had to face not only Chinese obstructiveness but the opposition of the Portuguese, who had established a trading-post at Macao in 1537 and wished to preserve their lucrative monopoly. The result of China's remoteness and general inaccessibility was that until the nineteenth century the only botanical investigations were those made by keen amateurs among the few Europeans – missionaries, merchants or diplomats – whose business took them into the country, or by go-and-return collectors who sailed in some trading-vessel and stayed in China only for the time necessary to unload and reload cargo.

The first of the amateurs was **James Cuninghame**, about whom little is known except that he was a Scot, and had probably lived some time in London before sailing to China in 1698 as a surgeon in the service of the East India Company. After a sufficiently adventurous voyage, during which the ship was impounded for a breach of regulations by the Spanish authorities of Palma in the Canaries, and the crew imprisoned for some weeks, they reached the port of Amoy in the Strait of Formosa. Their stay there cannot have been long, but Cuninghame managed to procure 'the Paintings of near eight hundred Plants in their Natural Colours, with their Names to all, and Vertues to many of them'.[1] From every port of call he sent home to the botanist James Petiver collections not only of dried plants and seeds, but of shells, insects 'and, indeed, whatever came his way'. He was back in England in 1699.

Next year he made a second voyage, this time to Chusan, where the Chinese had recently 'granted us a Settlement and Liberty of Trade'.[2] He stayed there for more than two years, and sent to the *Philosophical Transactions* an account of the island's agriculture, which included the first description of the cultivation of tea. He hoped, when he had learned Chinese, to see something of the inland towns, and to be able to satisfy the 'longing Expectations' of his correspondents at home, who included Leonard Plukenet and Sir Hans Sloane as well as Petiver; but 'the Chineses' put too many obstructions in his way. 'Had I the Libertie I could wish for, I might have made greater Collections, but the Jealousie of these People among whom we live, restrains so much that we have no Freedom of rambling.'[3]

Early in 1703 the factory[4] at Chusan was given up, and Cuninghame was transferred to the station on Pulo Condore (an island east of Cape Cambodia) from which the Company was trying to open up trade with Cochin-China against the opposition of the Chinese. But in 1705 some of the Macassar servants of the company rebelled, fired the fort during the

[1] Petiver *Museum Petiverianum* 1699
[2] Cuninghame *Philosophical Transactions* 1702
[3] Letter to Petiver, quoted by Britten and Dandy *The Sloane Herbarium* 1958
[4] In the old sense, an agency or trading-station

night, and killed all but a handful of the Europeans. The remainder were murdered a week later by their so-called allies the Cochin-Chinese, who had come to the rescue and had been very active in the pursuit of the Macassars, but were unable to resist the opportunity of looting what remained of the fort, including a large quantity of money which the English had so far managed to preserve. Cuninghame, wounded in the arm and side, was the only survivor of this second massacre. Three weeks later he was taken as a captive to Barrea on the Cochin-China mainland, and after another three weeks was brought before the governor to answer three charges, directed against the Company rather than himself; the principal grievance being that no Embassy with presents had been sent that season to the King of Cochin-China. In spite of his capable defence, Cuninghame was kept prisoner in the country for nearly two years. After his release in 1707 the Honourable East India Company[1] appointed him chief of their settlement of Banjermassin in Southern Borneo; but here he was no more fortunate. Ten days after his arrival the settlement was attacked by natives (again incited by the jealous Chinese) though not this time with so much loss of life, and had to be abandoned. Shortly afterwards Cuninghame embarked to return to England; he wrote to Sloane and Petiver from Calcutta on 4 January 1709, but after that no more is heard of him, and it is thought he died on the voyage home.

Cuninghame sent home at least 600 specimens of oriental plants, often with full descriptions and notes on localities, uses, and native names; among them were the first examples of the camellia ever to be seen in Europe. A few plants were raised from his seeds, but nothing of great garden importance.

Traders were not the only Europeans to experience Chinese obstructiveness. Missionaries also found that the Government would condescend to accept the best that European civilization had to offer, while granting very little in return. The Jesuit colleges sent out very able men, highly trained in acceptable skills such as clock-making and astronomy, but even they found that difficulties were many and converts few. The Emperor Ch'ien Lung made use of some of these Jesuits as architects and hydraulic engineers for the embellishment of his celebrated 'Garden of Perfect Brightness' (Yuan-ming-yuan), and the introduction and cultivation of European flowers was entrusted to Father d'Incarville.

Pierre Nicholas le Chéron d'Incarville (1706–1757) joined the Compagnie de Jésus in Paris in 1726, and underwent the usual long training, including five years' teaching in Quebec (1730–5). He received no instruction in natural science until after he had been chosen for the China mission in the summer of 1739, so he had only six months for his studies, which included an intensive course of botany under Bernard de Jussieu. (He ever afterwards regretted that he had only a 'superficial tincture' of scientific knowledge.) He sailed on 19 January 1740, and after spending some months in Macao and in the East Indies, reached Pekin in February 1742.

[1] The East India Company, founded on the last day of 1600, became the Honourable East India Company on amalgamation with a rival company in 1701

He was appointed Master Glass-maker to the Court, and was kept hard at work, which, with the restrictions imposed by the Chinese, gave him very little opportunity for botanizing. In November he wrote to ask de Jussieu to send him bulbs and seeds (though they would take two years to reach him); he had found that the Emperor was fond of flowers and hoped to lead him to religion through natural history. He also hoped to get Chinese plants in exchange for European ones; but it was ten years before these plans were realized.

When the Emperor was absent from Pekin the missionaries were allowed to go a little farther afield, and in 1743 d'Incarville was able to take two trips into the nearby mountains – both, unfortunately, at the wrong season for plants; on the second, the snows of winter were already beginning to fall. A collection he sent home in 1743 was captured by the British and that of 1745 was lost by shipwreck. In 1746 an anti-Christian persecution began, which further limited excursions and also damped hopes of getting collections made for him in other provinces – 'our missionaries have enough to do, to hide themselves'.[1] Meantime he made a study of the language and compiled a French-Chinese dictionary, besides writing several memoirs on the arts of China.

Bernard de Jussieu was an exceptionally bad correspondent, and seldom answered letters; in 1755 d'Incarville complained that he had not heard from him for six years. But meantime a flourishing intercourse had sprung up with the secretary of the Royal Society (Cromwell Mortimer), and other correspondents in Britain, including Philip Miller and Peter Collinson. Latterly, many of d'Incarville's letters to France were sent via England or overland by the three-yearly caravan to St Petersburg.

In 1753 d'Incarville presented the Emperor with two specimens of the Sensitive Plant (*Mimosa sensitiva*), which he had grown from seed sent to him by de Jussieu; and this at last won him the coveted Imperial favour. Ch'ien Lung was 'greatly diverted' by the plant and 'laughed heartily' at its performance; he asked for more European plants and gave d'Incarville free admission to all the Imperial gardens. The Father wrote home for more bulbs and seeds, a book on propagation and instructions for putting up a small greenhouse; for he was charged with the difficult task of keeping the mimosas alive through the bitter Pekin winter. He was promised all the Chinese plants he wanted, but he did not long enjoy his late-won privileges. He died in May 1757, of an illness caught from a sick man he had tended, which was not at first considered serious.

Although no Chinese plants had previously been gathered so far north, those sent by d'Incarville were ungratefully neglected. A few of his 300 herbarium specimens were described and published (including a plant of a new genus named after him *Incarvillea sinensis*[2]), but the majority were not properly examined until 140 years later, when some of the seeds he had sent were found still unsown. The Chinese Tree of Heaven (*Ailanthus altissima*)[3] was, however, raised from his seeds both in England and

[1] Letter to de Jussieu, quoted by Bernard-Maitre
[2] Even this was not done till 1789, by Bernard's nephew, Antoine Laurent de Jussieu
[3] Called by d'Incarville *Frêne puant*, Stinking Ash

France; his introductions to Paris included *Albizzia julibrissin, Cedrella sinensis, Sophora japonica* and *Thuya orientalis,* and to St Petersburg, *Koelreuteria paniculata.* He is credited with the introduction of the China Aster (*Callistephus chinensis*), but seeds of this popular flower had already been sent to Antoine de Jussieu by another Jesuit in 1728.

Peter Collinson, at least, appreciated d'Incarville's value, and urged Linnaeus to send a botanical expedition to China to follow up his work. Linnaeus needed no urging; he had already made arrangements for the Swedish East India Company to give a free passage every year for one of his pupils, to wherever the ship might be trading. Both of his emissaries to China were clergymen – the Rev. **Christopher Ternstroem**, who sailed about 1735 but died on the homeward voyage in 1738; and the Rev. **Peter Osbeck**, who left Sweden in November 1750 and was back by June 1752. Many of their finds were included in the first edition of Linnaeus' *Species Plantarum* (1753) – some rather loosely labelled '*indica*'. In China Osbeck had 'made a number of useful and interesting observations, at the expense of his whole salary',[1] and these were incorporated in a book about his voyage which appeared in 1757, the cost of production being defrayed by his equally public-spirited parish.

Three years after Osbeck's visit, the doors of China were closed to foreigners, except for some concessions at Canton and Macao. As time went on the European traders were subjected by the local officials to so many abuses and impositions, and their privileges so much eroded, that their situation became intolerable. The British merchants in particular were at a disadvantage, compared to the long-established Portuguese, and it was at last decided to send an impressive embassy to the court at Pekin, to demand redress for their wrongs and, if possible, to obtain improved trading conditions. Lord Macartney, previously Governor of Madras, was appointed as Ambassador.

Scientists saw in this embassy an excellent opportunity for investigating the almost-unknown country, and needless to say, Sir Joseph Banks had a hand in the affair. It was at one time proposed to appoint an official naturalist, and the name of Afzelius was suggested, but he had already contracted to go to Sierra Leone. In the end, the party had to be content with two 'botanic gardeners', one official, and one engaged 'at the expence of an individual of the Embassy' – probably Staunton. The distinguished personnel not being under Banks' authority, he could not issue his usual copious instructions; but he addressed some 'Hints on the Subject of Gardening' to the gentlemen of the Embassy, in which he reminded them that there were only two plants of the yulan (*Magnolia denudata*) in England, and that more would be desirable; and that although the tree-paeony had twice reached England alive, both plants had succumbed to their first winter.

The function of naturalist was to a certain extent fulfilled by Sir George Staunton (1737–1801), Macartney's second-in-command; he was well-versed in botany and was a Fellow of both the Linnean and Royal

[1] Kalm *Travels* 1772

Societies. He was the author of the official account of the Embassy with which Jane Austen's heroine Fanny Price improved her mind at Mansfield Park.[1] Unfortunately, in two folio volumes he hardly mentions the gardeners, and never by name; it is from other sources that we learn that they were called **Stronach** and **John Haxton**. All the introductions and botanical discoveries resulting from the trip are credited to Staunton alone.

The Embassy – with military escort, band, and artificers of various sorts it numbered nearly one hundred strong – left Portsmouth on 26 September 1792 in a warship, the *Lion*, and an East-Indiaman, the *Hindostan*, with the *Jackall* as tender; later another tender, the *Clarence*, was added to the little fleet. It was necessary to have a small vessel of shallow draught to go before and take soundings, for the party was to go by sea to Tientsin, and the Yellow Sea had never been charted by European navigators. There is much both of human and botanical interest in Staunton's account of the voyage, but it is too long for inclusion here. By the time they reached the East Indies so many were ill with scurvy or dysentery that a month's stop had to be made at Turane Bay in Indo-China to give the invalids a chance to recover. A Chinese pilot was sent to meet them at Chusan, and shortly afterwards they were joined by the East India Company's brig *Endeavour*. The harbour at Tientsin being shallow, the two big ships, *Lion* and *Hindostan*, were sent to make a tour of the China seas, and the company transferred to the three smaller vessels for the journey up the River Pei-ho. They disembarked on 16 August 1793 at Tong-chow-ju, for the twelve-mile overland journey to Pekin. Here the greater part of the Embassy remained, to restore the invalids and to unpack and arrange the bulkier presents in a pavilion in the garden of Yuan-ming-yuan, while the principals went on to the Emperor's summer residence at Jehol.

The chief obstacle to the success of the Embassy was the ceremony of 'kow-tow' – the nine prostrations that everybody was supposed to perform, not only before the Emperor, but before 'every piece of yellow rag which they might choose to consider as emblematical of his Chinese majesty'.[2] This the English steadily refused to do, unless a Chinese of similar rank would perform the same ceremony before a portrait of George III. The consternation caused among high Chinese officials was extreme; it was almost impossible to find anyone who would dare to convey so blasphemous a message to the Emperor. The throne, however, was still occupied by the same Ch'ien Lung who had employed d'Incarville and the other Jesuits. Aged,[3] wise and conversant with the peculiarities of Europeans, he announced that he would be satisfied if the Embassy would pay him the same homage as they would to their own King – that is, to kneel on one knee; and this the stiff-necked English graciously consented to do. The audience was given, presents exchanged, and the visitors invited to attend the Emperor's birthday celebrations; and afterwards it was kindly but firmly indicated that they might now take their leave.

[1] 'You in the meantime will be taking a trip to China, I suppose. How does Lord Macartney go on?'

[2] Abel *Narrative of a Journey* 1818

[3] He was now eighty-three

Macartney felt that his mission was but half accomplished, but for various reasons felt obliged to comply; one of them being that news had arrived of the outbreak of war with France, and the British merchants at Canton were anxious to secure the escort of the warship *Lion*, now lying at Chusan. On 7 October the party set out on the 1400-mile overland journey to Canton.

'Overland' gives perhaps a wrong impression, as nine-tenths of the journey was made by China's wonderful system of inland waterways – lake, river and canal. Only two sections of the route were traversed by chair or on horseback; a short stage after Chen-San-Shen, and a longer one over the mountains between Nan-gan-fu on the River Kan-Kiang, and Nan-shu-fu, where they joined the Pe-kiang river flowing south to Canton. This travel by water made botanizing difficult, and so did the lateness of the season; it was mid-December by the time they arrived. It is tantalizing, however, to read nothing of the activities of the gardeners. John Haxton, at least, was a man of ability; after his return he became an associate of the Linnean Society and attained some prominence as an entomologist. He kept a journal of the expedition, but it does not seem to have survived.

At Canton the Embassy parted 'not without tears', from the two mandarins who had escorted it from the start, and removed to Macao, where its members might live without further cost to the Chinese, who had supported them all the time they were on Chinese territory. They sailed on 17 March 1794, escorting fifteen[1] East India Co. merchantmen, one Spanish ship and one Portuguese; a French squadron was cruising in the Banca Straits but did not dare to attack so large a force. By 6 September all were safe at Portsmouth.

Considering the dangers and the limited opportunities, Staunton reaped quite a creditable botanical harvest, bringing home specimens of some 220 plants and introducing seven or eight novelties, the most important of which were *Macleaya (Bocconia) cordata* and the lovely *Rosa bracteata*, still known as Macartney's Rose. James Main says that Haxton brought home plants of *Camellia sasanqua*, but they were more probably *C. oleifera*, the two species at that time not having been clearly distinguished.

Staunton's eleven-year-old son accompanied the Embassy as page to Lord Macartney. He learned Chinese more easily than the older members of the party, and in Pekin was able to render valuable service as an interpreter and intermediary. When he grew up, he returned to China, and his father having died, became in turn Sir George. His services as interpreter were called upon when the second embassy was sent out, under Lord Amherst.

A high Chinese dignitary had urged Lord Macartney to arrange for another embassy in a year or two – not necessarily at so much expense – to consolidate what had been gained; and the Emperor himself had indicated that a return would be welcome before his intended retirement in 1796, after sixty years of reign. The proposal was often discussed, but England

[1] or twenty-eight, according to Main

was much occupied with the Napoleonic Wars; it was not till peace was restored that Lord Amherst's embassy was launched – and was a failure.

The start was auspicious enough. The precedent of the earlier Embassy was closely followed except in numbers, the new party consisting only of seventy-two members. The physician appointed, Dr **Clarke Abel** (1780–1826), was interested in all branches of natural history and anxious to make collections, and Banks provided him with every facility. He supplied a 'plant cabin' and a trained gardener from Kew, James Hooper (at the usual £100 a year) whose instructions were to help in the naming and preparation of specimens, to collect seeds and to plant in pots and boxes all living plants collected by Dr Abel, for transport to England. They sailed in February 1816, were joined at the Lemna Islands off Hong Kong by the younger Sir George Staunton and some others, and reached Tientsin on 28 July.

A very different Emperor now occupied the Chinese throne. Ch'ia Chang was haughty, vacillating, and much under the influence of venal officials, who had their own reasons for wishing the embassy to be a failure. Fierce arguments about the 'kow-tow' raged from the start, the Chinese blandly asserting, as a precedent, that Lord Macartney had performed it, and even trying to persuade Staunton to bear witness to that effect. It leaked out accidentally that the day of their audience had been fixed in advance for 22 August, before the Emperor's projected removal to Jehol; which left very little time for the remainder of the journey. The Manchu emperors were accustomed to hold their audiences at the unholy hour of six or seven a.m., and Amherst, exhausted and unwell, was summoned to attend before he had had time to change his travel-stained garments, after a gruelling journey which had terminated only in the small hours of the same morning. He therefore presented his apologies and said he was unable to attend. The Chinese intermediaries chose to regard his illness as a pretext, and suppressed the more acceptable excuse of his inappropriate dress; and this, with the refusal to kow-tow, so incensed the Emperor that he broke off further intercourse. Later on, when he discovered the true facts of the case, he was placated and sent conciliatory messages and orders for good treatment; but by this time the Embassy had been dismissed in disgrace, and was on its way home.

The party followed the same watery route, and at the same season of the year; and conditions were even more unfavourable for Abel than for Staunton, for he was ill for part of the time, and their Chinese escort was hostile and allowed them little liberty. Nevertheless, with the help of his colleagues he amassed a considerable collection of plant-specimens, and Hooper had in his care a number of living plants and 300 packets of seeds, most of them believed to be of plants new to science. Before leaving Canton in the *Alceste* for the voyage home, Abel was given charge of some natural history collections made by others – plants from other regions, zoophytes, madrepores and geological specimens. But on 18 February 1817 – a fine day with a fair wind – the ship ran on an uncharted reef in the pirate-infested Banca Straits and was badly holed. There was no loss of life, but it was necessary to take to the boats and make for the nearest island; and Abel had the mortification of seeing the chests containing his seeds brought up and emptied overboard 'to make room for some of the linen of one of the gentlemen of the Embassy'.[1] The next day, the wreck being still afloat, Abel returned and found one of his collections still relatively unharmed; he managed to get it put aboard a raft – which was promptly burnt to the water by the hovering Malay pirates. The casualties included a very large pink azalea, which the day before the wreck had been so covered with flowers that not a leaf was visible.

The principals of the party set out next day on a hazardous five-day boat-voyage to Batavia, and from there sent help to the stranded remainder, just in time to save them from a native attack. Six weeks later, they resumed their voyage (in the *Caesar*); and one hopes it was some consolation to Abel, that on the way home they were granted an interview with Napoleon on St Helena. All that remained to show for the doctor's laborious endeavours was a small collection of duplicate plant-specimens which he had given to Sir George Staunton before leaving Canton, and which Staunton generously restored to him; among them was the plant afterwards named *Abelia chinensis*.

Long before the dawn of the nineteenth century, gardeners as well as scientific botanists were casting covetous eyes to the East; they longed to obtain the magnificent flowers portrayed in Chinese paintings, textiles and ceramics. Employees of the East India Company at Canton were pestered by correspondents to send home plants, and relatives besieged the captains of the tea-clippers. One of the first to respond to the demand was Benjamin Torin, who sailed to China in 1753, and rose to become one of the seven resident supercargoes maintained by the Company at Canton. He is seldom remembered, but of a consignment of plants he sent to Kew in 1770, five were new and valuable additions to horticulture – *Daphne odora*, *Osmanthus fragrans*, *Saxifraga sarmentosa*, and two showy greenhouse species, *Cordyline terminalis* and *Murraya exotica*. (After some twenty years in China, Torin served for another thirty years in India, but he does not seem to have sent home any more plants.) A little later, Banks brought his ponderous influence to bear, and between 1782 and 1796 received a number of plants from the brothers John and Alexander Duncan, successively surgeons to the Company at Canton. The Horticultural Society enlisted several Chinese contributors, soon after its foundation in 1804.

Foreign officials were permitted to live in Canton only while the merchant fleet was in port; the rest of the year they spent on the island of Macao, where there were fewer restrictions and more space. On Macao it was possible to have a garden, and there were some notable gardeners among the English who made their homes there. John Livingstone, Thomas Beale[2] and John Reeves all lived in China for twenty years or more, so there was time for their gardens to mature. Reeves was particularly zealous. He was sent out as a tea-inspector to the East India Company in 1812, the last of the gardening fraternity to arrive; correspondence with Banks began at once, and plants followed shortly after. On his first leave

[1] Abel *Narrative of a Journey*
[2] Beale stayed for fifty years

in 1816, Reeves brought 100 living plants to the Horticultural Society, which subsequently commissioned him to obtain a series of plant-drawings executed by native artists under his supervision – the celebrated Reeves Collection. All his plants were carefully established in pots in his own garden for a long time before despatch, and he was more than usually successful in persuading ships' captains to carry and care for them.

Nevertheless, the patrons at home were still unsatisfied, chiefly because so few of the plants sent, survived the voyage. The 'plant cabins', or miniature greenhouses of the day, were large and heavy; to keep them out of the spray they had to be carried high on the poop, and captains complained that their weight at that level unbalanced the ship and impeded her sailing. Many refused to take them at all, and in any emergency they were the first things to be thrown overboard. Even when captains were sympathetic they were not necessarily sufficiently skilful to maintain vegetation in good health through a six-month voyage that entailed two crossings of the Equator and the rounding of the Cape of Good Hope. Obviously it was more satisfactory to send a gardener to fetch what was wanted; so Slater sent James Main, Banks sent William Kerr, and the Horticultural Society sent Potts and Parks.

Gilbert Slater was very anxious to obtain Chinese plants for his Essex garden. He had connections with the East India Company and a controlling interest in several of their ships. By tracing the characters from Chinese pictures of flowers, he compiled a catalogue which he distributed to ship's captains and other travellers, asking them to purchase such living plants as corresponded with the characters; but only a tithe of the plants survived the voyages, and among these were many duplicates. In the end, he sent out three gardeners; two of whom, unfortunately, did not return.[1] The third was **James Main** (*c.* 1770–1846), a young Scot who had risen in his employ to be foreman of the greenhouses and flower-garden.

Delighted and full of enthusiasm, Main embarked on the *Triton*, one of Slater's own ships, in the autumn of 1792. He was armed with instructions from Banks 'written with his own hand, as to self-government in the ship and when on shore'[2] as well as on the management of seeds and plants, and was entrusted by the firm of Loddiges with four experimental boxes of plants, differently packed, to see which best withstood the voyage. He was to return by the same ship, so his stay in China was not long; but the trip was full of incident.

The *Triton* was carrying despatches, and made all speed to the Cape, without the usual stops at Madeira and Rio de Janeiro; these despatches were to inform the Dutch governor of the outbreak of war between Britain and France. At Cape Town, Main had an interesting meeting with Francis Masson and saw his garden and collections. He arrived at Madras on 3 April 1793 – with his boxes of plants 'in a sad mutilated condition' – and the *Triton* was shortly afterwards commandeered to take part in the siege of the French settlement of Pondicherry. This being concluded, she continued her voyage via Pulo Penang (Prince of Wales Island) and

[1] One was drowned on the outward voyage, in the Straits of Malacca

[2] Main, in the *Horticultural Register* 1836

Malacca, in both of which Main collected seeds and specimens. On the Malayan side of the 'Straits of Sincapore' they encountered a French frigate with six or seven prizes; the *Triton* and two other ships gave chase, and the Frenchman fled, leaving his prizes to the British. The typhoon season was now beginning, and the ship had to lie to for eight days in a storm, in the neighbourhood of Manila.

When at last they reached Canton, they found the *Lion* and *Hindostan* at anchor at Whampoa, and Lord Macartney's Embassy just returned from the north. Main gives the impression that he thoroughly enjoyed his stay. He had letters of introduction to Dr Alexander Duncan and others; nearly threescore Indiamen of different nations were lying in the bay, their officers and part of their crews ashore; and among them and the personnel of the Embassy the sociable gardener found several old school-fellows and others that he had previously known in England – notably the Embassy botanists, Stronach and Haxton. But he worked hard, too. Before leaving the *Triton* he had measured all the places where plant-boxes might stand, and had cases made to fit;[1] he had calculated how many plants the cases would hold, and had mixed special compost. He was not allowed to visit the famous Fa Te nurseries, though he was taken on conducted tours of certain celebrated gardens; but through a Chinese agent called Samay he made numerous purchases. He bought few camellias, for he thought the Chinese varieties already surpassed by those being bred in Europe; but he got half a dozen tree-paeonies, several magnolias, *Chimonanthus præcox* and *Clerodendron fragrans*, *Spiraea crenata*, and a plant of *Chaenomeles speciosa*, 'introduced several years before by Mr. Slater, but repeatedly lost by being treated as a stove-plant'. There was much else; but owing to subsequent events, the subject is chiefly of academic interest. Duncan thought it worthy of remark that he transferred his plants to ordinary earthenware pots, instead of the glazed china pots in which they were received.

The *Triton* sailed early in March 1794, as part of the convoy escorted by the *Lion* and the *Argo* which brought home the Macartney Embassy. Main had persuaded the captains of several other vessels to take consignments of plants, and he looked after his own with the most solicitous care. Often he sat up with them at night, for he believed the night air to be refreshing to them after the heat of the day, but they had to be covered at the approach of squalls with a platform on which the men stood when making or reducing sail. He was too liberal with the water, and this and the heat of the tropics stimulated rapid growth which the plants could not maintain; by the time they reached Sarawak many were in a 'feverish state of excitement, evidently much exhausted'. Some of the sickly ones recovered during the passage of the Indian Ocean, but they met with two tremendous gales when rounding the Cape, when the plants had to be closely covered for two successive days, and Main began to despair of getting more than half his collection to England alive. Fortunately they then had fair weather to St Helena, where he was disconcerted to find that some of

[1] Slater being the owner or part-owner, the captain could hardly refuse to take them

the plants carried by other captains, which had been neglected, had survived much better than his own, almost killed by kindness.

At St Helena, Main received the devastating news that his employer, Gilbert Slater, whom he liked and admired, had died. Much disheartened, he resumed the voyage – without incident till they reached the Channel, when, on the threshold of home, the *Triton* was involved in a collision with an outward-bound frigate. Her foremast, main- and mizzen-masts went by the board, and the two latter in falling crushed all the plant-boxes to the deck. She was towed into the Thames, and the shattered remains of Slater's plants were sold to George Hibbert Esq., a keen gardener of Clapham. According to some authorities Main went with them; he, too, must have been shattered, for he had had no arrangement in writing with Slater, so had no claim on the administrators of the estate. He received no remuneration for his arduous voyage, and never was able to find out what had happened to the plants left in charge of other captains, and what, if anything, he had actually introduced. Fortunately he had no difficulty in finding other employment, and later became a prolific writer on gardening topics and assistant to Loudon on the *Gardener's Magazine*.

Sir Joseph Banks was hoping to send out a collector, as early as 1796; but no promising opportunity occurred till 1803, when David Lance, a superintendent of the Company's factory at Canton, was home in England and about to return again to China. He was interested in natural history, and applied to Banks for instructions as to what he should bring or send from the East; and Banks suggested that he should take out with him a gardener from Kew, who should be under his direction and help him to collect and despatch seeds and plants. Within a month[1] everything had been settled. Banks had an interview with George III, who expressed great interest in the project and assented to the arrangement that all expenses should be borne by the East India Company, except the gardener's salary of £100 a year, to be paid by Kew. The candidate selected was **William Kerr** (d. 1814). Known to Banks for some years as a 'considerate and well-behaved man', Kerr was given an advance on his salary to purchase his equipment, and was well-briefed both by Banks and Lance; and very soon afterwards was on his way. He was the first resident, professional collector to go to China, and he stayed for more than eight years, though he never got farther than Canton and Macao.

Lance and Kerr sailed in the *Coutts* about the end of April 1803, and the collector sent home several consignments of plants by different ships in 1804, those carried by the *Henry Addington* in a plant-cabin, designed by Banks, being particularly successful. When the fleet for that season had sailed, he withdrew, as was the custom, to Macao, and explored the vicinity and the neighbouring islands, though much impeded by the activities of pirates. David Lance returned to England that year, but acting on his written instructions, Kerr went from December 1804 to September 1805 on a long-planned journey to Cochin-China and Manila. For some reason Banks disapproved of this expedition, although he had given Kerr

[1] April 1803

instructions to enquire particularly into plants producing fibre and cordage – Manila's major industry. Afterwards he had admitted that it had been justified by the results. Kerr had collected more than 700 living plants, and though many were lost in a gale on the way back to Macao, he hoped to get replacements from friends he had made among Manila's Spanish priests.

The usual shipping difficulties were experienced, in both directions. European plants were sent out from Kew in 1805 and 1806, but in spite of every care, nearly all of them were dead on arrival. Kerr himself was often unsuccessful; a consignment he sent by the *Winchelsea* was almost a total failure. Banks did not blame him for this; he had come to the conclusion that it was hopeless to send living plants except under the charge of a competent man, and suggested that next time a Chinese gardener should be sent, to look after the cases on the voyage. This seems to have been done, for Kerr later refers to one A-hey who 'since his return from England' had been employed by Roberts on Macao, and treated better than he deserved. John Roberts, a chairman of the Company at the time of Kerr's appointment, became superintendent of supercargoes at Canton in the spring of 1807, and had a fine garden with a large staff. It was he who advised Kerr to have his plant-cases made smaller and lighter, in order to deprive ship-masters of one of their excuses for refusing to carry them.

1808 was a year of trouble for the English residents. Portugal was England's ally in the Napoleonic wars, and Admiral Drury was despatched with a naval force to 'protect' Macao against France. Learning that a French force was operating off Java, he landed British troops on Macao on 18 September, thus violating a pledge previously made by the Portuguese, that no foreign troops should be admitted without Chinese authority; and the Emperor issued an edict suspending all trade with the foreigners till the troops should be withdrawn. Roberts, then President of the Select Committee, refused to make any concession to the Chinese officials and threatened to withdraw the whole British community if the garrison were re-embarked. Drury then made a half-hearted attempt to force a passage to Canton; but it all ended with an agreement and the retirement of garrison and navy, with much loss of 'face'. During the disturbances all the Chinese servants on Macao deserted, including Roberts's ten gardeners, led by A-hey; and Kerr was kept busy trying to maintain the garden, the weather being very hot and much watering necessary. The merchant fleet, too, was detained much beyond its usual sailing-date. Nevertheless Kerr sent by each of four vessels a plant-cabin containing eighteen boxes of plants, all of which had been established in the boxes for six months at least before despatch. Nearly all of them were garden varieties of trees and shrubs, but some were wild plants, identified only by their Chinese names and descriptions such as 'small wild tree'. He also sent more than 238 parcels of seeds – 129 of wild species collected by himself near Macao, thirty of Bengal plants from Roberts' garden, and others purchased from Chinese sources. Besides these consignments Kerr had also sent plants to the botanic gardens of Calcutta and St Helena, had supervised the production of plant-drawings by native artists, and had endeavoured to establish on Macao the European fruit-trees and other plants sent out from Kew.

It is surprising, therefore, to find that some people (notably John Livingstone) regarded Kerr as a failure. Livingstone put this down to the inadequacy of his salary, which was 'almost too small for his necessary wants, and he consequently lost respect and consideration in the eyes even of the Chinese assistants whom he was obliged to employ. . . . I have not the slightest doubt that his failure is to be attributed chiefly to the necessity he was under of associating with inferior persons, from his deficiency of means to support himself more respectably.' Bretschneider quotes a mysterious statement from the *Chinese Repository* (written, however, twenty years after Kerr's death) to the effect that he was at first very active but after three or four years became greatly changed – 'He was then unable to prosecute his work, in consequence of some evil habits he had contracted, as unfortunate as they were new to him.'

Up till 1810, however, Banks seems to have been perfectly satisfied with his work, and always refers to him with approbation. In June of that year he wrote to inform Kerr that in consideration of his good conduct as a plant-collector, he had been appointed Superintendent of the Botanic Garden which was to be established in Ceylon; but he did not actually take up his new post till August 1812; and it is possible that during the long tedious interval he took to opium-smoking, as the quotation hints. There is no indication that he ever came home on leave nor made excursions to other parts of the East; he must long ago have exhausted the resources of the Canton nurseries, and probably had not enough to do.

The reason for the two-year delay is uncertain. A botanic garden had been founded near Colombo in 1799 by a Mr North, under the curatorship of M. Joinville; in 1810 it was transferred to a seven-acre site on Slave Island, thereafter known as 'Kew'. It is possible that Joinville was unwilling to resign his charge, and that Kerr was unable to take over until he left. It was not till a year after his arrival in 1812 that negotiations were concluded for the purchase of 560 acres of land at Caltura, and a new official garden started; and he died only fifteen months later, on 24 November 1814, soon after his return from an expedition to the extreme north of the island.

Chinese plants introduced by this 'failure' included (besides garden varieties) *Begonia evansiana*, *Juniperus chinensis*, *Lilium japonicum* and *L. tigrinum*, *Nandina domestica*, *Pieris japonica*, *Pittosporum tobira* and *Rosa banksiae*. In Aiton's *Hortus Kewensis*, however, all Kerr's plants are credited either to Lance or to the Court of Directors of the Hon. East India Co. – a form of courtesy to the powerful body from which further favours were hoped. Lance was given as the introducer of *Corchorus japonica*, which we now call kerria; it was renamed in honour of its collector by de Candolle in 1816.

Except for the care of the plants in transit, it would seem hardly worth while for the Horticultural Society to have sent out two go-and-return collectors in 1821 and 1823, when they already had such active correspondents as Beale, Reeves and Livingstone on the spot. It was only nine years since Kerr had left, and little new could be expected; moreover the Chinese restrictions were even more severe, and the Fa-Te nurseries could only be visited on three days a month, at a charge of eight dollars. It is difficult not to lump Potts and Parks together, in the absence of detailed information, especially on the former; their collections were similar, though not their personalities.

John Potts, a gardener in the Society's employ, sailed on the *General Kyd* in 1821 and returned on the same ship the following year. He brought back several new plants, including *Camellia maliflora*, *Callicarpa rubella* and *C. japonica augustata*, and a large quantity of seed of *Primula prœnitans* (=*sinensis*). He also procured forty varieties of chrysanthemums, all of which were lost 'in consequence of an accident which befel the ship on her voyage home'.[1] He reached home in August, and died of consumption barely two months later. The next season, 1823–4, **John Damper Parks** sailed and returned on the *Lowther Castle*; he was rather more fortunate with his collections and successfully introduced twenty or thirty chrysanthemums, the double yellow Banksia rose, a valuable consignment of *Camellia reticulata* varieties, and the first aspidistra (*A. lurida*). He touched at Java on the way home, and from there got several greenhouse plants and orchids. Both gardeners, Potts and Parks, spoke gratefully of the kindness and help they were given by Beale and Reeves on Macao; Parks in particular had 'received more favours as a stranger from Mr. Reeves than I ever did from any gentleman in my life'. No further attempts to obtain Chinese plants were made for twenty years.

At the conclusion of the infamous Opium War of 1840–2, the unwilling oyster was prised partly open by the sword; Hong Kong was ceded to the British, Chusan temporarily occupied, and four new treaty-ports, Amoy, Foo-chow, Ning-po and Shanghai, opened to European trade. John Reeves had retired to England in 1831; he now rendered his ultimate service to gardening by urging the Horticultural Society to take this opportunity of sending out a collector. No time at all was wasted; a volunteer, **Robert Fortune** (1812–1880) was ready to go, and less than eleven months after the signing of the Treaty of Nankin on 26 August 1842, he was on the spot.

'When the news of the peace with China first reached England in the autumn of 1842', wrote Fortune, 'I obtained the appointment of botanical collector to the Horticultural Society of London, and proceeded to China in that capacity early in the spring of the following year.' It was not quite so simple as it sounds. The Government was notified of the intention to send out a collector, and Lord Canning advised delay until the new officials were established at the treaty-ports and able to afford some protection. The Chinese Committee of the Society, headed by Reeves, discussed equipment, and drew up a list of instructions unsurpassed by any of Banks' efforts in the same line. It ran to some twenty-four paragraphs and contained a list of twenty-two sorts of plants about which he was to en-

[1] *Transactions of the Horticultural Society* VOL III

quire, including blue paeonies and camellias with yellow flowers 'if such exist'. At Fortune's particular request he was furnished with firearms – a fowling-piece and pistols – the Society having considered that a life-preserver would afford a sufficient defence. He was to take out with him European seeds, and live plants in some of the new Wardian cases, and to observe their behaviour on the voyage. And his salary was to be the usual £100 a year – less adequate than ever; but when, after a year, he asked for a rise, he was coldly informed that 'the mere pecuniary returns of your mission ought to be but a secondary consideration to you'.

The man to whom this tough assignment was allotted was a Scot, who had been employed by the Society for no more than a year, in the capacity of superintendent of the hothouse department at Chiswick. Before that, he had received a thorough training at the Royal Botanic Gardens, Edinburgh; but he had no previous experience of foreign travel nor of the work of a collector. It was subsequently amply demonstrated that the Society could not have chosen a better man.

It had been suggested that Fortune should establish some sort of base at Hong Kong; he arrived there on 6 July 1843, and stayed for seven weeks, probably arranging his lines of communication. There was as yet no regular steamer service from the new port of Shanghai, either to England or to India; and on his first three journeys to the East Fortune returned to Hong Kong at least once a year, to organize and oversee the despatch of his collections. It is pleasant to record that he was helped during his travels by the sons of two of the Macao enthusiasts – by John Russel Reeves, junior, after whom he named *Skimmia reevesiana*, and by the younger Thomas Beale (commemorated in *Mahonia bealei*), who became a great friend, and had a fine garden at Shanghai.

On leaving Hong Kong on 23 August, Fortune went first to the port of Amoy, for he had been asked to investigate the British-held island of Ku-lang-su in Amoy harbour before it was handed back to the Chinese. He stayed for about a month, pursuing his botanical researches so far as it was safe to do so; but the local flora was poor, and the garden-flowers were already well-known. He left his departure too late – at the end of September the monsoon changed, and adverse winds and currents came swooping down the Formosa straits. The first ship on which he took a passage sprang her bowsprit and had to put into the bay of Chin-chow for repairs. Fortune transferred to another vessel, but this time met with a gale in which two glazed cases of plants from Amoy were dashed to pieces, and the ship received so much damage that when after three days the storm abated, she put back into Chimoo Bay – south of the point from which she had started. This time Fortune waited while the repairs were being done, botanizing meanwhile in the vicinity; but the local inhabitants were a wild lot – as Fortune himself remarked, the worst elements in any country tend to congregate at the ports – and on one occasion his pocket was picked and the servant carrying his plants attacked and beaten. At the end of October he sailed once more, and this time got safely to Ting-hai on the island of Chusan – the first collector to land there since Cuninghame in 1701.

Hong Kong had proved barren, except high in the hills, Amoy almost equally so; it was in Chusan that Fortune got his first glimpse of the glories of China's temperate flora. He arrived in late autumn, but he returned many times at other seasons, and in later years formed the habit of retiring to Chusan to avoid the summer heats in Shanghai. He thought the island a paradise – as would most of us if translated to a place where wisteria grew wild in the hedges and azaleas covered the hills. There were gardens, too; it was in a mandarin's garden at Ting-hai that Fortune first saw one of his favourite introductions, *Diervilla florida* (*Wiegela rosea*).

On this first occasion Fortune soon left Chusan for Ning-po and Shanghai. Late though it was in the season, he found it worth while to comb the Shanghai nurseries – when he could get into them. The proprietors, mistrusting his intentions, set little boys to watch for his coming, and everywhere he was met by locked and guarded doors. His command of the language being still imperfect, he had to take a Chinese official along, to explain that any plants he took would be paid for; this little misunderstanding having been cleared up, he was welcomed and shown everything he wanted. But Shanghai was an uncomfortable place in the winter of 1843; no European-style houses had yet been built, and even the best of the Chinese houses were far from weatherproof and had no provision for heating; it was not unusual to find snow on the bedroom floor in the morning. Doubtless he was glad to return to Hong Kong and a warmer climate, and to despatch from there his first consignments of plants, in January 1844. The Wardian cases had proved their worth on the outward voyage, and he afterwards used them exclusively.

Before returning to the north, Fortune paid a visit to Canton and Macao, chiefly from curiosity, as that was the one part of China where no new finds were to be expected. It nearly cost him his life; on the outskirts of Canton he was surrounded by a hostile crowd bent on robbery, if not murder; he was pursued and stoned, and escaped with considerable difficulty. In April he was back in Shanghai – and saw the tree-paeonies in flower. The previous autumn the nurserymen had told him that these plants were expensive, because they were brought from a great distance; now he found by chance that the region where they were raised was quite close at hand, and he could go and see for himself what varieties were available. They were different from anything previously seen at Canton, which was supplied from another district; there was no communication between them.

Early in May Fortune went to Ning-po, and from there rode out with three compatriots to the temple of Tien-tung, about twenty miles to the south-west. The others returned to the port, but Fortune stayed on. The scenery was beautiful, the Buddhist priests hospitable and kind – but it was the first time he had been alone in China, and for a time he felt isolated and apprehensive. He liked the place, however, and afterwards returned several times, making it his headquarters for long periods. Moreover, it was in a tea-producing district, and he began to acquire the information on tea and its culture which afterwards served him so well. Again he had a mishap which almost proved fatal, when he nearly fell into a pit dug in a remote and lonely spot to trap wild boar. As he clung to the 'twig' which enabled him to scramble to safety, he thought of the fate that had overtaken David Douglas ten years before[1] – 'which made me doubly thankful for my escape'.

Europeans were not then allowed to travel more than twenty or thirty miles from a treaty-port. Fortune wanted to go to Soochow, which was beyond the prescribed limits, but which had the reputation of producing all that was most beautiful, from women to flowers. He set off by boat from Shanghai, 'of course' in Chinese costume complete with shaved head and pigtail, without at first telling his boatmen where he wanted to go; when he reached the thirty-mile limit he was able by a mixture of guile and bribery to persuade them to take him on. They tied up for the night at the town of Kia-ting-hsien; Fortune slept on board, but nevertheless his cabin was entered and robbed during the night and all his clothes taken. His money having been under his pillow, he was able to send his servant to buy new clothes, and proceeded undaunted on his way. He reached Soochow on 23 June, and was probably the first Englishman to go there. He was very nervous, but did not seem to attract any notice; the nurseries, however, were disappointing, though he got some new plants including a white wisteria and a double yellow rose mentioned in his instructions. He stayed for a few days, and when he returned to Shanghai, still in his Chinese dress, several of his acquaintances failed to recognize him.

During the summer, Fortune returned to his beloved Chusan islands, and this time was nearly drowned trying to make the crossing back to Chusan after a visit to Ning-po, in unfavourable weather. Back at Shanghai, he was delayed for a fortnight by a bout of fever; but eventually got himself and his second season's collections to Hong Kong in November. For extra safety he sent his plants by a number of different ships, the last leaving on 12 December.

Having had sufficient experience of a Shanghai winter, Fortune filled in the time both usefully and in a better climate, by a visit to Manila, from January to March 1845. For this it was necessary to get no less than four permits from the Spanish authorities – one to land, one to stay, one to travel in the interior and one to leave again – and even then, there were difficulties. His main object was to obtain plants of an orchid, *Phalaenopsis amabilis*, which had been introduced to England but was still extremely rare; in this he succeeded so well that the Horticultural Society was able to distribute forty-five plants of this species among their Fellows.

After returning to Shanghai on 14 March, Fortune was anxious to make a particularly fine collection, for this time the plants would be taken home by himself. All his old haunts were revisited, some several times, in order to see the different varieties in flower. Foo-chow was the only treaty-port he had not yet visited, and in July he set out to remedy this omission. Here he found the crowds insulting and turbulent, and the mandarins obstructive; it was hard even to locate the nurseries, but he eventually managed

[1] See p. 314

to go a short distance into the 'tea-hills', and made the important discovery that the plant producing the local 'black' tea was the same botanically as that producing the 'green' teas – the difference lying only in the method of manufacture.

Few European vessels used the port (which was never to prove much of an asset), so Fortune took a passage on a Chinese timber-junk, one of a fleet sailing regularly to Shanghai; he was sickening for a fever and retired to bed as soon as he embarked. The junk sailed in good (Chinese) time – that is, a fortnight later. The South China coast was then infested with pirates, over whom the authorities had little control; they took regular toll of the timber-boats, like wolves harrying a flock. A few days after sailing Fortune, still fevered, was summoned from his bed by the news that five of these pirate junks were approaching. The Chinese crew and passengers vanished below at the sound of the first broadside from the leading junk; the two helmsmen on duty would have followed suit, had not Fortune forced them at pistol-point to remain. Two more broadsides followed, but Fortune held his fire; this, he remarks, 'was a moment of intense interest'. He warned the helmsmen to lie down, while the fourth broadside whistled overhead, and then stood up and let fly at the crowded decks of the pirate with his double-barrelled gun, at a distance of twenty yards, causing great execution. Now it was the pirates who vanished below, and the boat fell back; but another junk moved in to the attack; Fortune repeated the same manoeuvre, this time killing the enemy helmsman and casting all into confusion. The remainder of the fleet sheered off; and afterwards, he reports, the Captain, pilot, crew and passengers 'actually came and knelt before me'. Three days later, six more pirates appeared; Fortune, still suffering from fever, repeated his former tactics with equal success, and went back to bed. In spite of these services, the ungrateful Captain tried to take him to Ning-po instead of to Chusan, as he had promised, and Fortune had again to resort to threats before he was landed 'in a most deplorable condition' at the place of his destination. Soon afterwards he was able to get a European boat to Shanghai and some much-needed medical attention.

At Shanghai he had a grand final pack-up of all his booty, and sailed for Hong Kong on 10 October. From there he sent off eight cases of duplicates, and reserved eighteen cases 'filled with the most beautiful plants of North China' which he transferred to Canton to be brought home on the *John Cooper* under his own care. They contained 250 plants, only thirty-five of which died on the voyage – a small proportion for those days. On 6 May 1846 he reached home with his harvest – and what a harvest! Besides many varieties of familiar flowers such as azaleas, camellias, chrysanthemums, paeonies and roses, he introduced from this first voyage the 'Japanese' anemone (*A. hupehensis*), two little chrysanthemums from Chusan which became the parents of the 'pompom' strains; *Dicentra spectabilis* and *Platycodon grandiflorus*, both of which had been introduced before but had been lost to cultivation; the white wisteria and the Winter Jasmine (*J. nudiflorum*). But his richest finds were among the flowering shrubs; he got more than a dozen good new species, including the first

Forsythia (*F. viridissima*) and the first weigela (*Diervilla florida*); the winter-flowering honeysuckles, *Lonicera fragrantissima* and *L. standishii*; *Spiraea prunifolia fl. pl.*, *Caryopteris mastacanthus*, and three fine viburnums, *V. dilatatum*, *V. macrocephalum* and *V. tomentosum plicatum*. There were also a number of plants not quite hardy enough for general cultivation out-of-doors but valuable for the greenhouse – *Chirita sinensis*, *Indigofera decora* and *Trachelospermum jasminoides*, besides a couple of conifers (but most of these came later) and *Ilex cornuta*.

It seems to have been expected that on his return Fortune would go doucely back to his job in the Chiswick greenhouses; but instead, he accepted a post as Curator of the Chelsea Physic Garden (where he 'exerted himself most energetically in rearranging the plants'[1] and erecting new buildings), and settled down to the production of his first book. It appeared in 1847 under the romantic title of *Three Years' Wanderings in the Northern Provinces of China*; actually his base at Shanghai is only half-way up the map, but it was so much further north than any part of China previously explored that it was regarded as practically at the Pole. The book contains two chapters on the cultivation and manufacture of tea, with some suggestions as to the possibilities of its development in India; and the next time Fortune sailed to China, it was as an employee of the East India Company, with a commission to procure seeds and plants of the best tea varieties and transfer them to India, along with expert Chinese tea-manufacturers and their apparatus. Tea-plantations had already been established in India, from plants procured with difficulty from Canton; but Canton is a long way from the main tea-growing districts and the varieties obtained had been poor. He set out on this new assignment on 2 June 1848.

Except for his surreptitious visit to Soochow, Fortune on his first trip had never been able to go beyond the permitted thirty miles or so from a treaty-port. On the second visit he was much more venturesome, though the same regulations were still nominally in force; and though he had some bad moments, he was never again attacked, and his alarms were mostly due to quarrels between his servants and villagers whom they had tried to defraud. Actually, the farther he went, the safer he was; for in remote country places the inhabitants were too ignorant to recognize a European, and accepted the explanation that he came from a distant province, beyond the Great Wall.[2] He avoided as much as possible any encounter with the more sophisticated Chinese, who would know him for what he was.

The difficulty was, that the best green-tea districts were a considerable distance from the coast, and the conscientious Fortune, knowing that it was impossible to rely on the good faith of Chinese emissaries, felt it was necessary to go to them himself. Emboldened, perhaps, by the success of his visit to Soochow, he again assumed Chinese costume, and submitted to having his head painfully shaved; and left Shanghai in October 1848 for the Hwuy-chow district, some 200 miles inland. He had with him two

[1] Field and Semple *Memoirs of the Botanic Garden at Chelsea* 1878

[2] Yet Fortune had markedly European features, with a strong nose and piercing eyes that might have been blue

servants – a fool of a coolie, who could not resist blabbing the delicious secret that his master was actually a foreigner, and a knave of an interpreter, always causing trouble by trying to impose on the locals. Perhaps the most dangerous part of his journey was at the start, when he had to cross the city of Hang-chow-fu from the canal by which he had come, to the cargo-boat on the river Tsien-tang in which he was to continue his journey. He dared not eat any dinner at the inn that night, as he was out of practice with chopsticks and feared to give himself away.

Slowly the boat with its cargo and its score of Chinese passengers worked westwards upstream to Yen-chow-fu where the river divides, and then turned up its north-westerly branch, where progress was impeded by many rapids. 'This', said Fortune, 'suited my purpose exactly, and enabled me to explore the botanical riches of the country with convenience and ease.' He would go ashore with his men in the morning, and climb the nearest hill in order to observe the course of the river ahead. If it was full of rapids, he knew that the boat would be slow and he could wander far afield; if it seemed smooth, he must keep close to the shore. 'Thus day after day passed quietly by; the weather was delightful, the natives quiet and inoffensive, and the scenery picturesque to the highest degree.' He was greatly excited by his first sight of the weeping Funeral Cypress, *Cupressus funebris*, and was able to collect seed, which germinated successfully in England; he also found an almost-hardy palm, and sent some young plants to Sir William Hooker at Kew, with the request 'that he would forward one of them to the garden of His Royal Highness Prince Albert, at Osborne House, Isle of Wight'.

When, after three weeks, they neared Whei-chow, the river became very shallow; a second boat was chartered to take some of the cargo and reduce the draught, and Fortune saw with surprise two coffins complete with contents removed from under his bunk – he had been quite unconscious of their presence. Navigation ceased at the village of Tung-che, and it was here that he made one of his most important finds – *Mahonia* (*Berberis*) *bealei*, growing in an old neglected garden; he obtained some cuttings on the homeward journey.

From Tung-che he took a chair westward to the celebrated hill of Sung-lo Shan, where the green-tea shrub was supposed first to have been discovered. At the foot of the mountain was a farm belonging to the father of his servant, Wang; and here Fortune took up residence for about a week. For the first three days it never stopped raining; four Chinese families lived in the house, each with its separate cooking-fire, and there was no outlet for the smoke, which got into his eyes until he was almost mad with pain. Then the weather cleared and he was able to roam the countryside, collecting tea-seeds, plants and information. He went back by much the same route, this time in a boat hired by himself, which gave him a little more freedom of movement; but instead of returning to Shanghai he left the river at Nechow and crossed by various canals and waterways to Ning-po. From there he sailed to Kiu-tang or Silver Island, another place where green tea was extensively cultivated, and which had been little explored. In due course he returned to Shanghai, and after the

Chinese New Year sailed for Hong Kong, where he saw his plants and seeds safely despatched to Calcutta. He then returned for more.

Hitherto he had been working in the green-tea districts, but now he wanted to investigate the areas where the black teas were made. His first attempt was by going up the river from Foo-chow; but on reaching the limit of navigation at Sui-kow he found that the overland journey was going to cost more than he could afford, so he sent his servants by the land-route to Ning-po – by which they could not avoid passing through the black-tea country – while he went back downstream and to the same place by sea. But although his two men duly arrived with seeds and plants, his conscience nagged him for not having supervised their work in person; and on 15 May 1849 he embarked on a new venture. It was a wet day, and he felt 'low spirited . . . knowing the journey I had undertaken was a long one and perhaps full of danger'.

He followed his previous route up the Tsien-tang to the junction at Yen-chow-fu, but this time took the more southerly branch, which he followed to the limit of navigation at Chang-shan. From there he took a chair over the Pu-ching mountains to Yu-shan and sailed down the Kwang-sin river to Kwang-sin (now Shang-jao) and Ho-chow. This brought him to the west of the Bohea range, and he transferred to a chair again to travel eastward over the steep mountain passes – but walked nearly all the way, in order to 'inspect the natural productions', a procedure highly approved by his chairmen. He found, and dug up, some new plants, but had difficulty in persuading the Chinese to carry what they regarded as worthless weeds; by a mixture of threats and promises he got the plants alive to Shanghai and thence to Europe. They included a hydrangea, *Abelia uniflora* and *Spiraea japonica fortunei*.

The road over the hills eastward to Chung-ngan-hsien, the centre of the black-tea district, was a busy one, and trains of loaded coolies were continually passed, but from Chung-ngan Fortune turned south-west by quieter ways to the sacred mountain of Wui-shan, an outlying spur at right angles to the main Bohea range. Here he stayed for a few days at two of its many temples before descending on its southern side to Tsin-tsun on the Min river. It would have been easy from here to go down the stream to Foo-chow; but that would have taken him through country he already knew; moreover, Foo-chow, he considered, was easier to enter than to leave. (Perhaps he was thinking of the pirates.) Instead, he turned northward by a new route to the east of the Bohea mountains – to Pu-cheng-hsien (where he had some bad moments at an inn frequented by gamblers and opium-smokers) and north to Ching-che and Ching-hoo, where he was again able to take to river-transport. He got back safely to Shanghai after an absence of three months.

Fortune is sometimes dismissed as a collector who procured nearly all his plants from nurseries, and who introduced novelties only because the places he worked in were new. It is apt to be forgotten that he made these two dangerous journeys of several hundred miles into the interior, in regions never before visited by a European, during which he collected in the wild. Although his main object was tea, he missed no opportunity of collecting ornamental plants also, which were housed temporarily in the garden of the handsome European house his friend Beale had built at Shanghai. He was no longer under any obligation to the Horticultural Society, and his finds were sent to various nurserymen.

Again, Fortune despatched his collections from Hong Kong, and again returned in the spring of 1850 to Shanghai for another season's work; but this time he entrusted his tea-collecting to emissaries and agents, and remained in the vicinity of Ning-po, Chusan and Shanghai. He revisited his old friends in the nurseries, with profitable results, and spent a two-month holiday in the islands of the archipelago, partly for the sake of his health. On 16 February 1851 he left for Calcutta, via Hong Kong, with sixteen glazed cases of tea-plants and germinating seeds, a team of expert tea-makers and their apparatus. After his arrival in India he was sent to inspect the plantations already established in Gurwhal and Kumaon; and got back to England in the autumn of the same year. The horticultural introductions from this journey included *Clematis lanuginosa*, *Exochorda racemosa*, *Forsythia suspensa fortunei*, *Poncirus trifoliata* and *Skimmia reevesiana*.

The Directors of the East India Company were so satisfied with Fortune's work, that they asked him to repeat the process, on very similar lines; and he left England again in December 1852, for an absence that was to last four years. This third trip, however, was much less interesting than the previous ones. The smouldering Tai-ping Rebellion had burst into flame, and he could not have made any long journeys, even if these had been necessary. On one of his visits to Shanghai a contingent of insurgents arrived on the same day as himself;[1] they captured the city and held it for the following eighteen months, but without any serious threat to the safety of the European community. Fortune revisited old haunts, such as the temple of Tien-tung, and went some short trips, but did not cover much new ground. (His boat was again robbed, but the courteous thieves came back an hour later and restored everything except his money.) His most important discovery was made almost by chance, at the half-past-eleventh hour. After seeing off the tea-plants and tea-makers from Hong Kong, Fortune returned to Shanghai to settle his affairs and pay off his collectors; and he made a final trip (at the end of October 1855) to Poo-in-che, to get seeds of *Pseudolarix amabilis*, which he had not been able to gather before. Here he was told of a handsome cephalotaxus (*C. fortunei*) which grew ten or fifteen miles further on; and on the way to its locality he noticed a fine-looking rhododendron in seed, which the Chinese assured him bore handsome flowers. He collected the seed and sent it to Messrs Glendinning of Chiswick, and in that way made one of his most valuable introductions – *R. fortunei*. A week or two later he sailed for Hong Kong and Calcutta, arriving on 10 February 1856; he spent eight months in India and was home on 20 December.

In 1858–9 Fortune was again in China in quest of tea, this time on behalf of the American Government, who wished to experiment with tea-growing in the Southern States; but this time he wrote no book about his

[1] 7 September 1853

travels – probably because they only covered the same ground as before – and we do not even know whether he went straight from China to Japan in 1860, or whether he had been home in the interval. It was almost as if the infection of the Taiping Rebellion was transmitted along with his tea-seeds. His second visit to India was followed within a few months by the outbreak of the Indian Mutiny[1] and the winding up of the East India Company in November 1858. The Company's plantations were neglected and Fortune's Chinese teas were eventually superseded by the indigenous Assam variety, which proved more suitable both for the Indian climate and the English taste. Soon after his introductions arrived in America, events began to build up towards the outbreak in April 1861 of the Civil War; so here, too, his work came to nothing.

In some ways Fortune was very much the conventional Englishman of his day. He shook his head over the heathen darkness of the Buddhist temples and longed for the time when the missionaries should have converted all China to sweetness and light. His sense of humour was simple and elementary, and so was his devotion to 'our beloved Queen'. His courage was phenomenal, but it sprang partly from his conviction that the Chinese could not be his superiors in anything but numbers – and even numbers he was prepared to outface, unless they were actively hostile. During nearly nineteen years of oriental travels he was in England for a total of only five years; yet he never once hints in his books that he had left a wife and family at home, or mentions the arrival of wished-for news or letters. He made considerable sums of money from the sale of Chinese and Japanese *objets d'art*, which he collected in the East, and enjoyed eighteen years of peaceable retirement, spending part of each year on his eldest son's farm in East Lothian. Bretschneider credits him with the introduction of about 190 species and varieties of plants, more than 120 of which were new – modest figures compared with those of some later plant-hunters, but they probably take no account of the numerous garden-cultivars which necessarily formed a large part of his collections.

Only a year after Fortune's third visit, war again broke out between China and four European powers (England, France, Russia and America), due chiefly to Chinese bad faith, in persistently violating existing treaties.[2] After its conclusion in 1860, several more ports and some inland cities were opened to foreign commerce, and subjects of the four powers were at last permitted to travel in the interior. It is remarkable how quickly the Europeans – French missionaries, Russian explorers, English administrators – penetrated into the furthest marches, and even over the border into Tibet. It began with Père David's journeys to Mongolia in 1866 and to eastern Szechuan in 1868; Tatsien-lu on the Tibet border was visited by travellers from 1875 onwards; Delavay was in N.W. Yunnan in 1881, Potanin at Sining in 1884, and Henry, who had been at Ichang since 1881, was transferred to Szemao in South Yunnan in 1897. These were not gardeners or indeed professional botanists, but their discoveries were of the utmost importance in stimulating the despatch of horticultural collectors;

[1] 10 May 1857
[2] During this conflict, the celebrated garden of Yuan-ming-yuan was destroyed

it was through them that the extraordinary and hitherto unsuspected richness of the flora of Western China was revealed.

Armand David (1826–1900) was an all-round naturalist with a preference for zoology, who went to Pekin as a missionary in 1862. The specimens he sent from there to the Natural History Museum in Paris were so interesting that the authorities obtained permission from his Order to send him on a scientific mission to Mongolia, and nearly all the remainder of his stay in China was spent on scientific research. He made three journeys, with a stoic disregard of local insurrections – like Michaux in Persia, he remarked that if he took notice of such things he would never get anywhere – and his own frail constitution. (He had five severe illnesses and several minor ones, in eight years.) His first journey, accompanied by a missionary colleague, took him north-east from Pekin into country where Chinese settlers were just beginning to penetrate among nomad Mongol tribes. He established his headquarters in the summer of 1866 at Saratsi, a large village with a famous lamasery near the Hwang-ho river, where Europeans had never before been seen; and from there made three extensive excursions, often under conditions of great hardship. Transport was difficult to obtain; for his first trip he hired a donkey, but at one of his camps wolves were so prevalent that at night the donkey had to be taken into the tent. For the other two trips all he could get was a camel – useful in some areas, but not adapted to steep mountain slopes. Natives were hostile, weather conditions severe, and water and provisions scarce, but he made an extensive survey of the flora, fauna and geology of an entirely unknown region, where he found silkworms genuinely wild, as well as *Rosa xanthina* and the shrub *Xanthoceras sorbifolium*, which he had already introduced as a cultivated plant from Pekin. They set out for the return on 27 August, David so ill and exhausted that he fainted several times before arriving at the comparatively civilized post of Erh-shih-san-hao. He was back in Pekin on 26 October.

The Mongol plateau was not very rich in new animals and plants, but his next journey made up for all deficiencies. He left Pekin on 26 May 1868, feeling 'as serene as if it were a question of a short walk', for a trip that was to take him to the borders of Tibet. He went by sea to Shanghai, arriving on 13 June; but at that season the upper reaches of the Yang-tse-kiang in spate were not navigable for native boats, and he had to wait till the river subsided to its winter level before he could reach his objective. He therefore went by steamer as far as Kiu-kiang, and explored the neighbouring country for four months before resuming his journey on 14 October. Then came the long, long haul up the river to Chung-king, which took till 17 December. 'I begin to feel very tired, morally and physically,' he wrote in his journal, 'and it seems as if this tyrannical navigation will never end.' The most dangerous section was in the rapids and gorges above Ichang, where it took up to eighty hauliers to get through the current, where the tow-rope broke several times, and where the reaches were studded with depressing frequency with the remains of previous wrecks.

David spent Christmas at Chung-king and then went overland to

Cheng-tu, the capital of the province of Szechuan. When the Chinese New Year celebrations were over, he set out on 22 February 1869 for Muping (now Pao-hing) the capital of a small native principality under the Prince of the Man-tzu, where there was already a well-established theological college. He arrived at this, his intended destination, on 28 February – nine months after his start.

He now found himself in a region with a remarkably rich flora, which he was unable to investigate fully because of his preoccupation with zoology. He found the giant panda and the rare musk-deer; but he also found fifteen rhododendrons (and procured a number of young plants of the handsomest kind, which he hoped to send to France) and many other new and beautiful plants, including his namesake tree, the davidia – as remarkable in its way as the panda, and much easier to acclimatize. He had intended to stay for a year, but was twice seriously ill – first with fever and an 'intestinal irritation', which laid him low for the whole of June and which he cured by a diet of goosefoot (*Chenopodium album*), and in September and October, with symptoms of typhus accompanied with acute pain and swelling in the legs; the Chinese called it 'bone-typhus' and poulticed him with ginger and onions moistened with brandy. He was obliged to cut short his visit and leave for more civilized regions on 22 November. Seven months later he returned to France for a nice rest – during the Franco-Prussian war.

In 1872 David returned to China and embarked on the investigation of two little-known mountain ranges. This time he left Pekin with a cart and two servants on 2 October, and travelled by slow stages south-west to Hoai-king and across the Yellow River to Honan; then turned westward into Shensi province, presently striking south to Inkia-po in the Tsin-ling mountains. This was a winter journey; he arrived in early December and stayed till the middle of February at villages in and around the range, hunting for rare birds and animals. On 18 February 1873 he left Yenkia, one of his principal centres, and travelled south-west to a village near Ch'eng-ku on the Han river, where he embarked on 17 April on a native boat for Hankow. (Six days later the boat was wrecked in a rapid and part of his baggage lost.) After a short rest at Hankow David continued down the Yang-tse to his old quarters at Kiu-kiang, and having packed and despatched the remainder of his collections, embarked with hardly a pause on what was virtually a fourth journey – south-west to the Bohea mountains in Kiang-si.

He left on 22 May, his transport this time being a sedan chair, for Nanchang and Fu-chou[1] on the River Wu-yang-shui, somewhat further south than the point previously reached by Fortune. He stayed for nearly three months at Tsi-tsu, a Christian seminary near Kien-chang; then crossed the range into the province of Fo-kien, arriving on 9 October at Koa-te, a Christian village in the tea-district. But again his body betrayed him. Already in August he had suffered from 'intermittent fever' (malaria) and at Kao-ten he developed inflammation of the lungs and was so ill that

[1] Sometimes spelt Foo-chow, but not the same as the port of that name

on 11 November he was given the last sacraments. He recovered, however, and on 5 December began the 'wearisome journey on foot' back to Tsi-tsu – with his collections in four bamboo hampers. Here he rested for six weeks before embarking at Kien-chang, still burning with fever, to return by river and lake to Kiu-kiang, where he arrived on 2 February 1874. Two months afterwards he sailed from Shanghai to France; some years later he managed to visit Tunisia, but he had regretfully to abandon plans to botanize in Japan, the Philippines and America. It is pleasant to record that this tall, thin, gently-indomitable naturalist lived happily almost ever after, establishing a museum of his own in Paris, where he gave lectures to prospective missionaries, and dying in 1900 at the age of seventy-six.

On his return to France, David found over eighty species of plants growing in the garden of the Natural History Museum, from seeds he had sent home; but not all of these became permanently established, or were of value as ornamental plants. His discoveries – one might say his revelations – were a great deal more important than his introductions, although these included *Prunus davidiana* – at one time thought to be the original wild ancestor of the peach – and *Clematis heracleifolia davidiana*. He is commemorated also in *Clematis armandii* and in the celebrated Père David's deer, extinct now except for the descendants of three specimens he managed with difficulty to obtain from the Imperial Hunting Park south of Pekin.

Five years after David's first arrival in Pekin, another missionary, Père **Jean Marie Delavay** (1838–1895), was sent to Hui-chou in the southern province of Kwan-tung, not far from Canton. He was not, like David, an all-round naturalist, nor was he released from his apostolic duties for the purposes of science, but he was a very keen botanist and collected many plant-specimens for Dr Henry Fletcher Hance,[1] a member of the British Consular Service in Canton and Hong Kong. In 1881 Delavay went home on leave; in Paris he met David, who introduced him to Adrien René Franchet, the Director of the Natural History Museum, to whom his own discoveries had been consigned. Delavay agreed to send his future collections to Franchet.

It was like the breaking of the broomstick by the sorcerer's apprentice. Franchet had dealt well enough with David's finds, which after losses by various accidents numbered about 1577, including more than 250 new species and ten or eleven new genera; he published a two-volume *Plantae Davidianae* in 1884, and two years earlier had examined d'Incarville's long-neglected collection of about 300 plants. But after his return to China Delavay sent him some 200,000 beautifully prepared specimens, representing more than 4000 species, about 1500 of them new; and these were presently augmented by contributions from two other missionaries, Farges (*c.* 4000; 1892–1903), and Soulié (*c.* 7000; 1890–1905). Despite his best efforts, Franchet was almost swept away by the flood he had invoked; he published a short list of Delavay's plants in 1885, began a catalogue which never got further than the letter A in 1886, and started a *Plantae*

[1] Hance did not himself travel in China, but was a great promotor and purchaser of the collections of others

Delavayi in 1889 which was still unfinished at the time of his death in 1900. The result was that many of the discoveries of Delavay (some of whose boxes were not even opened) remained unknown and unappreciated till long afterwards.

On his second visit to China Delavay was stationed for nearly ten years in the hills to the north-east of Tali-fu in Yunnan. His range of operations was limited, but he combed his area with great thoroughness; he is said to have climbed Mt Tsemei Shan 'which he called at the same time his garden and the Mont Blanc of Yunnan',[1] sixty times from all sides and at all seasons, to be sure of missing nothing it produced. In spite of this, he himself maintained that he had hardly tapped the resources of the rich alpine flora, and that it would be worth while sending a scientific expedition to the area. In 1888 he contracted bubonic plague, and though he recovered, he suffered from its effects for the short remainder of his life, eventually losing the use of his right arm. Undefeated, he returned to China after visiting France in 1892–3, and after a few months in Long-ki, reached Yun-nan-sen in February 1895. He died in December of the same year.

It has been said that more worth-while garden-plants were discovered by Delavay than by any other botanist. He sent seeds as well as specimens to the Paris museum, but many were lost by mismanagement; of thirty-four species of rhododendron only three were raised. Among the successes were *Deutzia purpurascens*, *Incarvillea delavayi*, *Iris delavayi*, *Paeonia lutea* and *Rhododendron racemosum*; but as with David, his introductions were of small significance compared to his discoveries. One need only mention *Meconopsis betonicifolia*, *Primula malacoides*, *Thalictrum dipterocarpum* and the genus Nomocharis, among hundreds of species which were left for other collectors to bring home.

Paul Guillaume Farges (1844–1912), who collected at Cheng-kan-ting near the north-eastern border of Szechuan from 1892–1903, was more fortunate with his seeds, for besides his consignments to the museum (including *Decaisneya fargesii*) he sent seeds to the nursery firm of Vilmorin, where they received much more skilled and careful attention. Maurice de Vilmorin raised *Deutzia vilmorinae*, *Incarvillea grandiflora*, *Rhododendron augustinii* and a single, prized seedling of *Davidia involucrata*. He also received seeds from Delavay and from **Jean André Soulié** (1858–1905). The latter was a medical missionary who made many dangerous journeys in the troubled Tibetan borderland; his introductions were few, but included one item of great garden importance – *Buddleia davidi*, now omnipresent and even naturalized in Britain. Soulié fell into bad hands, and was tortured and shot in 1905.

During this period (1860–1900) several scientific expeditions were sent out by Russia, and these, too, made their botanical discoveries. Those of Przewalski and Komarof took place beyond the boundaries of China proper – in Turkestan, Tibet and Manchuria; and so did the Mongolian journeys of Potanin. In 1884 and 1893, however, Potanin led two impor-

[1] Cox *Plant-Hunting in China* 1945

tant expeditions to Western China, of which some mention at least should be made.

Grigori Nicolaevich Potanin (1835–1920) had a stormy youth, and was sent for a time to forced labour in Siberia; but by the date of his Chinese travels he was high in repute, and his third expedition, 'fitted out on a pretty large scale',[1] was financed by the Imperial Russian Geographical Society and sent out in a Russian man-of-war. He is almost unique among botanical explorers in that he took his wife on each of his four journeys.[2] (Most plant-hunters were bachelors, or, if married, left their wives at home.) Alexandra Victorovna was herself a botanist, and helped to collect and look after the party's botanical specimens.

Having voyaged to Pekin by sea and stayed there about six weeks, they set out on 25 May 1884 in a westerly direction, crossed the Great Wall just beyond So-ping and proceeded north-west to Kui-hua-cheng. Here they were well on the way to David's Saratsi, but they turned west before reaching it, crossed the Yellow River, and skirted round the barren Ordos country to Boro-algasu. Crossing the Great Wall again at Hua-ma-chi, they entered Kansu from its north-east corner, and after visiting its capital Lanchow, continued to Si-ning in the extreme north-west, where China meets Mongolia and Tibet. They then turned southward, working through the mountains to Min-chow, Siku, the Satani river and the Chago-la pass – the country afterwards explored by Reginald Farrer. Potanin's expedition passed beyond Farrer's most southerly point of Weng-hsien to Lungan in Szechuan, before looping back by a slightly more easterly route to Lanchow and Si-ning. The winter of 1885–6 was spent at a place with the grand name of Gumbum, and in the spring they started for home – almost due north from Lake Koko Nor across the Gobi desert to the River Orkhon, and along it to Kiahkta on the Russian border. They brought back about 12,000 specimens of plants, representing some 4000 species, gathered and prepared by Potanin and his wife.

On his Chinese journeys Potanin was accompanied by three or four scientific colleagues, who kept splitting off and reuniting like so many amoebas, after the usual pattern of Russian exploration. On the second trip they arranged to meet at Sian (Si-ngan) in Shensi, some coming from Pekin and some already in the field. They left this town in February 1893 – a little early for plants, but they began to gather them at Feng-hien, further west, and afterwards collected almost daily on the way to Cheng-fu, which they reached on 9 March. In April they visited Mt Omei, but their principal objective for the summer was Tatsien-lu, which they reached in May. In July, they arranged to go north to Li-fan-fu. The party divided; a colleague, Kashkarov (just back from a thirty-nine-day trip across to Batang), was to take Mme Potanin by the easier eastern route, while Potanin and another colleague, Rhabdanov, took the mountainous northerly direction They had an exhausting but rewarding journey, through country very rich in plants; but on arriving on 17 August at

[1] Bretschneider

[2] Another Russian, Alexei Fedstchenko, did the same, taking his botanist wife Olga to Turkestan in 1868–71

Li-fan-fu, Potanin found that his wife had fallen very ill. He abandoned his further plans and hastened back by the shortest road to civilization and medical aid. But the shortest roads are long in China and Tibet; it took a month to get to Pao-ning on the Kialin river where they could transfer to the easier water-transport, and Alexandra died on 1st October, two days before they reached Chung-king, at the junction with the Yang-tse. Even after death, the unfortunate woman had not concluded her travels; her body was taken down-river to Shanghai and by sea to Tientsin, and from there was despatched by caravan across Mongolia and the Gobi desert to Kiahkta – the nearest place where she could be buried in Russian soil. (A fine species of rhubarb was named *Rheum alexandrae* in her honour.) Her husband meantime returned from Tientsin to Shanghai, to bring his party's collections home the long way round, by sea. This time the botanical harvest amounted to some 10,000 specimens, plus a great many seeds and drug plants. Nothing of great garden importance is recorded to have been grown from his seeds, though *Arundinaria nitida* was raised at Kew from seed supplied by the zoologist Berezovski, a member of his party.

The botanical achievements of these dedicated priests, these intrepid explorers, were equalled if not surpassed by an insignificant-looking little Irishman, who came to China as a Medical Officer and Assistant Inspector of Customs in 1881, and lived for seven years in one place. The Chinese Government had made the discovery that Europeans were less corrupt than their own officials, and had arranged that their Imperial Customs Service should be staffed by the British. **Augustine Henry** (1857-1930), aged twenty-four and just finishing his medical training at Belfast, was among those recruited for the service. After a few months in Shanghai he was transferred in March 1882 to Ichang, at the limit of steam-navigation on the Yang-tse, where Maries had spent a brief fortnight three years before.

Here he was in the very heart of China; on every side the smiling inscrutable country stretched for hundreds of miles. (A circle drawn on the map with Ichang as its centre and Tientsin as its radius neatly encloses almost the whole of China, except parts of Chi-li in the north-east and Yunnan in the south-west.) There were few resident Europeans, and Henry, whose office hours were from 10 a.m. to 4 p.m., found time hanging heavily on his hands. One of his duties was to compile a report on plants used as drugs by the Chinese; this started him botanizing, and the difficulty of identifying some of the plants was the occasion of his first letter to Kew, in 1885, in which he offered to collect specimens and asked for instructions. An encouraging letter and a leaflet of routine instructions were sent in reply, and he despatched his first collections in November of the same year – a thousand plant-specimens containing ten new genera, and including examples of the coffin-juniper, which the then Director, William Thistelton-Dyer, for years had vainly tried to obtain. The cordial relations thus established were ever afterwards maintained, and Henry turned more and more to collecting. A few miles to the west began the fantastic Ichang Gorges, with their precipitous limestone cliffs and rich flora; he explored a part of these during his brief vacations, but in 1888

Thistelton-Dyer obtained for him a long leave and a small grant towards his expenses, which enabled him to go further afield.

He made two excursions; the first was in a narrow arc to the south of the river – to Chang-yang on the tributary Tsing-kiang, then west to Pa-tung and Wu-shan, both on the Yang-tse itself, and back by boat. After a short rest at Ichang he went due north to Pao-kang and Fang-hsien, and then worked his way westward through the Tapa Shan mountains, over the border to Szechuan and down to Wu-shan again. This was wild and difficult country, though rich in animals and plants, but Henry merely remarks that 'I had a very pleasant trip, being on excellent terms with the people'. He had no great opinion of the Chinese, but got on with them well, and instituted the novel practice of training and sending out native collectors. Wilson was afterwards welcomed wherever Henry had been.

Henry's finds on these two journeys and in the Ichang neighbourhood are too numerous to list here; to get some idea of their value one needs only to mention a few of his first-class namesakes – *Acer henryi*, *Lilium henryi*, *Parthenocissus henryana*, *Rhododendron augustinii*[1] and *Spiraea henryi*. The lily was the only one of these desirable plants that he introduced himself; he sent bulbs to Kew in 1889.

The years 1889–95 were important in Henry's private life, but less botanically productive. He was transferred to the island of Hainan; after three months he contracted pernicious malaria; he went home on leave; he married, and lost, a charming, consumptive wife; and he spent a couple of years on Formosa. After another spell of leave he was appointed in May 1896 to Meng-tse, in southern Yunnan, near the borders of French Indo-China.

Here he again found himself in a region with a rich and little-known flora, of more interest to botanists than to gardeners in temperate regions, as the climate was semi-tropical. When his duties permitted he went on expeditions to the mountains; at other times a trusted Chinaman called Old Ho collected for him. He became very interested in a non-Chinese mountain tribe called the Lolos, and studied their language and customs, much as J. F. Rock later studied the Na-khi people, further to the north. In January 1898 he was moved to Szemao, eighteen days' journey to the west; a remote place with five resident Europeans and a wet climate. He despatched thirty-two cases of specimens to Kew and set out.

For some time his duties prevented him from going far afield, but when he did so, he was dismayed to find how rapidly the surrounding country was being deforested by the Chinese, with consequent destruction of the flora. He had already suggested to Thistelton-Dyer that it would be worth while to send a collector to the area; he now represented this as urgent, for whole genera, of great importance to the study of geographic botany, might be exterminated within the next fifty years. Unfortunately Kew had not the funds to finance such an expedition. Professor Sargent of the Arnold Arboretum in America, with whom Henry had corresponded since 1894, offered to pay for an expedition if Henry would lead it; but

[1] *Rhododendron henryi* was named after an earlier Henry, a missionary

after some hesitation he refused. He was tired of China, and wanted to go home; it was a job for a younger man.

Fortunately the younger man had already been found – not by a learned institution but by a commercial nurseryman. Harry Veitch[1] of Chelsea had seen Henry's specimens, especially those of the davidia, of which Henry had also written a glowing description; and decided that it would be worth the firm's while to send out a collector. On the recommendation of Thistelton-Dyer he despatched in April 1899 the last and greatest of the Veitch collectors – E. H. Wilson.

Wilson was instructed to go straight to Szemao, and learn from Henry everything he could – including the locality of the one davidia tree that Henry had found. But although he arrived at Hong Kong on 3 June, he was held up for a long time at Lao-kai on the Red River by local insurrections and transport difficulties, and did not reach Szemao till October. Henry, who had been waiting for him for months, was inclined to be disappointed when he arrived; possibly finding him more of a gardener and less of an academic botanist than he had expected. But on closer acquaintance he began to think that Wilson would do very well; he had the most important of all qualifications – the art of getting on with the Chinese.

Very soon after Wilson's arrival Henry was summoned back, to take charge at Meng-tze. Wilson went with him, and then left for the coast, en route for Hong Kong, Shanghai and Ichang. After his departure Henry spent a troubled year in the turmoils of the Boxer Rising; he eventually had to evacuate Meng-tze and withdraw his staff to Ho-kow, just across the river from Lao-kai and in safer territory. He left China for good on 31 December 1900[2] – as he himself pointed out, the last day of the nineteenth century. It was the end of an era. The seminal period of botanical discovery was over; the age of the horticultural collector had begun.

There are no university courses for plant-hunters, but **Ernest Henry Wilson** (1876–1930) had a more thorough training than most. Born at Chipping Camden, Gloucestershire, he went, on leaving school, to Hewitt's nursery at Solihull; at the age of sixteen he transferred to the Birmingham Botanic Gardens and attended classes at the Technical College, where he won the Queen's Prize for botany. In 1897 he went to Kew, and about a year later was contemplating taking a post as a teacher of botany, when Veitch applied for a collector and Thistelton-Dyer recommended Wilson for the job. He then worked for six months in Veitch's nursery at Coombe Wood in order to learn their trade and requirements, and was sent to China by way of the Arnold Arboretum at Boston, Mass., where he was to study the latest techniques of plant-collection and carriage. (He was there only for five days, but in this short time laid the foundations of a friendship with the Arboretum's celebrated Director, Professor Sargent, which was to be of great importance to his subsequent career.) He then crossed the continent to San Francisco and sailed from there to China – the first collector to approach from the east

[1] Younger brother of John Gould Veitch, and now head of the firm

[2] After twenty years in China, Henry started a new career in forestry, which lasted for thirty distinguished years

instead of from the west. From Hong Kong he made the difficult journey through French Indo-China to Henry at Szemao, whose advice about Chinese conditions and collecting localities was probably worth all that went before.

The particular object of this expedition was to obtain seed of the davidia, and Veitch's instructions had been that he should 'stick to the one thing you are after, and do not spend time and money wandering about. Probably almost every worthwhile plant in China has now been introduced to Europe'. Fortunately Wilson did not take these instructions too literally; his first consignment of plants was collected on the way back from Meng-tze to the coast, and included *Jasminum primulinum*. From then on, he sent home numbers of valuable plants, which must far have exceeded Veitch's most sanguine expectations.

From Hai-phong Wilson went by sea to Shanghai, and there equipped himself with a houseboat, on which he lived for his first season at Ichang. This was a very practical arrangement; in the unsettled state of the country a boat would afford greater independence, mobility and security than a lodging ashore. Not all the Chinese provinces supported the Boxer[1] Rising, which reached its climax that summer of 1900 with the siege from 13 June to 14 August of the foreign legations in Pekin; and Ichang was comparatively unaffected. Henry had supplied Wilson with a rough sketch-map showing the location of the only davidia tree he had seen, twelve years before; and Wilson's first task was to find this single tree, in a mountainous area about the size of New York State. From his arrival in February he waited till April, when the tree should be in flower, and with the help of some of the collectors that Henry had trained was then able to pinpoint the very spot where it had, till recently, stood; it had been cut down to build a Chinese house. No other specimen was known to exist except at Mupin, a thousand miles to the west, where the species was originally discovered by David. About a fortnight later, however, Wilson chanced to find another davidia much nearer to his headquarters, and a search of the neighbourhood revealed a grove of about a score of trees; these were carefully watched until the seed was ripe and a good quantity was obtained. In subsequent journeys Wilson saw plenty of davidias, but never again did he find them bearing seed.

In his second season (1901) Wilson went further afield, and followed Henry's footsteps both south and north of the Yang-tse. The mountainous region of north-west Hupeh was wild, inhospitable and inaccessible, its half-obliterated tracks representing, as he afterwards said, the worst going in China; but it was particularly rich in trees and shrubs, all of which proved reliably hardy in cultivation. In addition to the consignments he had previously sent, he brought home with him thirty-five cases of bulbs and roots, seed of 305 species and over 900 herbarium specimens. He reached England in April 1902, and returned to the Veitch nurseries at Coombe Wood.

The summer of that year must have been a happy time for Wilson. He

[1] Literally, the Society of Patriotic Harmonious Fists

was twenty-six years old, newly married, with his difficult first mission successfully accomplished; and he had the pleasure, not given to every collector, of seeing his Chinese seeds germinate and his Chinese plants establish themselves on British soil. They included (besides the davidia), *Acer griseum*, *Actinidia chinensis*, *Clematis armandii* and *C. montana rubens*, *Ligularia clivorum* and *Rodgersia pinnata*. But his honeymoon was brief; nine months after his return to England he was off again on his second Chinese journey, on behalf of the same firm.

This time he was commissioned to obtain seeds of *Meconopsis integrifolia*, which involved going much further to the west. For the following two years he made his headquarters at the town of Kiating-fu, on the Min river half-way between the Yang-tse and the provincial capital of Szechuan, Cheng-tu. This was an excellent centre of communications, of which Wilson took full advantage. He went twice to Sung-pan, some 300 miles to the north, and three times north-west to Tatsien-lu, varying his route every time in order to cover more country. West of Kiating-fu lie three remarkable limestone mountains, Omei Shan, Wa Shan, and Wa-wu Shan, enclosing in their triangle a wilderness called the Laolin which Wilson was to cross on a later occasion. Omei, being a sacred mountain, was well-wooded; the others had been deforested but were covered with flowering shrubs, including many rhododendrons. Between his northern journeys Wilson explored Omei Shan, already famed as a rich botanical area, and the less-known Wa Shan. On this journey much of his collecting was done above the tree-line, for Veitch had demanded hardy herbaceous plants, and Wilson obligingly produced *Aconitum wilsonii* and three fine primulas (*P.P. cockburniana*, *polyneura* and *pulverulenta*) besides the meconopsis he had been sent to obtain, and poor Mme Potanin's *Rheum alexandrae*. But there were woody plants, too, including *Berberis verruculosa* and *B. wilsonae*, and rhododendrons *calophytum*, *intricatum*, *lutescens* and *souliei*.

Unfortunately the great firm of Veitch was on the verge of decline. Three of the family died tragically young, and after the death of the third (the James H. Veitch who had collected in Japan) in 1907, there was no one to carry on the firm except his uncle Sir Harry, now approaching his seventies. When the lease of the Coombe Wood nursery expired in 1914, the stock was sold and the firm wound up. Many of Wilson's tree and shrub introductions were neglected just as they were coming to maturity, and were never commercially propagated and distributed; some of them in consequence long remained rare.

After his return to England in the spring of 1905 Wilson took a post as botanical assistant at the Imperial Institute – the sort of work he had been about to undertake when his career as a collector began. But he did not hold it for long. He was invited by Professor Sargent to make another expedition to China on behalf of Harvard University (the owners of the Arnold Arboretum) and some private subscribers. Sargent, like Wilson himself, was chiefly interested in woody plants, and he had personal experience of the work of a collector in the field. It was an opportunity not to be declined; Wilson sailed to Boston for his instructions in December 1906, and by San Francisco to China in January 1907.

Unfortunately Wilson's books are singularly uninformative about his travels. In the two volumes of *A Naturalist in Western China* (1913) he tells us much about Chinese plants and where he found them, but the examples are picked out at random from his last three journeys, and no dates or itineraries are given. It seems that on this third trip he revisited both Ichang and Cheng-tu; explored Mupin where David had worked and got seeds of many of his plants; and went to Tatsien-lu again by a more northerly route, over the difficult pass of Ta-p'ao Shan. He revisited Wa Shan, finding *Paeonia veitchii* and *Ribes laurifolium*, climbed the third of the trio of mountains, Wa-Wu Shan, and crossed the Laolin from north to south, probably the only European to have done so. A welter of crags and gullies, the Laolin had a particularly wet climate, and was avoided even by the Chinese; in seventy miles of hunting for storm-obliterated trails and wading along the margins of streams, Wilson never saw more than fifty yards ahead, owing to mist and rain. Among the shrubby plants he brought back were *Magnolia wilsonii*, *Staphylea holocarpa rosea*, *Hydrangea sargentiana* and *Lonicera nitida*. That great but egocentric gardener, Ellen Willmott, was one of his subscribers, and *Ceratostigma willmottiae*, *Corylopsis willmottianum*, *Rosa willmottiae* and *Lilium davidii* var. *willmottiae*, were raised by her from Wilson's seed.

In all his years in China, Wilson made no attempt to learn Chinese; not for him Fortune's plucky adventures in disguise, though his rather heavy features had more of an oriental cast than Fortune could command. He preferred to rely on the services of a competent interpreter, and travelled always with a train of twenty-five or thirty coolies and attendants. Some of these he had trained as collectors, and re-engaged when possible from year to year; but they were kept more or less under his own eye, and not sent off, as Forrest's were, to work independently. He also took everywhere a bulky whole-plate camera with its supply of glass plates – and a sedan chair; the latter essential for prestige even if it had to be dismantled and carried.

Wilson returned to Britain in May 1909, but in September of the same year he emigrated to America, as a permanent member of the staff of the Arnold Arboretum, of which he ultimately became Keeper. He did not take out American citizenship, for he intended to spend in England the retirement that never came. Meantime, there was still work to do. 1908 had been a bad season for conifers, and many species had borne little or no seed. Moreover, he had discovered a beautiful new lily in the Min valley, which he wished to obtain in quantity. A new expedition was therefore launched, early in 1910.

This time Wilson travelled across country from Ichang to Cheng-tu, by a route parallel to the Yang-tse but approximately fifty miles north of it, from the limestone mountains of Hupeh to the fertile sandstone plains of Szechuan. The going was exceedingly tough; there were main roads – some six to eight feet wide – but few main roads led in the direction he wanted to go. He and his men walked the 200 miles of rough tracks in twenty-two days – and that was considered good going. He then revisited Sung-pan, far to the north.

The road from Cheng-tu to Sung-pan was an important trade-route to Tibet. It passed through the Min river gorges, where the cliffs consist of mud shales which tend to disintegrate after heavy rain, causing dangerous rock-falls. It was in this region that *Lilium regale* grew. In late summer, Wilson camped seven days' journey from Sung-pan while he made arrangements for six or seven thousand of the bulbs to be dug up in October and packed according to instructions for transport to America. He then set out to return to Cheng-tu, but shortly afterwards was caught in a rock-avalanche, which swept him down the hillside; he struggled out of his overturned sedan-chair, but a boulder hit him in the leg, breaking it below the knee in two places. Under his instructions his men devised a splint from the legs of his camera-tripod; while this was being adjusted, another mule-train came by. There was no room for it to pass, and it was too dangerous a place for it to stop; there was nothing for it but to lay the patient across the path and let the train walk over him. Nearly fifty mules stepped over his prostrate body and not a hoof touched him, though, as he feelingly remarked, each one looked as big as a plate. Three days later he reached the comparative civilization of Cheng-tu, but the lacerated leg had become infected, and for six weeks the bone showed no sign of uniting. He was threatened with amputation, but after three months it healed sufficiently to allow him to start on crutches for the long homeward journey. In America the twisted limb was broken and re-set in a Boston hospital, and it afterwards served him well for many thousands of miles, though he limped slightly for the rest of his life. The lily-bulbs arrived safely, and it is said to be from this one introduction that all the Regal Lilies in our gardens are derived.

Wilson never returned to China, although he made several other journeys to the East. As we have seen, he did not escape all the tribulations of the plant-hunter; he was almost drowned in the Yang-tse on his first journey, and almost starved on a trip by an unfrequented route to Tatsien-lu, on the second. But compared with that of earlier collectors, his work presents an almost monotonous record of success. His relations with the Chinese were amicable; transport had improved, and we hear no more of plants dying or collections being lost on the way home. Professor Sargent was nobly generous in distributing his seeds to other botanic and private gardens, to make sure that some of them at least should be successfully grown; and the regions in which he worked ensured that his plants would be hardy enough for general outdoor cultivation. It would be impossible to mention a tithe of the garden-worthy plants he introduced, a hundred of which received horticultural awards; the lowest estimate puts his new introductions at more than a thousand. (They included sixty rhododendrons.) But he missed by a narrow margin the restricted area of survival of that wonderful 'fossil' tree, the metasequoia, found by Chinese botanists in 1944. He was spared the knowledge of what he had missed, for he and his wife were killed in 1930, in a car accident in the United States.

The private patron now becomes of importance, perhaps for the last time; it seems unlikely that there will ever again be an employer of collectors of the scope and capacity of Arthur Kilpin Bulley (1861–1942), a wealthy Liverpool cotton-broker, who started to build a house and lay out a garden near Neston on the estuary of the Dee, in 1898. He had already been in correspondence with Augustine Henry, who visited the Bulleys in 1901, soon after his retirement from China. Henry's account of the Chinese flora was probably the direct cause of Bulley's decision to send out a collector – his previous attempts to get seeds from missionaries having only resulted in 'the best international collection of dandelions to be seen anywhere[1]'. In 1904 he asked Professor Isaac Bayley Balfour of the Royal Botanic Gardens, Edinburgh, if he knew of a suitable man for the purpose; and the professor recommended Forrest.

George Forrest (1873–1932) had already been something of a rolling stone. He was born at Falkirk, and after leaving Kilmarnock Academy had started life as a pharmaceutical chemist. He gave this up, and for some years led a roving life in Australia; then he returned to Scotland, and at the time of Bulley's enquiry had been working for two years in the herbarium of the Edinburgh Botanic Garden. All of these seemingly unrelated experiences were afterwards of great service to him, but the most important was his close association with Bayley Balfour, to whom he sent henceforward all his new species for identification.

The arrangement with Bulley was not concluded till May 1904, and it was August before Forrest arrived, by way of Burma, at Teng-yueh in Henry's Yunnan, which was to become the headquarters for all his subsequent travels. It was too late in the season to do much collecting, and Forrest spent the first months surveying his field and learning his way about. The British Consul at Teng-yueh, G. L. Litton, was very helpful, and soon after his arrival Forrest accompanied Litton on a journey by way of Likiang, the Yang-tse valley and the Karu pass to the French missionary station at Tsekou on the Mekong, where the sight of dried plant-specimens collected by the fathers caused Forrest to make the fateful decision to return there the following summer. After his return to Teng-yueh he attempted, in December, to reach Atuntse, but only got as far as Chung-tien before all the passes were blocked by snow.

The year of 1905 was one of trouble along the Chinese-Tibetan borders. The Tibetans had been upset by Colonel Younghusband's unauthorized entry into Lhasa the year before, and by a Chinese attempt to take over the town of Batang. A rising of the Batang lamas spread all along the borders and served as an excuse for a general massacre of the missionaries.[2] In July, when Forrest was established at Tsekou, word came that Atuntse, only two-and-a-half days' march away, had been taken by the Tibetans and its Chinese garrison massacred. It was decided to retreat to a place of greater safety thirty miles down the valley, where Chinese troops were stationed.

Towards the end of July a party of about eighty set out by night – Forrest and his troop of seventeen servants and collectors, and the two aged French missionaries with their Chinese converts and their families. It is difficult for an undisciplined party of this size to move silently, and on

[1] Mrs Bulley, quoted in the *RHS Journal* VOL LXXXV, 218
[2] It was at this time that Père Soulié was murdered

passing the lamasery of Patong a noise made by some member of the cortège betrayed them to their enemies. The next day revealed a column of smoke over the spot where Tsekou had stood, a party of Tibetans in hot pursuit, and another sent in advance to cut them off. When the attack came, the unfortunate party panicked and scattered; but even if they had made a stand, the result would probably have been the same. All but fourteen were slain or captured; one of the priests was killed on the spot, the other caught after a day or two in hiding and slowly tortured to death; of Forrest's own party only one survived. At a bend in the path where he was momentarily hidden from both parties, Forrest fell, rolled and plunged steeply downhill into the jungle, and so escaped.

For the next eight days he was hunted like a wild beast. He discarded his boots on the second day, to avoid leaving tell-tale tracks, and he had no food except a few ears of wheat and a handful of parched peas which someone had providentially dropped. He still had his arms, and at the end of the ninth day, desperate and at times slightly delirious, he decided to 'hold up' a Lissu village of some half-a-dozen huts, to obtain food. Fortunately the headman proved friendly, and supplied a meal of parched barley-flour (the only food they had) of which the famished man ate so unwisely that he suffered from gastric inflammation for weeks afterwards. The Tibetans were still watching the roads to the south, and the headman, after concealing him for four days at great risk, passed him on to the headman of the next village, who agreed to supply him with guides to take him by little-known paths over the dividing-range to the east and then southwards to the Mekong valley, thus making a detour round the most dangerous area.

It was a miserable journey. It rained without cease, they dared not light a fire, they had no shelter and very little food. After hacking a way through dripping jungle, they had to struggle through the snows, rocks and glaciers of the 17,000-ft ridge – and Forrest, be it remembered, had no European boots – and then to make an equally steep and difficult descent on the further side. At last they approached cultivated ground. The villagers of the region were accustomed to protect their crops from marauders by sharp slivers of bamboo, concealed in the paths; Forrest trod on one of these and it penetrated right through his unprotected foot, projecting several inches beyond, and causing a painful wound that took long to heal. Fortunately he soon afterwards reached a village where he was known, and the worst of his troubles were over. Four more days took him to a respectable garrison-town (Hsias-wei-hsi) and 200 Chinese troops were supplied to escort him and a surviving French missionary from another district for the further nineteen days' march to Tali-fu. He had, of course, lost everything except his rifle, revolver, and the tattered clothes on his back.

At Tali-fu Forrest met his friend, Consul Litton, who re-equipped him to the best of his ability, and they returned together to Teng-yueh in the last week in September. Forrest then had eight days in which to recover from his unnerving experiences, before setting off again to accompany Litton on an exploring expedition up the Salween river, to a point not hitherto visited by any European. This was further west and nearer to

Tibet than the dangerous Mekong; but the valley was inhabited by a tribe called the Black Lissu, and though they had a bad reputation,[1] they were peaceable compared to the murderous Tibetans. Unfortunately there were other dangers in the Salween valley; it was notoriously unhealthy and even the natives of the party were affected by illness. Forrest and Litton got back to Teng-yueh, much exhausted, on 13 December; Litton succumbed to blackwater-fever a month later, and Forrest had in his blood the poison of the dreaded Salween malaria, which flared into serious illness the following summer.

After the Yang-tse parts company with the parallel courses of the Salween and Mekong, it makes a great loop to the northward round the Likiang range before turning east across China to the sea, with the town of Likiang-fu at the base of the loop. These mountains became one of Forrest's principal hunting-grounds; he described them as one huge natural flower-garden, fifty miles in extent, from the base to the limit of vegetation at 17,500 ft. He worked in this paradise from March 1906 to August, when he was overtaken by the Salween malaria and had to return to Tali-fu, leaving his men, now well-trained, to complete the harvest on his behalf. He was so ill that he was advised to return at once to Europe, but he feared to lose the results of yet another season's work and stayed on, eventually making a slow and partial recovery. Sometime in 1906 he managed to make two 'short and hasty' trips to Delavay's locality, the Tali range; this, too, was to become one of his special regions. He was home before the end of the year.

The main lines of Forrest's collecting were all laid down during this eventful first journey – his areas chosen, his methods established, the first of his 'large and competent' staff of helpers trained, and his relations with the Chinese put on a friendly footing. He had found, among much else, a number of fine primulas, including *P.P. bulleyana, beesiana,*[2] *forrestii*, and *littonii*; primulas and rhododendrons were henceforth to be his specialities.

Four years later, Forrest was back in Teng-yueh. According to Bulley, he had 'gone over' to J. C. Williams of Caerhays in Cornwall, and it seems that this second trip was financed by a syndicate of which Williams was the principal member, but in which Bulley nevertheless had a share, for two of Forrest's most valuable introductions were raised in his garden from seed collected during this short and little-known expedition of 1910–11 – *Pieris formosa forrestii* and *Gentiana sino-ornata.*

It was Williams, at any rate, who was the principal patron of Forrest's important third journey, 1912–14. Fighting was still in progress between China and Tibet, and in addition Southern Yunnan was ablaze with revolution. On arrival, Forrest was prohibited from going beyond Teng-yueh, and in mid-September, as the political trouble spread, he was ordered by the British Consul to withdraw for a time to Burma. He was able to return to Teng-yueh at the end of October, and the following spring (1913) moved on, first to Tali-fu and then to his base for the season at U-lu-kay, about fifteen miles north of Likiang. He increased both the number of his

[1] They massacred a party of Germans a few months later

[2] In 1905 Bulley had founded the firm of Bees' Seeds Ltd

staff and the radius of his operations, ranging across the bends of the Yang-tse to Yung-ning in the north-east and to the Haba mountains and the pass of Bei-ma Shan in the north-west. In December he returned to Tali-fu; and the day he entered it the local soldiery (some 3000 strong) mutinied, shot their officers and captured the city. Less lucky than Fortune on a similar occasion, Forrest was kept a prisoner for three weeks and forced to act as medical attendant to the wounded rebels. (He later remarked that living in China was like camping alongside an active volcano.) The city was stormed and re-taken with much bloodshed by loyal troops from Yunnan-fu on 24 December; after which, Forrest writes, 'Order being once more restored, I proceeded on my journey to Teng-yueh, packed up my collections there, and despatched the whole to England'. Then he returned for more.

He made his base this year (1914) a little further north, and worked north and west till he reached the point where he had been in 1905, in the vicinity of Tsekou; but the seed-harvest was poor, owing to the unusually wet season. That of 1913 had been copious; Forrest at times employed as many as twenty men, whom he sent in pairs to regions he was unable to cover himself, and with their help was able to send home fully 200 lbs of seeds of 400–600 different species, and about 3000 herbarium specimens. Among the finds of this journey were *Berberis jamesiana*, named after his brother, and *Dracocephalum isabellae*, named after his sister.

About this time the horticultural world was swept by a mania for rhododendrons; it seemed impossible to have too many. The Rhododendron Society was founded in 1915, with twenty-five members, and Forrest made his fourth and subsequent journeys largely on their behalf. J. C. Williams was prominent among them, and he offered Forrest a bonus for every new species he introduced. Over the years he made more than 5000 gatherings of rhododendron seeds, and they included so many new kinds that the whole genus had to be revised;[1] it was no longer possible to maintain the previous distinction between rhododendron and azalea after the discovery of so many intermediate forms. Forrest's own species, *R. forrestii* (=*reptans*), was found in that fateful 1905 season when all his collections were lost; later he was able to collect and introduce it. The 1917–19 gatherings were made chiefly on the Mekong-Salween divide, and yielded *R.R. giganteum* and *griersonianum*. On his next journey, 1921–22, Forrest found an even richer field in the Salween-Taron watershed; the rhododendrons there were so varied and so fine that he believed he was approaching the centre of distribution of the genus, somewhere to the north, in Tibet; but this postulated rhododendron-paradise he never attained.

Forrest had developed a sense of territory as strong as that of a robin, and was as jealous of intruders. He was 'irked' that summer of 1922, to find that the American J. F. Rock was collecting in what he considered his preserves; and Kingdon-Ward was also in the area, for the third and last time.[2] (Forrest might well look a little askance at Kingdon-Ward,

[1] By Sir Isaac Bayley Balfour
[2] Forrest met them both that summer in the Likiang range

who had been sent by his own employer to his own area while he was still in it.) Farrer and Kingdon-Ward courteously avoided as far as possible any encroachment on Forrest's territories, but he showed no compunction about invading theirs; collecting both in the vicinity of the latter's base at Atuntse, and in the region of Farrer's second and last trip on the frontiers of Burma.[1] This was on Forrest's sixth journey (1924–5) when he worked north-west from the Schweli-Salween divide to the Htawgaw, Hpimaw and Chimili passes. In one small valley in the Chimili area which had been visited three times (May, August and October) a fortnight at a time, by Farrer and Cox in 1919, Forrest found a primula and a rhododendron they had missed, and two colonies of the rare *Codonopsis farreri* to their one; there was very little that he overlooked. He made this season one of his most important discoveries – *Camellia saluensis* (parent of the valuable hardy hybrids raised by J. C. Williams) and also two other good camellias, *C. cuspidata*[2] and the wild single form of *C. reticulata*.

By this time Forrest was approaching middle-age, and announcing every journey as the last he intended to make. In 1930 he set off for a final expedition, with the object of revisiting all his best collecting areas and obtaining, in particular, seeds of species which had been introduced but which had failed in cultivation. He was accompanied at the outset by Major Lawrence Johnstone of Hidcote, but Johnstone fell ill at 'almost the last post of civilization' and had to return home – taking with him, however, *Jasminum polyanthemum*[3] and *Mahonia lomariifolia*. Forrest had a good season, and returned satisfied to Teng-yueh. He had 'of seed such abundance, that I scarcely know where to commence, nearly everything I wished for, and that means a lot. Primulas in profusion, seed of some of them as much as 3–5 lb, the same with Meconopsis, Nomocharis, Lilium, as well as bulbs of the latter. When all are dealt with and packed I expect to have nearly if not more than two mule-loads of good clean seed, representing some 400–500 species, and a mule-load means 130–150 lb. . . . If all goes well I shall have made a rather glorious and satisfactory finish to all my past years of labour'. Fate took him at his word. On 5 January 1932, his packing completed, he was out with his gun, when he suddenly cried out, collapsed, and was dead before anyone could reach him. He was buried at Teng-yueh, close to his friend, Litton.

Forrest's journeys in China covered a span of 28 years; he introduced an enormous number of plants and worked on a wholesale scale. His were plants for the specialist – for the rock-garden and alpine-house enthusiast, for the amateur of primulas and of rhododendrons; many of them are hardy only in the most favoured areas of Britain. Some of his introductions overlapped with those of Wilson; both are credited with *Meconopsis integrifolia*, *Iris chrysographes* and *Thalictrum dipterocarpum*. Forrest also established in cultivation a number of plants previously known, but still rare.

Having lost Forrest to J. C. Williams and his syndicate, A. K. Bulley applied to Sir Isaac Bayley Balfour for another collector for his own ex-

[1] See p. 175
[2] Previously introduced by Wilson
[3] Previously introduced, but lost to cultivation

clusive use; and Bayley Balfour this time recommended the son of one of his friends, a young man called **Frank Kingdon-Ward**[1] (1885–1958). On leaving college in 1907, Kingdon-Ward had accepted the first post that offered an opportunity of seeing the world – that of a teacher in a school at Shanghai. In 1909 he accompanied the American, Malcolm P. Anderson, on an expedition to Tatsien-lu; the object was zoological, not botanical, but young Frank collected a few plant-specimens and learned the elements of Chinese travel. When Bulley's offer arrived in January 1911 he decided to accept before he had finished reading the letter, and was off within three weeks. He remained a collector for the rest of his life, and was planning a new journey when he died at the age of seventy-three; but only three of his many expeditions took place in China – the rest were over the border in Upper Assam, Burma and Tibet.

As on nearly all his journeys, he went by sea to Rangoon and approached his objective through Burma. There were a good many hitches by the way, before he arrived on 6 March at Teng-yueh, and found the entire European population of six having tea with Litton's successor, the British Acting-Consul Archibald Rose. Anxious not to encroach on Forrest's territories, Kingdon-Ward asked for advice as to a good collecting centre, and Rose recommended Atuntse. This proved an excellent choice, at least from the floral point of view, and Kingdon-Ward returned there on two subsequent occasions. (See map on page 159.)

Atuntse lies far to the north on the watershed that divides the Mekong from the Yang-tse. This is the country where the great rivers, bursting out of Tibet, are gripped at the waist by the iron corsets of the mountains before spreading east, south and west over Asia. East of the Yang-tse are its tributaries, the Litang and the Yalung; west of the Mekong are the Salween, the Taron, which joins the Nmai Kha and ultimately becomes the Irriwaddi, and finally the Lohit, turning westward to join the great Brahmaputra.

It took two months even to get there. He went first to Tali-fu, then northwards to Kien Ch'uan and over the watershed to Wei-hsi-ting, and north again up the arid Mekong valley to Tse-kou, where Forrest had so nearly lost his life six years before. Here he rested for ten days at the French mission-station before going on to Atuntse, which he reached in mid-May. It was still too early for the flowers on the surrounding heights, so after a fortnight's survey he returned to Tse-kou and from there embarked on a trip westwards across the divide to the Salween, up that river as far as Min-kong and back by another route; during which time he suffered a good deal from insufficient food and extremes of climate. By 27 June he was established at Atuntse.

A month later he received a message from one of the French priests at Tsekou, saying that the English were in Lhasa, that the Chinese had sworn to exterminate every Englishman, and that he had better leave at once. Feeling that it would be useless to return to Yunnan (which would in any case mean the loss of a year's work) Kingdon-Ward decided to go north

[1] He preferred this form of his name, though he was christened Francis, and was not entitled to the hyphen

to Batang, where there were a number of Europeans. This meant a journey of 180 miles with slow yak transport, east and north to the Gartok and Yang-tse rivers and then upstream to Batang. Having covered the distance in six instead of the usual eight days, he found the alarm had been completely false, and after a short stay returned by a different route. He regained Atuntse on 19 August, after three weeks of hard and quite unnecessary travel, and settled down for the remainder of the season, making only short excursions in the immediate neighbourhood.

In October, rumours began to be heard of revolution in the south; and presently news arrived that all Europeans had left Batang. There were at this time eleven foreigners at Atuntse; they could not all leave together as there were not sufficient transport-coolies available. Kingdon-Ward and his party set out on 1 November; but even then he was not content to go straight down to Tse-kou. With one servant, a soldier-interpreter (who promptly deserted) and four porters, he made a detour to the west to collect seeds on the Salween – where the Chinese authorities had forbidden him to go. He rejoined his main baggage-train at Tse-kou and proceeded to Wei-hsi (22 November) where he learned that South Yunnan was 'ablaze with revolution' and several towns were in the hands of the rebels.[1] Nevertheless, having sent his caravan, under escort, by the main road to Teng-yueh, he himself made another detour, in the course of which he encountered the vanguard of the revolutionary army; they proved perfectly civil and harmless, and their captain gave 'Mr Wha' a passport which assured him a safe-conduct from village to village. In mid-December, bearded, ragged and dirty, he arrived safe at Teng-yueh. His caravan did not turn up till the 27th, and two days later he set out on his homeward journey, via Bhamo, Mandalay and Rangoon. Considering his lack of experience in horticulture, botany or Chinese languages, he had made a very creditable beginning; he had gathered specimens of 200 plant-species, twenty-two of them new, and seeds of seventy-six species, which included many beautiful plants, but, apparently, no outstanding novelties.

In 1913 he returned to the same area by a more easterly route. From Teng-yueh he went to Likiang instead of Wei-hsi, crossed the Yang-tse loop by ferry and travelled north to Chung-tien; then up the Yang-tse to Pang-tzu-la and north-west across the mountains to Atuntse on 4 June. This season was less troubled, and he made several excursions, including two or three to the Do-kar-la, a pass on the pilgrim road round the sacred mountain of Kar-kar-po, which he made two unsuccessful attempts to climb. It was on the Do-kar-la that he found the beautiful yellow-flowered *Rhododendron wardii*. Sometime in July he met Forrest on the Bei-ma Shan, but neither has left any account of the meeting.

Fond though he was of flowers, Kingdon-Ward was at heart an explorer; on most of his journeys he made maps and surveys as well as collections, and he was prouder of his medals from the Royal Geographical Society than of all his horticultural honours. On 30 October, when the collecting-season was over at Atuntse and his seeds packed up and des-

[1] Kingdon-Ward in 1911, Forrest in 1912–13, Farrer in 1914 and Meyer in 1918 all suffered from this political upheaval

patched to Tali-fu, instead of retiring thankfully to winter quarters for a well-earned rest, he started on a long journey of exploration in Tibet. Hundreds of miles of the Upper Yang-tse were still unmapped, and so were the sources of the Mekong; but on this occasion he was unable to attain his objectives. China and Tibet were still at war, and the Chinese authorities, concerned for his safety, quietly put obstacle after obstacle in his way. He reached Pitu, but could not get permission to go on to Lhasa; he was welcomed back at Jana on the Salween (where he had been in 1911) but was not allowed to go up to Min-kong without a passport, which 'of course' he had not got; he went down the river to Tramutang, but was refused porters to cross the watershed westwards to the Taron; he attempted to explore lower down the Salween but was hampered by bad weather, fever and rheumatism, surly and truculent porters and a treacherous interpreter. At last he was obliged to return to Tramutang, and from there by his outward route to Atuntse, which he reached about the end of January 1914. When the Chinese New Year ceremonies were over he returned as he came to Teng-yueh (10 March) and thence went westward to the railhead of Myitkina in Burma. But he did not go home to England; after a month in Rangoon he was off again on what proved a disastrous expedition to Upper Burma. During the Great War he served in the Indian Army and rose to the rank of Captain; in 1919 he made another Burmese journey, so it was not till 1921 that he returned to China for the third and last time.

His base for the season was to be Muli, about the same latitude as Atuntse but further to the east, between the rivers Sholo and Litang. As usual, it took a long time to get there. Railways now ran to Lashio in the west and Yunnan-fu in the east; but in between, the only transport was still the age-old mule caravan, or in the mountains the slow and stumpy yak. Kingdon-Ward left Lashio on 11 April 1921, and made the now-familiar journey to Tali-fu and Likiang, and after some hard travel through the Likiang range reached the Yang-tse ferry at Feng-kow. The Waha mountains had next to be crossed, to the valley and town of Yung-ning. Muli now lay directly north, the capital of a local principality about the size of Yorkshire. Its king was accustomed to spend a year in each of three towns in turn – Muli, Kong and Warichen; Kingdon-Ward visited all of them during the course of the summer.

He arrived at Muli early in June, and at the end of the month established what he called Glacier Lake Camp on one of the adjacent mountains. He returned to this camp four times during the season, and in between explored Muli cliff, a precipitous limestone crag fifteen miles in circumference between the Litang and Rongtu rivers – hair-raising to climb but supporting a fine flora. It was also rich in animal life, and he twice encountered black bears, who fortunately showed no hostility. Besides visiting Warichen and Kong across the Litang, he also explored west to Penong on the Sholo. His final visit to Glacier Lake Camp was from 30 October–10 November for the last of the seed-harvest, and he left Muli for Yung-ning and the south on 15 November.

He returned to Bhamo, but can hardly have had time to go home to

England, for he was back at Yung-ning in the following April (1922) and again visited Muli in May and June. He had conceived an ambition to cross all the great river-divides from the Yang-tse in the east to the Irrawaddi in the west, and in July he was at Likiang and ready to start; but the rainy season was approaching and he was in poor health (his rheumatism was again troublesome) so he decided to postpone his journey till October and meantime make another trip to the north. He went first to Atuntse, and found the place on the whole little changed since his last visit nearly ten years before. An uneasy truce now reigned over the borders and he could go where he wished; but the land was full of brigands. He spent some time in the neighbourhood of Tsakalo on the Mekong-Salween divide, and he also climbed the Tibetan peak of Damyou; some of the roads he travelled being such that his head-muleteer remarked, 'My heart was a little tiny thing all the way!'

Early in October Kingdon-Ward returned to Atuntse and south to Tsekou; and from there sent his heavy luggage with some of his men to Tali-fu, while he set off with seven Tibetan porters on his cross-country journey. They first of all went north to Batang, and then struck steadily south-west. From Tramutang on the Salween they crossed the Gompa-la pass, through wonderful rhododendron-country, to the hitherto unattainable Taron river, which they descended to its junction with the Nam Tamai. Then they crossed another range to the Nam Tisang – each pass lower than the one before, and each valley revealing a more tropical luxuriance of vegetation. On 11 November they came to a real village – six or eight huts at least, the first for a month; and within another week were at the British station of Fort Herz (Putao). On 29 November the indefatigable explorer started on a trip to the source of the Mali Kha river, but was attacked by fever and had to return to Fort Herz on a stretcher and go straight to hospital. When he recovered he left for Myitkina, where he took train to Rangoon at the end of January 1923; and this time he really *did* go home to Britain.

Even on this strenuous winter journey Kingdon-Ward did not cease to collect. Nothing, of course, was in bloom, but he was now sufficiently experienced to recognize when a plant in seed might be worth introducing; *Rhododendron taronense* and *R. aperantum* were among the plants he collected 'blind' – without seeing the flower. In Muli he had found *Primula melanops* and *Gentiana trichotoma.*

While Forrest and Kingdon-Ward were sharing the flowers of Yunnan and Szechuan, other collectors were working further to the north. The achievements of one of them, William Purdom, are often overlooked. He was one of the taciturn sort, and wrote little about his plants and nothing about himself; and unfortunately his collecting-career was short. His first journey was made in 1909–12 on behalf of Messrs Veitch and the Arnold Arboretum, so he had to contend at the start with the established reputation of Wilson, and the dissolution only two years later of the firm for which he had worked. On his second journey (1913–14) his self-effacing personality was overshadowed by that of his companion, the voluble Farrer.

William Purdom (1880–1921) was born in Westmorland and trained in the gardens of Brathay Hall. He worked in two noted London nurseries, Low and Sons and H. J. Veitch and Co. and for six years at Kew (1902–8) where he became sub-foreman of the arboretum nurseries. Leaving England early in 1909, he must have had at least two full seasons, if not three, in Chihli, Inner Mongolia, Shansi, Shensi and Kansu, but hardly anything is recorded of his activities except in second-hand reports from Farrer, most of which refer to 1911. It was in that year that Purdom and his three Chinese helpers were attacked near Shan-chow on the Hwang-ho by a band of brigands – according to Farrer, they numbered 200. Two of his horses were shot, but Purdom (armed with a rifle) and his men made a stand, and eventually managed to break away, though not without loss of life.

Westward of Shan-chow the Yellow River bends a fantastic elbow, and having flowed for hundreds of miles from north to south, continues its course from west to east; on the north-south part of its course it forms the boundary between Shensi and Shansi. On its western side, in the neighbourhood of Han ch'eng, stands Mou-tan Shan, Paeony Mountain, where according to a seventeenth-century manuscript tree-paeonies grew wild in such profusion that they scented the air, and were used locally for fuel. Sometime during his travels – but which year is uncertain – Purdom visited this mountain (never before seen by Europeans) and explored it thoroughly, but no paeonies were to be found. In 1911, however, he found a dark-red, wild tree-paeony in the foothills of the Min Shan, near the borders of Kansu and Tibet, and considerably west of the region where Farrer later found the white form. Farrer's paeony was named *P. suffruticosa*, and Purdom's given the minor rank of *P. suffruticosa* sub-species *spontanea*, although his was the first to be found, and the Chinese have always regarded the tree-paeony as typically a red flower.

The Min Shan is in the country to which Purdom afterwards escorted Farrer; it was Purdom who discovered the excellent centre of Cho-ni, where he stayed for some time before going further north, to the 'grass-country' and the mountains south-west of Lanchow, where he found *Ligularia purdomi* and primulas *conspersa*, *purdomii* and *woodwardii*. On the sacred mountain of Lien Hwa Shan he had a bad fall, and lay stunned until rescued by a 'grubby little old Daoist nun' who dragged him to her cell and nursed him till he recovered. On his second visit he went back to see her, but found that she had been murdered by the White Wolves.

Purdom's introductions included *Clematis macropetala* and *C. tangutica* var. *obtusiuscula*, *Deutzia hypoglauca*, *Syringa microphylla*, and (from Chinese gardens) *Viburnum farreri* (=*fragrans*). Later, he and Farrer together found this invaluable shrub growing truly wild; but it is always Farrer who is given the credit for its discovery and introduction.

Reginald Farrer (1880–1920) met Purdom in London in 1913, and at once recognized that here was the opportunity to realize his long-standing dream of exploring the alps of Tibet, a journey he had not felt capable of making alone. He asked Purdom if he would be willing to share and direct such an expedition, for his mere expenses, and such was Purdom's enthusiasm that he promptly accepted. Farrer's object was to obtain alpine plants from a more northerly region than that being combed by Forrest and Kingdon-Ward, which would be likely to prove hardier in cultivation. He was his own patron, but some of his gardening friends took a small share in financing the expedition. Without the experienced Purdom the journey would not have been possible; it was he who engaged the staff, arranged the transport, pacified rebellious coolies and interviewed local mandarins; but Farrer himself, already a Buddhist, quickly learnt the elements of the language and also the sympathetic comprehension of the Chinese viewpoint which was the basis of good relations.

So the incongruous pair set out – Farrer harsh-voiced, exciteable, heavily moustached and tending to fat; Purdom tall and lean, of magnificent Nordic physique; equable and reticent, but to Farrer 'an absolutely perfect friend and helper'. (Farrer was the elder – by two months.) The new Trans-Siberian Railway[1] took them to Pekin, which they left on 5 March 1914, also by train, for Honan and the terminus at Mien-chi Hsien. Here at once they ran into difficulties. Their intention was approximately to follow the frontier between Kansu and Tibet from south to north, where only Potanin had been before them. But trouble on the borders had been more than usually acute, and the region to which they were bound was something of a no-man's-land, giving allegiance only to its own local chieftains. Moreover, rebellion had broken out in the south, and a disaffected army headed by a general called the White Wolf, was sweeping up into southern Shensi and threatening the town of Sin-ning-fu. No contractor would allow his mules to be taken into the danger area, in case they should be commandeered; and Farrer and Purdom were hampered by having to carry sufficient silver in bar-ingots to cover two years' expenses. Eventually they obtained three oxwains to take them to Sin-ning-fu, where they were detained for another three weeks (during which both were ill) before they could obtain permission to go further west. At last they reached the foothills of southern Kansu, and there, within three days of each other, they made two of the most important finds of the whole two-year expedition – the wild *Viburnum farreri* (*fragrans*) on 16 April, and the magnificent wild white tree-paeony on 19 April. Two days later they reached Kiai Chow on the Hei-Shui-Jang or Blackwater River, and were at the beginning of their chosen country.

They followed the arid gorge of the river southward, and then struck west over the Feng-Shan-Ling pass to Wen-hsien on the Pei-shui-jang or Whitewater – and by so doing escaped the ravages of the White Wolf, who charged up the Blackwater immediately afterwards, burning and devastating everything in his path. By this time, however, Farrer and his party were following the tributary Dung-lu Ho north-west into the high mountains, and by 6 May were crossing the pass of Chago-ling in the Sha-Tan alps. This was Tibetan territory, and the lamas were unfriendly. Outside the village of Chago the inexperienced Farrer unwittingly violated a local taboo by trying to ride along a path that the lamas had declared closed to traffic; and this was enough to bring the whole hornet's

[1] Completed in 1904

nest about their ears. Having narrowly escaped being murdered, they were obliged to withdraw north to the Chinese village of Ga-hoba, and after a few days, west again to the friendly little village of Sha-tan-yu. But here, too, the hostility of the lamas pursued them. An excursion they made into the mountains chanced to be followed by thunder and a hailstorm which destroyed the sprouting crops further down the valley; obviously the Celestial Powers who lived on the mountain-tops were angry at the intrusion, and the lamas roused the whole countryside against the impious foreigners. To save the friendly villagers from reprisals a retreat had to be made, but the only roads led still further into Tibet, or to Siku, which was reported to have been destroyed by the White Wolf. They chose the latter course and found Siku quite intact, though in a great state of ferment, being threatened by the Wolves from one direction and by a local tribe, the Black Tepos, from another.

Siku was their base for the next six weeks, and when the countryside had calmed a little, they were able to explore the nearby mountain of Lei-go Shan or 'Thundercrown'. But they still yearned to get into the Tibetan alps, and on hearing that the roads to Min-chow were quiet, they set out on 6 July, went about twenty miles down the Blackwater to its junction with the Nan Hor, and then up that tributary to Min-chow, through a high cold country of 'vast and rolling green dish-covery grass downs'. From there they went west, now on the Chinese, now on the Tibetan side of the boundary river of Tao-Hor to the little town of Cho-ni (on Chinese territory, though the stronghold of a Tibetan prince) which had been Purdom's headquarters in 1911. They were greeted by the resident missionary (who was 'quite unconscious of having been murdered' as rumour had reported) and were housed in the mission, making such local excursions as circumstances permitted; but they were not allowed to go far without a military escort. They were able, however, to camp for a time in the foothills of the Min Shan – with an escort of forty and a Union Jack. On their return to Cho-ni on 19 August they learned of the outbreak of the European war.

After this they separated for the seed-harvest, Purdom staying in the Cho-ni district while Farrer returned to Siku, and their headman Mafu went right back to Weh Hsien. It was the first time that Farrer had been alone with his Chinese staff and his scanty command of the language; but the country that had been so turbulent two months earlier, was now perfectly calm, and when Purdom rejoined him they both went back to Sha-tan-yu without the slightest opposition. It was still dangerous to cross the Tibetan border to Chago, but Purdom made a flying visit there, disguised as a coolie, to obtain seed of a beautiful dipelta they had seen nowhere else. By 18 October winter had shut down on the alps, and even sheltered Siku became bitterly cold; a month later they set out to the north for the capital, Lanchow, where they passed the winter. Farrer spent much of the time correcting the proofs of his great two-volume work, *The English Rock Garden*; but first of all they had to clean, sort and despatch some two mule-loads of seed-packets. Introductions from this first season included *Buddleia alternifolia*, *Daphne tangutica*, *Gentiana hexaphylla* and *Rosa farreri*.

Their second season was more peaceful, but less productive. Misled by the enthusiastic accounts of the Russian travellers Przewalski and Potanin who alone had visited the area, they had decided to explore the Da-tung alps to the north-west of Lanchow. But the further north they went, the poorer grew the flora; the granite mountains were 'too high, too bleak, too cold, too lonely' and even the few limestone areas yielded nothing of interest. Disappointment followed disappointment, until Farrer vowed 'to forswear alps altogether, and henceforth cultivate no more difficult eminence than Brighton Esplanade'.

They left Lanchow on 28 March 1915, and established themselves at Sining, six days' journey to the north-west. But it was still too early for the mountains[1]; Sining, though set in a wide, rolling plain, lies higher than the summit of Mt Cenis. Farrer stayed at Sining for a month, while Purdom made a quick journey of reconnaissance to find a suitable site for their summer quarters. On 3 May they moved off – in a snowstorm; travelling for a short distance east down the Sining river and then due north up its tributary the Wei Yuan Pu. Two days' travel and a hard pull to the north-east brought them to the pass of Lang Shih Tang ('Wolfstone Dene'), on a regular mule-route between Sining and Ping-fan, with inns at intervals. Farrer and his party took over the whole of one of these inns ('Wolvesden House', at about 11,000 ft) at a rent of half-a-crown for six months, and made themselves comfortable for the summer.

Summer, however, had not yet come; and while waiting for the heights to become more accessible, they went off to visit some of the big local monasteries, since the success or failure of the expedition depended on the goodwill of the ecclesiastical authorities. Leaving their base on 18 May, they followed the valley north to the Da-tung Hor and downstream to the bustling abbey of Tien Tang Ssû, the Halls of Heaven. Here the monks had at first been hostile, but they had been mollified by the receipt of viceregal and other official letters of introduction, and were quite melted by the visitors' accomplishments and charms. After a very happy stay Farrer and Purdom made a circuit up the Da-tung Hor and across the mountains by a more northerly pass to Gan Chang Ssû ('Chebson Abbey') a large establishment, very dignified and serene; and here, too, they were well received.

Early in June they returned to Wolvesden for their disappointing summer, making local excursions which mounted higher as the season advanced, but never rewarded their hopes. At length it was decided that Purdom should make a quick trip to the Koko Nor lake and the Kweite Salar ranges south of Sining, while Farrer remained at Wolvesden for the seed-harvest and paid return visits to Tien-Tang and Gan Chang. On the heights above the latter the greatest floral display at 13–14,000 ft was made by the common dandelion; 'It takes a granitic alp of the Da-tung', wrote the disgusted Farrer, 'to achieve so grotesque a violation of Alpine proprieties'. Purdom returned almost empty-handed; the southern ranges were as unproductive as the northern ones.

[1] Farrer maintained that Purdom was in a hurry to remove him from the temptations of Lanchow's antique-shops

Reunited at Wolvesden, the companions planned one last camp in the mountains before the season ended; but this was delayed for a time owing to bad weather and an attack of influenza by which Farrer was prostrated. They were not ready to quit the heights and return to their base until 30 August, and it was only then, when the homeward journey had in fact begun, that the greatest find of the season revealed itself – *Gentiana farreri*, which its namesake could not at first believe to be a new species, but which he considered 'worth all the trouble, anxiety and expense of the whole two years' tour'. It was only beginning to bloom, and seed could not be hoped for; but living plants were collected, and every effort made to bring them safely home.

On 13 September they turned regretful backs on 'Wolvesden House', proceeding first to Tien Tang, and thence making the five-day transit to Ping-fan, with everybody tired and food beginning to run short. At Ping-fan a cart was hired and travel was a little easier – south to the Hwang Ho river and eastward to Lanchow. Then followed a round of farewell visits and a grand packing-up; equipment, purchases, specimens and all heavy baggage being sent by cart and rail to Pekin, while Purdom and Farrer, travelling light, went south by different routes – Purdom by way of Ardjeri and Cho-ni for a last snatch at certain seeds that had been scarce the previous year; and Farrer, by mule-litter or sedan-chair, almost due south through Tsin-chow to Hwei Hsien and Lo-yang, where he waited for Purdom to rejoin him. He had hoped for some new plants en route in the Tapa Shan hills, but at a mere 8000 ft they were not high enough for alpines.

Lo-yang was on the Ja-ling-Jang river, but the weary botanists had to descend as far as Pao-ning-fu in Szechuan before it was possible to transfer to the relative ease of boat-travel. Even then there were many rapids to be negotiated, and brigandage in the guise of civil war beset the lower reaches of the river. With the help of a rifle rigged up on a camera-tripod in the prow of the boat with a cloth thrown over it to look like a hooded machine-gun, they reached Chung-king on the Yang-tse without incident on 16 November – two months and two days after leaving Tien Tang. They were now back in civilization; a steamer took them to Ichang, and they were in Pekin by 6 December. In Pekin their ways parted; Farrer returned home, to work in the Ministry of Information for the remainder of the war, and Purdom entered the forestry service of the Chinese Government, where he remained till his untimely death in 1921.

The journey was hard for the men, but it was harder still for the plants. Most collectors in Western China made their introductions only by seed; but Farrer brought with him a number of living plants of which seed, for one reason or another, could not be obtained – albino forms of *Iris goniocarpa* and *Trollius pumilus*, his late-flowering gentian, *Androsace mucronifolia* and other treasures. Some, brought from Gan Chang, were temporarily established in a piece of ground at Wolvesden; then, planted in biscuit tins, they were transported across the 'blazing distances' of China to Pekin, where they were wintered in the grounds of the British legation. In spring the journey was resumed by steam-heated train across Siberia and Russia to Petrograd; then through Finland and Sweden and across Norway and the mine-strewn North Sea to Hull. Quite a number of the plants survived – but not the gentians; and Farrer thought the species lost for ever, until it turned up unexpectedly in Edinburgh Botanic Garden, from seed that had inadvertantly been gathered by native helpers along with that of *Gentiana hexaphylla*, in the Ardjeri alps of the Min Shan.

Towards the end of his first season, on his return to Siku after a brief absence, Farrer had heard 'in a tempest of surprise, by no means wholly pleasureable', of the arrival of another plant-collector – an extraordinary encounter in so tiny and remote a place, where the appearance of two foreigners at a time was quite unprecedented. He was relieved to find that the visitor was **Frank Meyer** (1875–1918) collecting economic plants for the American Department of Agriculture, and therefore no rival in the introduction of alpines.

This was not Meyer's first visit to China. Born in Amsterdam, and trained in the Amsterdam Botanic Garden under Hugo de Vries, he went to Washington in 1900 to work in the greenhouses of the Department of Agriculture, but he wandered from there to other parts of America, and was in the Missouri Botanic Garden at St Louis when Dr David Fairchild of the Office of Seed and Plant Introduction (himself a much-travelled collector of economic plants) was looking for a suitable emissary to send out to China. Meyer seemed ideal for Fairchild's purpose. He had a thorough knowledge of botany and gardening; he loved travel, especially on foot, and in his early days had walked by compass from Holland to Italy to see the orange-groves, and had nearly died when crossing the Alps in a snow-storm. He was an entertaining talker and a good linguist, fluent in Dutch, German and English and with a smattering of French, Italian, Spanish and Russian; and like Farrer, he was a Buddhist. But there was a darker side to the picture. He was subject to recurrent periods of 'nervous prostration', sometimes accompanied by complete amnesia; and he seems never to have made the slightest attempt to accommodate himself to oriental ways and customs. Farrer had found that, given a share in the work, the Chinese would respond with enthusiasm, and that an uneducated mule-boy from Shansi would learn to mount plant-specimens with exquisite patience and skill; Meyer kept aloof from his men and then complained that they took no interest in his enterprise. He never learnt Chinese, but took with him to Siku a highly-sophisticated interpreter from Pekin; and when this man refused to go further into what he considered a barbarous wilderness, Meyer threw him and a coolie downstairs. This antagonized the whole neighbourhood, and it was only through the mediation of Farrer that he was able to continue his journey. He had similar trouble with his servants on other occasions.

Meyer's mission was to introduce economic plants that could be used in the breeding of specially hardy or disease-resisting strains; his journeys consequently took place in cultivated areas, and in more northern regions than those of the majority of collectors in China. He did, however, introduce to the United States a number of ornamental plants not in cultivation in that country, though many were already known in Britain. He completed three sojourns in the East, each lasting three years, and died in the course of the fourth.

Late in 1905, Meyer arrived in Pekin for his first journey, and in the following year made an early spring excursion south to Soo-chow, inland from Shanghai. It was probably on this trip that he discovered the maiden-hair-tree (*Ginkgo biloba*) growing near Changhua Hsien, about seventy miles west of Hang-chow; it had long been known in cultivation but had never before been found in a truly wild state. Returning to Pekin, he then proceeded across Manchuria to Harbin (Pinkiang) and into North Korea, where he tramped through primitive forest from Kang-Ko to Hoinyong. (It is said that in Manchuria he walked 1800 miles.) 1907 saw him at Tsin-chau fu in Shantung and exploring the little-known Lan Shan mountains, and in the following February he was in the Wu-tai Shan west of Pekin, where he found a new lilac, *Syringa meyeri*.

In 1909, after an interval at home, Meyer travelled via the Caucasus and Persia to Merv and Samarkand. Here he was deserted by his interpreter and assistant, but with compass and map and a smattering of Russian he made his way north-east across Turkestan to Tchungutchuk, on the borders of Dzungaria and Semipalatinsk. Of the fourteen nights of this journey only four were spent under cover; one of the others was disturbed by rain, one by a wolf and four by prowling robbers, 'and the remaining three, we made the most of'.[1] Meyer did not consider this at all a bad trip; the supposedly-ferocious Kalmucks were quite helpful, and milk was abundant and only froze overnight on a few occasions. On the way home he is said to have visited Japan.

On his third journey Meyer came back to northern China, and spent part of 1913 in the vicinity of Kalgan in Inner Mongolia. Early in 1914 he moved to Sian-fu in Shensi, and from there made trips into the Tsin-ling mountains; then west again to Kansu, and the meeting with Farrer at Siku on 21 October. After a short visit to Ga-hoba for certain nuts and peach-kernels, Meyer returned to Siku and thence went north to Lanchow, where, like Farrer, he stayed over Christmas and the New Year. A wild-cat scheme for a mid-winter visit to the Da-tung alps having fallen through, he returned in 1915 by Sian-fu and Honan to Pekin, and so home.

In the autumn of 1916 he returned to Pekin, and searched about Jehol for the forests of the wild Pekin Pear, which will thrive in almost pure sand. In the spring (1917) he went overland to Hankow, and thence by boat to Ichang, where he stayed for over a year, working when possible in various parts of Hupeh. Since the revolution in 1912, China had been in a state of anarchy, with one would-be ruler fighting against another; Ichang was involved in these civil wars, and for several months Meyer was confined there, unable to travel and suffering from the discomforts of food-shortage, cheerless and uncomfortable quarters, and an atmosphere of suspicion and fear. At last, in May 1918, he broke through the lines, and walked through burnt and looted villages to King-chow, where he picked up baggage and seeds previously deposited, and continued, still on foot, to

[1] Quoted by Fairchild *National Geographical Magazine* 1919

Sha-si, where he got a steamer to Hankow. He planned to visit some tung-oil plantations further down the Yang-tse at Kiu-kiang, but he never got so far. He disappeared from the steamer on which he was travelling on the night of 1–2 June, and his body was later recovered from the river about thirty miles above Wu-hu. It was never discovered whether he fell or was pushed, or perhaps committed suicide during one of his periodic fits of depression. He is said to have introduced more than 2000 species and varieties of plants – mainly cereals, fruits, timber-trees and other useful products, but including ornamental trees and shrubs, such as *Aesculus chinensis*, *Rosa xanthina*, and *Juniperus squamata* var. *meyeri*.

Two years later, Dr Fairchild was responsible for sending another remarkable collector to the East. **Joseph Rock** (1884–1962) was an Austrian of good family and brilliant attainments, who had made an erratic progress across Europe and America, and at the time of his engagement by Fairchild in 1920 held the Professorships of Botany and Chinese at the University of Hawai. He was sent by the Department of Agriculture to Indo-China, Siam and Burma to obtain seeds of the chaulmoogra tree, whose oil was then used in the treatment of leprosy; but in 1922 he went on to China, and remained there, on and off, for twenty-seven years. Three times he had to flee the country, and twice returned within a year or two; the third time, in 1949, he retired to Hawaii for the remainder of his life. For the first part of his stay in China he was engaged in making botanical and ornithological collections for various scientific societies, but the last fifteen years were devoted entirely to the study of one of the non-Chinese aboriginal tribes, the Na-Khis, who lived in the neighbourhood of Likiang. By 1944 he had published six articles and two books on the subject, and had translated the key volumes of over 8000 books of Na-Khi literature.

Rock's collecting years can be divided into four periods – 1922–4, 1925–7, 1928–30 and 1930–33, based mostly on Likiang. In 1922 he worked the Schweli-Salween divide and the Likiang range – to the disgust of Forrest, who, however, could hardly expect to keep so rich a region entirely to himself. Indeed, it was all Forrest country; Teng-yueh, where Rock spent the winter; Likiang; Wei-hsi to the west, the Bei-ma Shan to the north, and the Salween-Irrawaddi and Mekong-Salween divides. (The area was vast and the flora rich; but the choice of approach-routes and centres of supply was very limited.) Before leaving in 1924, Rock visited Muli, where Kingdon-Ward had worked two years before, and became friendly with the king of the tiny principality, an association that was to become of great importance on a later journey.

In 1924 he returned for a short time to America, visiting Washington and the Arnold Arboretum, and when he came back to China it was with a commission from Professor Sargent and the Harvard Museum of Comparative Zoology to explore two little-known ranges on the Chinese-Tibetan border. Accordingly the spring of 1925 saw him established at Cho-ni, which Purdom and Farrer had used as a base ten years before. He established cordial relations with the Prince of Cho-ni, thus ensuring permits and military escorts – a necessity, as the regions he was about to visit were unsettled and full of brigands, and the tribes were hostile. That season he worked northwards to the Richthofen Range and the Koko Nor. He found the Richthofen mountains very barren from a botanical point of view, though rich in bird and animal life; but two conifers he collected there (*Picea asperata* and *P. likiangiensis* var. *purpurea*) afterwards turned out to be of great importance, as they were hardy in very severe climates. He returned to Cho-ni for the winter of 1925–6, and in the spring went to explore the Amne Machin range, which lies tucked into a loop of the infant Hwangho well over the western border in Tibet. Here, too, the flora was poor, but later in the year he visited the rich region of lower Tebbu-land (the region of Farrer's dreaded 'black-faced Tepos'), never before seen by a European, and here the scenery and flora were magnificent. Again he wintered in Cho-ni, but in 1927 political troubles and difficulties of travel obliged him to quit the country, and he returned to America, with 2000 botanical specimens and much propagative material.

Very soon afterwards he was back again – this time working for syndicates of garden-owners in Britain and the USA, as well as for the National Geographical Society. He returned to Yung-ning and Muli, and spent the winter of 1928–9 at Nu-lu-ko[1] in Yunnan. In 1929 the friendship of the King of Muli enabled him to penetrate to Mt Jambeyang in the Konkaling district – the territory of a local robber-chieftain. Unfortunately, as so often happened, a thunderstorm that destroyed the local crops was attributed to his sacrilegious violation of the mountain-tops; he was asked to leave and warned that if he returned he would be murdered. He turned instead to the 24,900 ft peak of Minya Konka, between the Yalung and Tatu rivers, seen from afar but never previously explored; but it was granite, not limestone, and the flora was less good. He also revisited the valleys of the Yalung, Yang-tse, Mekong and Salween, and after another winter at Nu-lu-ko, returned with thousands of seeds and specimens and 1700 birds. The details of his 1930–3 travels are not recorded; seeds from them were distributed by the Botanic Garden of the University of California. After 1933 Rock, then nearly fifty, thought himself too old for further exploration, and settled down, still in China, to his ethnological studies.

Rock sent home large quantities of seeds (he collected, for example, nearly 500 species of rhododendron), but they yielded little that was new. The unexplored regions that he visited were barren, and the better ones had previously been searched by others, except perhaps for Konkaling, where he was forbidden to return for the seed-harvest. One plant, however, worthily commemorates his name. In 1926 he sent to the Arnold Arboretum seeds of a tree-paeony which had been cultivated for many years in the courtyards of the lamasery of Cho-ni, and which strongly resembled the wild white form that Farrer had found, but had not been able to introduce. Sargent distributed the seeds to growers in several countries, and this magnificent flower is now in cultivation under the name of *Paeonia suffruticosa* 'Rock's variety'. In 1938 the lamasery was burnt to the

[1] Could this be Forrest's U-lu-kay, about fifteen miles north of Likiang?

ground in one of the periodic Mohammedan uprisings, the lamas killed and the paeonies destroyed; when it was rebuilt some years later Rock was able to send the restored establishment seeds of the Cho-ni paeony, which had in the interval been grown on the other side of the world.

Since 1939, politics have again interfered disastrously with plants; the second Sino-Japanese War of 1937–45 and the Chinese Civil War of 1946–50 effectively closed the frontiers to foreigners, and the 'bamboo curtain' of the Communist government has kept them closed. What is worse, much of what we had has been lost; only a fraction of the plants introduced have survived in cultivation. Some were unsuited to our climate, but many of the losses were due to the arrival of too many novelties all at once, and to the war-time conditions that followed shortly after. There is still unattainable treasure in the East.

The Indies

INDIA, BURMA AND TIBET

1696	Browne	1847	Hooker
1776	Roxburgh	1848	Lobb
1787	Hove	1849	Booth
1809	Wallich	1870	Elwes
1819	Royle	1913	Cooper
1826	Amherst	1914	Kingdon-Ward
1829	Jacquemont	1919	Farrer and Cox
1832	Griffith	1933	Ludlow and Sherriff
1835	Gibson		

THE EAST INDIES

1653	Rumpf	1803	Leschenault
1688	Kamel	1843	Low
1735	Loureiro	1843	Lobb
1754	Poivre	1877	Burbidge and Veitch
1775	Thunberg	1880	Curtis and Burke
1794	Smith		

'... having Sayl'd along the Coast of *Ceylon* (famous for Cinnamon-trees and well-scented Gums) ... the Wind, that then chanc'd to blow from the Shoar, brought them a manifestly odoriferous Air from the Island, though they kept off many miles (perhaps twenty or twenty-five) from the Shoar.' ROBERT BOYLE *Essays of Effluviums* 1673

'... a trick was attempted on the passengers, which is on such occasions not unusual, by sprinkling the rail of the entrance-port with some fragrant substance, and then asking them if they do not perceive the spicy gales of Ceylon?' REGINALD HEBER *Narrative of a Journey through the Upper Provinces of India* 1828

The Indies

India, Burma, and Tibet

It may fairly be said that the first naturalist to study the plants of India was Alexander the Great, on his Asiatic campaign of 331–323 BC. He wanted his former tutor Aristotle to write a book on natural history, and he employed a special corps in hunting, fishing, hawking, and (presumably) plant-collecting, the results obtained being 'carefully transmitted' to the philosopher. Alexander crossed the Khyber Pass and the upper Indus to the river Thelum, which he descended to the Chenab and down to the Indus again, and so to the coast, building two towns (Bucephalus and Alexandria, now extinct) on the way. Among his discoveries was the banana, but he would not allow his soldiers to eat the fruit, which he thought unwholesome.

Aristotle never wrote the book, but after his death in 322 BC, his papers passed to Theophrastus, who used them in the preparation of his *Enquiry into Plants* to such good effect that in 1903 a German scientist published a work of 400 pages on the botanical results of Alexander's expedition, as recorded by Theophrastus. It is not impossible that seeds were also sent; the aromatic basil[1] and the everlasting amaranth,[2] both Indian plants, were cultivated in Greece.

Systematic botanical exploration began in the seventeenth century. The British East India Company was founded on the last day of the year 1600; and 'factories', or permanent trading-posts, were established with the permission of the local potentates, at Surat in 1612, at Madras in 1639, at Bombay in 1661 and at Calcutta in 1690. Other nations quickly followed suit; the Dutch East India Company was chartered in 1602, and a

[1] *Ocimum basilicum* [2] *Gomphrenia globosa*

Dutch governor stationed in 1669 on the Malabar Coast, H. A. van Rheede tot Draakestein, sent home to the botanist Commelin in Amsterdam the materials from which he compiled the *Hortus Indicus Malabaricus* (1678–1703) the first book to be devoted exclusively to the flora of India.

Madras began with six fishermen's huts under the shelter of Fort St George; by the end of the seventeenth century it had grown into a considerable settlement governed by a President and Council. In 1688 its hospital was 'in great want of an able Chirurgeon',[1] one practitioner having died and another gone home on leave; and 'Doct. Saml Brown late Chirurgeon of the Dragon being reputed so, and desirous of the employ'[2] was appointed to the post. **Samuel Browne**[1] (d. 1699) had a turbulent disposition and was frequently in trouble. In 1693 he carelessly poisoned an important patient, and sent an anguished little note to the President – 'Honble Sir: I have murthered Mr. Wheeler by giving him Arsnick. Please to execute Justice on me the malefactor as I deserve.'[3] (His servant had 'negligently powdered pearl in a stone mortar wherein arsnick had before been beaten'.)[4] He and his servant were tried and acquitted; but in 1695 he was again in trouble for challenging another doctor to a duel when drunk, and in 1696 he was imprisoned for an assault on a native (not without provocation), but was released after a week because his patients needed him.

Browne corresponded with the botanist James Petiver from about 1690, but we have no record of his plant-collecting till 1696. He seems to have taken to it with an enthusiasm in keeping with the general boisterousness of his character, for in five months (February–July) of that year he assembled specimens of 316 kinds of plants, gathered in six different localities. Most of them were within thirty miles of Madras, but some came from 'Trippetee' (Tirupatti, north-west in the Nagari hills?) which was seventy miles away. He may not have visited all these places himself, as he is known to have employed others to collect for him. He was allowed a plot in the Company's garden, where he established not only native plants, but a number of others, of medical or economic interest, which he obtained from acquaintances in other parts of the East. He also sent home seeds, which were distributed to several of the most celebrated gardeners in England, among them Jacob Bobart at the Oxford Botanic Garden and Samuel Doody at the Chelsea Physic Garden, besides private establishments. Some at least of the seeds were successfully raised, including probably the 'Hoary Willowherb Malabar yellow Rattle-Broom' (the annual *Crotalaria juncea*) which flowered in Bishop Compton's garden before 1700.

In 1697 the Company decided to abolish the post of second surgeon, which Browne had held for nine years, and he retired into private practice, and later, after a few months' tuition, took over the post of Assay Master. This left him little leisure for further collecting, but he sent Petiver all the books of plants he had by him, and these were probably the consignments acknowledged by the botanist in 1699. 'To the inde-

[1] Browne himself always spelt his name with an 'e'

[2,3] and [4] D. G. Crawford *A History of the Indian Medical Service*

fatigable Industry of this Worthy and Generous Person I am beyond expression obliged, for his large and frequent Performances . . . I have this year (besides several before) received near 20 volumes in Folio, filled with fair and perfect Specimens of Trees and Herbs, and amongst them some from China, the isle of Ceilon, etc.'[1] Browne was alive, and corresponding with Kamel, in 1699; he seems to have died between that date and 1703, when Petiver dedicated a botanical plate to his memory.

Nearly all the important botanists and collectors in India began their careers, like Browne, as surgeons in the service of the East India Company; the few exceptions had more than the usual difficulties to contend with. **Anton Pantaleon Hove** was not an employee of the Company; he was a Polish-born gardener sent out from Kew and was responsible only to Sir Joseph Banks. In consequence, he was treated with an indifference and lack of support amounting to inhumanity, and found the local Rajahs more helpful than his fellow-Europeans. He had already made one collecting journey, along the African coast, and in the interval seems to have qualified as a physician, as he is now styled 'Dr. Hove'.[2] He was sent to India chiefly to study the cultivation and preparation of cotton, and to obtain seeds and plants of the best varieties, for introduction to the West Indies; but he was also to get as many useful plants as possible, and any ornamental ones that came his way.

Hove sailed on the *Warren Hastings* on 14 April 1787, and reached Bombay on 29 July, where his services as physician were immediately requested – for the head-surgeon. He stayed in Bombay for two months, organising his next move and making local excursions. He joined a party to visit a hot-water cascade on the 'Mahratta continent', and found a tree of yellow blossoms which he thought was a magnolia (*Michelia champaca?*). On their return they passed 'innumerable guard-houses of the Mahrattas' and the first one 'would not let me pass my little Magnolia trees, which I dug up, but on giving them a bottle of spirits . . . nothing was said further about it'. They were warned against tigers, but those they saw Hove found 'not near so large as those I have seen on the coast of Africa', and at the sight of so formidable a party 'from us entirely every one of them flew'. He sent his first 'growing collection' from Bombay on 6 September; it was followed by a chest of plants, with some specimens and seeds, in charge of a Mr Pemberton, returning to England, which he 'obligingly promised his care on'.

On 1 October Hove set out for a tour of the cotton-growing centres to the north, and travelled by sea to the Company's oldest station, Surat. The country was unsettled, and when he left Surat on the 16th to go north to Broach, it was with an escort of seventeen soldiers for protection. He visited cotton-plantations in the Broach neighbourhood, but could get no help from the Governor towards further travels, and was obliged to return to Surat, being robbed by a native official on the way. (He complained to the Governor of Bombay, but no action was taken and he received no redress.) Again he set out by a different route, inspecting

[1] *Museum Petiverianum* 1699
[2] He is known to have studied under Dr John Hunter

cotton from village to village; again he sustained a robbery, but this time some of the property was recovered. A much more serious disaster occurred on 25 November when he was approaching Ahmood (Amod). During his absence in the fields his camp was attacked by a large force of dissident Grassia tribesmen. The sepoys left in charge, seeing 'several hundred' horsemen cross the river, hid what goods they could in the cotton-plantations, and prudently made themselves scarce, while the Grassias looted what was left in the camp and burnt the tent and its contents. On his return Hove retrieved what was left of his possessions, which included papers, seeds and specimens; but a second attack from a smaller band of Grassias followed, which was with difficulty beaten off by well-directed fire from his own few men. From this time till his return to Bombay eight weeks later, he was obliged to wear native dress, as his European clothes had gone; and his medicines having been destroyed, he was unable to treat the many sick who besieged him for aid. It does not seem to have occurred to him, however, to abandon his expedition on account of his losses, serious though they were. At Amod he hired a strong convoy of coolies to see him safe to Dewan (Degan?) and thence across the river Mahli to Cambay.

The next part of his journey northwards was made in the company of an officer on a mission to buy horses for the cavalry. At Lymree a young Rajah who had never seen Europeans before was 'so *delighted* with our persons, that he sent for all his relations to come and look at us'. The two friends were lodged in an empty store-room that swarmed with rats, and having had no rest for three previous nights, slept heavily. In the morning, Captain Torin, who had paid his visit to the Rajah the day before in full dress, 'had the misfortune of finding all his hair eat up [by the rats] wherever there was an ointment of pomatum, that he was under the necessity of cutting it close to his head'. Their farthest point seems to have been a place called Mittimpur on the river Sabermati, which they reached on 21 December; from there they turned back to Dholka and Cambay.

Hove was now anxious to get back to Bombay in time to despatch his collections in the care of a Mr Boddam, an official of the East India Company, who would be returning to Europe on the last ship of the season, and to obtain from the same source a reimbursement of his expenses. He started on New Year's Day, 1788, and went more or less directly by Jambusar, Amod and Broach to Surat; and finding that there was no vessel due to sail, continued overland down the coast, pausing only to collect a few new plants and much seed of the teak tree, which he hoped would be useful in the West Indies. He reached Bombay on 19 January – and found that Boddam had left on the ninth. He applied for funds to his successor, but was refused; the East India Company, he was told, was required only to give such help as was necessary to secure the safety of his person. He was obliged to raise a loan from a friend, to cover his present expenses and renew his wardrobe, until reimbursements could be procured from Europe. He sent off his collections on 16 February by the *Princess of Wales*, and his report and accounts probably went by the same boat; but when they were received they provoked from Sir Joseph Banks an explosion of wrath. Not realizing, perhaps, the dire effects of three successive robberies, he considered Hove's expenses 'most unjustifiably enormous', and said that if he could not keep within the stipulated financial limits, he should have abandoned his mission; as it was, he was to return at once to Bombay and take the first ship for England, where it was to be hoped he would be able to justify his conduct. This letter, however, was not written till September 1788, and probably Hove did not receive it until he was arranging his return.

Actually Hove was a most conscientious collector, and does not give the impression of undue extravagance. He left Bombay on 17 March for another tour in the north, covering much the same country as before; he made a point of visiting several of the plantations three or four times at different seasons in order to observe the whole cotton-industry from the seed-sowing to the weaving of the cloth, and collected not only seeds and samples but spinning and weaving utensils and plants used in dyeing. It says much for him (or perhaps for his lavish expenditure?) that whenever he returned to a place where he was known, he was given a warm welcome. The Rajah of Amod received him with 'transports of joy'; on one occasion he was entertained with dancing-girls (whom he did not appreciate), on another with bowls of hashish, the very steam of which overcame him to such a degree that he was obliged to leave the room. He was troubled that summer with a complaint of the liver, and in August he became so ill that he decided to leave Cambay for Surat and the services of the Company's physician. Even so, he collected by the way, and had to leave some plants in the care of the Rajah's headman at Amod, as they could not be taken further owing to the bad state of the roads. Arrived at Surat on 11 September, he rejected the treatment proposed by the head-surgeon at the hospital – a blister to the head – and preferred to wait for his ailment to improve with the better weather. He then went off on another short cotton-tour, and retrieved his collections from Amod, returning to Surat on the first of November.

All the European shipping was taken up by the Surat military changing station, so Hove embarked his collections for Bombay on two native cotton- boats. They encountered gales, and one of the boats never arrived; either it was lost at sea or 'taken by the coolys'. Again he lost his personal possessions, though his collection of living plants was safe. He stayed in Bombay only long enough to pack his specimens and establish his plants in a friend's garden, before going off again on local excursions – chiefly to avoid the 'exorbitant tavern expenses' charged in the town. (Perhaps he had received Banks' letter!)

He had been promised a passage home on the *Prince William Henry*, the last East Indiaman of the season – provided that this should be at no expense to the Company – but when the ship came in, he found that the price he was expected to pay for his passage was excessively high, and his collections could only be stowed in the hold, where he would have no access to them; the captain would not even guarantee a dry place for his cotton-seed. Hove went on board, and found that there was actually plenty of room on deck for his plant-cases; but he was blandly informed that as Mr Ramsay, the late Governor, was going home on the ship, the captain 'could not think of disgracing him with such encumbrances'. Hove then appealed to the Governor, but was told that the Captain could not be enforced if he alleged a deficiency in the ship. Still far from well, and desperate to get back to England before his collections were spoiled, Hove after many difficulties secured a passage on a Danish ship, the *Norge*, for himself and five cases of plants.

They sailed on 1 February 1789, and after experiencing much bad weather, arrived at Cape Town on 27 April. Here Hove met Francis Masson and saw his garden; Masson gave him some seeds to take home, and asked him to take some plants also, but Hove dared not do so, having already lost many of his own through shortage of water. They also received news of the outbreak of war between Denmark and Sweden; and the Captain decided to wait at Madeira for a convoy. Delayed by this, and by head winds in the Bay of Biscay, they did not reach the Scillies till 6 July. Water and provisions were short, scurvy prevalent, and Hove's plants were dying almost daily; so when they encountered a pilot-cutter, the *Dove of Dover*, Hove engaged her for thirty guineas to land himself and his collections at the nearest English port. The transfer was made as soon as the weather was calm enough to permit it, and he disembarked at Plymouth on 8 August, where he was allowed to put his plants into 'one of the warmest spirit-warehouses', pending their removal to Kew. After so many struggles and vicissitudes, it is sad, but not surprising, that Banks' report on his collections was 'unfavourable'.

It is not recorded what ornamental plants, if any, Hove managed after all to introduce. In any case, having regard to the region in which he worked, they would require hothouse cultivation in Britain. The only Indian plants likely to be of value to outdoor gardeners in temperate climates are those that grow where the latitude is counteracted by the altitude – that is, in the Himalayas. But for a long period the heights were politically as well as geographically inaccessible. The lands ruled by the British extended only to the foothills; the southern slopes of the great ranges were occupied by a chain of independent native principalities where foreigners were not welcome and permits hard to obtain. Nepal, Sikkim and Bhutan lay along the Tibetan border like a string of sausages at the base of a high-breasted turkey; Assam came under British control only in 1826, Kashmir and the Punjab in the 1840s. The base for the assault on these botanical fortresses was Calcutta.

The Botanic Garden at Calcutta was founded in 1786 at the suggestion of Colonel Robert Kyd, and with the support of the Government and the East India Company. The initial object was economic – to form a nursery of useful and rare plants for the supply of those who would undertake to cultivate them; but it broadened under successive directors into a true botanic garden, and then into a renowned pleasure-park. Colonel Kyd remained in charge till his death in 1793, when he was succeeded by Roxburgh.

William Roxburgh (1751–1815) was a Scot who had come to India as an Assistant-Surgeon to the Company in 1776. For some years he was stationed at Samulcotta, where he had charge of a small garden of economic and medicinal plants. In 1789 he was appointed the Company's Botanist in the Carnatic, and had already begun to send home that flood of botanical drawings by native artists (amounting finally to more than 2000) a selection from which was published by Banks as *Plants of the Coast of Coromandel* (1795–1819). From Calcutta Roxburgh sent to Kew great quantities of material – specimens, drawings, seeds and living plants; he also sent plants to private gardeners – Lady Amelia Hume, Charles Greville, James Vere and others. They were not all Indian species, by any means, for Calcutta had long been a repository for plants from China and other parts of the East;[1] Roxburgh's *Crinum amabile* and his namesake *Roxburghia gloriosa* (*Stemona tuberosa*) both came from the East Indies. He did not himself travel or collect to any extent, and his introductions were all greenhouse plants; the day of the hardy Himalayan was not yet.

Roxburgh resigned his charge in 1813 on account of ill health, and died two years later. The catalogue he compiled of the plants in the Calcutta garden[2] listed 3240 species, including a number of homely English weeds – cultivated with one knows not what nostalgia; plantain, hemlock, sorrel, gorse, clover and dandelion. Perhaps it was he who planted the 'little wretched oak . . . kept alive with difficulty' seen by Bishop Heber in 1824.

In 1809, when Roxburgh was nearing sixty, the Company appointed a twenty-three-year-old Dane to assist him. **Nathaniel Wallich** (Nathan Wolff, 1786–1854) was born and educated in Copenhagen, and came to India in 1807 as surgeon to the Danish settlement at Serampore. During most of the Napoleonic Wars the settlement preserved a prosperous neutrality, but the bombardment of Copenhagen by the British in 1807 brought this happy state to an end, and in 1808 Serampore was captured by British forces. Wallich was imprisoned until released to help Roxburgh, 'but without any additional allowances' unless for travel in the interior. After three years together it was the younger man who broke down, and had to go on a voyage to Mauritius for the recovery of his health. On his return Wallich entered the service of the East India Company as an army surgeon, but was given temporary charge of the Calcutta garden in 1815, and appointed its permanent Director in 1817, a position that he held for thirty years.

His influence during this period was incalculable. By 1824 he had turned the Botanic Garden (locally known as 'Wallich's pet') into a place that, according to Bishop Heber, more perfectly resembled Milton's idea of Paradise than anything he ever saw, with an auxiliary garden for economic plants at Tittyghur. The magnitude of his contributions to botany cannot be estimated here. What concerns us are his travels, and the use he made of native collectors to bring him plants from regions inaccessible to Europeans. From the first he was particularly interested in the possi-

[1] The Bengal Rose (*R. chinensis* var. *semperflorens*), for example, though introduced from Calcutta, came originally from China

[2] *Hortus Benghalensis* 1814

bilities of Nepal, and sent two gardeners there in 1817, the year of his appointment as Director and three years before he was able to visit the region himself. The British Resident then in Khatmandu, the Hon. Edward Gardner, was a keen amateur botanist, and until his departure in March 1819 sent Wallich frequent consignments of seeds and specimens. Between January 1818 and March 1819 Wallich sent to England six consignments of Nepalese seeds, including those of *Rhododendron arboreum*, solicitously – and successfully – packed in tins of brown sugar. In December 1819 he notified Sir Joseph Banks that he was sending (for Kew) two chests of live plants, a chest of bulbs and tubers and a box of Nepal seeds on the *Rose* and *Minerva*, and similar consignments on each of the remaining nine ships of that season's fleet, besides packets for the Edinburgh Botanic Garden and various friends, and a live musk deer.

One of Wallich's earliest tasks, in conjunction with the missionary-botanist William Carey, was the editing of Roxburgh's posthumous *Flora Indica*, 'to which he added much original matter; but his zeal as a collector of new plants was greater than his patience in working up existing materials, so Carey was left to complete the task alone'[1] – the opportunity having arisen for Wallich to accompany a Government mission to Nepal. Taking with him a native draughtsman, Wallich travelled to Patna early in December, 1820, and by way of Parsua, Chessapani and Thankot reached Khatmandu on 21 December. He stayed till the following November, but does not seem to have been allowed to go much beyond the city, except perhaps for a visit to Naiacot in the north-west. But the pilgrim-route to the sacred mountain of Gossain Than, one of the giants of the Himalayas, passed through the valley, and Wallich persuaded some of the pilgrims to bring him plants from the heights on their return – learning thus what he should ask his collectors to procure. Some of his finest discoveries – *Cornus capitata, Gentiana ornata, Meconopsis napaulensis, Podophyllum emodi, Lilium giganteum*, and *L. wallichianum* – had to wait for later travellers to bring home, but he or his men managed successfully to introduce *Bergenia ligulata, Cotoneaster frigida, C. microphylla* and *C. rotundifolia, Geranium wallichianum, Pinus wallichiana* and *Rhododendron campanulatum*. On his return to the plains late in 1821, Wallich contracted a severe fever; two months in bed and a cruise in the Bay of Bengal having failed to restore him, he went on a voyage to the East Indies, visiting Penang, Singapore, Sumatra and Java, and returning on the last day of 1822 with renovated health and rich botanical collections.

Wallich's researches in the Himalayas were quickly followed by others. Besides the Calcutta establishment, the East India Company also maintained a garden at Saharanpur in the Punjab, far to the north-west. From 1823 to 1831 this garden was in charge of Dr **John Forbes Royle** (1799–1858). He was skilled in all the natural sciences, but his duties as medical officer of the station, with two hospitals and the garden in his charge, left little leisure for travelling and collecting. Like Wallich, there-

[1] *Dictionary of National Biography*

fore, he trained native helpers; the first lot ran away when asked to go as far as the mountains of Kunawar, but he later got a less pusillanimous team, and sent them with returning shawl-pedlars to distant Kashmir, where they obtained many new plants for the Saharanpur garden. When Royle was back in England and at work on his massive two-volume *Illustrations of the Botany . . . of the Himalayan Mountains* (1833–1840) his men continued to send him seeds, of which he made very generous donations to the Horticultural Society and others. He introduced in this way many valuable plants, such as *Primula denticulata, Polygonum vaccinii-folium, Leycesteria formosa, Jasminum officinale* var. *affine, Sorbaria tomentosa* and, of course, the Policeman's Helmet, *Impatiens roylei*, now widely naturalized in Britain.

Wallich was in the neighbourhood of Saharunpur in 1825, when he was sent on a Government mission to inspect timber in Oudh and the Northern Provinces. He went to Hardwar, a place of pilgrimage on the upper Ganges, and from there up the valley of the Dehra to Dehra Dun; but did not actually penetrate the mountains. It was probably from here that he introduced *Tulipa stellata*.

The Governor-General of India at this time was Lord Amherst, the same who had conducted the unsuccessful embassy to China in 1816.[1] He was still attended by the same physician, Dr Abel, and was also accompanied by his wife, daughter and eldest son. After the conclusion of the first Burmese War, Lord Amherst embarked in August 1826 on an official tour of the Northern Provinces of India. It was a sad party that set out. Lord Amherst had been censured for his conduct of the war (though his actions were afterwards vindicated) and had tendered his resignation; and his promising son had died of fever a few days before. Dr Abel, too, fell ill on the journey and died at Cawnpore.

Slowly the procession wound its way across the plains of the Ganges, with feasts and receptions at every sizable stopping-place; it took them eight months to get up to Simla. Amherst was the first ruler of India to visit this remote hill-station, where the first permanent European house had been built only five years before; and he stayed for ten weeks. **Lady Amherst** (d. 1838) and her daughter Lady Sarah, both ardent botanists, were in their element; early and late they scrambled about the surrounding mountains, looking for new plants. This charming, clever and courageous woman was among the pioneers of Himalayan botany; *Clematis montana* and *Anemone vitifolia* were among the plants raised from seeds she sent home. She was also a keen ornithologist, and when she returned to England in 1828, she brought with her not only large botanical and zoological collections, but two living 'Amherst' pheasants, which Sir Archibald Campbell[2] had brought her from Burma.

While Lord Amherst's party was making its way north-west, Wallich was travelling in the opposite direction. John Crawfurd was being sent to negotiate a commercial treaty with the Burmese King Hpagyidoa at Ava, and Wallich was appointed to go with him, to examine and report

[1] See p. 94
[2] The General who conducted the Burmese War

Lady Amherst in the park at Barrackpur. 'The large tree in the centre is a peepul. . . . In the distance, between that and the bamboo, is a banyan. In the foreground is an aloe, and over the elephant the cotton-tree, which at a certain season exchanges its leaves for flowers something like roses.' From Heber *Narrative of a Journey through the Upper Provinces of India* 1828
By Courtesy of the British Museum

on the resources of the forests of Pegu and Ava, as well as those of 'our recently acquired possessions to the south of the Saluen [Salween] River' in Tenasserim – with particular emphasis on teak. The large party, which included a military escort and some of Wallich's collectors and draughtsmen, left Rangoon on 1 September 1826, on the *Diane*, the first steamboat ever to be seen on the Irrawaddi, accompanied by five Burmese boats carrying baggage, presents and additional personnel. The steamer was towing one of the boats and made slow progress against the current; moreover, stops had to be made to take on firewood, and this gave Wallich and Crawfurd opportunities to go on shore and look for plants, minerals and fossils. It took a month to reach Ava; and while Crawford was engaged on long, long conferences with shifty Burmese, Wallich obtained permission to make a week's outing to the hills north-east of the city, never before visited by a botanist. The highest point that he reached was about 3500 ft, but in four days he found more than 300 new species, collected 'large roots' of Scitamineae and Orchideae (unidentified because not in flower) and was, he believed, the first European to see oak and teak-trees, 'the two greatest glories of the forests of Europe and Asia', growing together in the wild.

The mission left Ava on 12 December, but the return took longer than the outward journey. Water in the river was now low, and navigation exceedingly difficult; the *Diane* was stuck for a week on a sandbank and was only refloated by emptying her of all her contents, including the engines. Rangoon was reached on 17 January 1827, and a week later the party sailed again for the new settlement of Amherst, on the coast of Tenasserim. Crawfurd had to present his report to Sir Archibald Campbell, who was then at Moulmein, and he took the opportunity of exploring the River Ataran as far as time and steam-navigation would permit. He had then to return to Bengal, but Wallich remained for a time, to 'prosecute his botanical investigations'.

The most interesting botanical discovery of this expedition was the spectacular tree which Wallich named *Amherstia nobilis* in honour of Lady Amherst and her daughter. Crawfurd had found it on a previous visit (1826) growing in a neglected monastery garden near Kogum on the river Salween. He brought back specimens which he showed to Wallich; the botanist hoped to find the tree at Ava, but it was not seen again until he made a pilgrimage to the spot where Crawfurd had found it, and saw the same specimen in the spring of the following year. He was able to obtain some layers, which he established in the Calcutta garden; shortly afterwards he tried to bring two of them to England, but they perished early on the voyage.

After nearly twenty years of hard work, travel and a trying climate, Wallich came to England on long leave in the autumn of 1828. One can hardly say that he came 'home', for he does not appear to have revisited Copenhagen; by this time he had become thoroughly anglicized, and resided in London when not in India or elsewhere abroad. Besides living plants, he brought with him an immense accumulation of plant specimens from the collections of the Calcutta garden, which the East India Company authorized him to distribute to the Linnean Society and other learned institutions in Britain and Europe. The sorting and despatch of these collections, and the preparation and publication of his *Plantae Asiaticae Rariores* (1830–2) kept Wallich incessantly employed until his return to India in 1833.

The province of Assam was ceded to Britain at the end of the Burmese war in 1826, and from that time reports began to come in of an indigenous tea-shrub that was used by the native Assamese. The East India Company's monopoly of the tea-trade with China was about to come to an end, and urgent attempts were being made to establish a tea-industry in India. In 1835 it was decided to send a commission to investigate the Assam plant, and the climate and conditions under which it grew. The party consisted of Dr Wallich, Dr John McClelland (a geologist), and a brilliant young botanist Dr **William Griffith** (1810–1845) who had entered the Company's service three years before and had since been stationed at Mergui in Tenasserim.

They left Calcutta on the last day of August 1835, and spent more than a month navigating the intricate system of rivers and canals of the marshy Ganges basin, known as the jheels, to Chattuk at the base of the Khasia hills, where at last they left the waterways and proceeded first by elephant and then by ponies and mules to Cherrapunji, which they reached on 6 October. They stayed for a month in this botanist's paradise; the flora was reputed to be the richest in India, perhaps in all Asia, partly due to the great variety of habitats within a small area and partly to the excessively high rainfall. On 2 November they resumed their journey north-west over the hills to Now-gong, then down into the valley of the Brahmaputra and west to Gauhati, the administrative centre for Assam. (The country they passed through was considered fatal to Europeans for nine months of the year, and botanizing was impossible owing to the impenetrability of the jungles and the prevalence of wild elephants.) They then ascended the Brahmaputra to Tezpur, founded only the year before, and on 16 January 1836, reached Sadiya. Here they were joined by Charles Bruce, one of the original discoverers of the Assam tea, who took them to the various districts between the great river and the Patkai Hills where the shrub had been found. They visited Bisa and Tingrai on the river Bari Dihing east of Dibrugarh, and several other remote and uncomfortable localities, finishing in the Jorhat area farther down the Brahmaputra, where the party broke up – Griffith, with McClelland, to wait at Sadiya for the expedition to the Burmese frontier which he had orders to join, while Wallich and Bruce started down the river for Calcutta.

They parted, probably, without regret, for the association had not been harmonious. Wallich, normally a friendly and amiable character, was nearly fifty – his birthday occurred during the expedition – and in poor health; Griffith was an opiniated young man of twenty-five, and said to have an unfortunate temper. It seems to have been Wallich's province to bring the drying-paper, and he did not provide nearly enough to satisfy Griffith, who, new to the flora, avidly collected everything he saw. It is also said that Wallich was found removing Griffith's specimens from the plant-presses and substituting his own. This might seem a natural procedure in Wallich's eyes, in view of his seniority and the fact that his more selective collections were probably of greater value to science than those of the younger man; but to Griffith it was unforgivable. When the last tea-area had been visited, Wallich, who had not been sleeping and was in a low nervous state, beat a precipitous retreat from a country that he feared would cause his death; he arrived in Calcutta a physical wreck and took weeks to recover, but Griffith and McClelland attributed his haste to a wish to forestall them with his report to the authorities. Here, too, a divergence of opinion was apparent; Griffith and McClelland advising the cultivation of the Assam tea in its own region and Wallich the establishment of the Chinese plant in areas farther west. Griffith, of course, was right, but he was not vindicated until more than twenty years later, after many expensive attempts to cultivate the Chinese teas had been made.

Meanwhile, the Duke of Devonshire had read the description of the amherstia in Wallich's *Plantae Asiaticae Rariores*, and had set his horticultural heart upon it. He decided to send one of his gardeners, **John Gibson**, to India to obtain this and other plants, and despatched him on 1 October 1835 with Lord Auckland's suite in HMS *Jupiter*, in charge of a consignment of plants from Chatsworth in the new Wardian cases,[1] for presentation to the Calcutta garden. When Wallich returned from Assam, he found the young man already installed, under the care of his head gardener, J. W. Masters. Wallich advised Gibson to go to Cherrapunji, but to wait for the rainy season; it was at first intended that he should go afterwards to Martaban to get the amherstia, but it was later decided that this would not be necessary as Wallich could supply a plant or two from the Calcutta garden. Gibson seems to have been a biddable fellow, quite willing to go wherever Wallich and Lord Auckland advised; he could not

[1] This was six years before their use by Robert Fortune

expect orders from Chatsworth, for up to the time he left the hills in January 1837 he had not received one letter from home, and he arrived back at Plymouth late in July, still ignorant whether any of the numerous consignments he had sent home had reached their destination.

The rains were late that year (1836) and it was July before Gibson was able to leave Calcutta; he filled in the time by laying out a new garden at Barrackpore for Lord Auckland and his sisters, the Misses Eden. He travelled to Cherrapunji by much the same route as Wallich's party, and arrived on 2 August. He was a true collector, full of enthusiasm and quite undeterred by the incessant rain. '. . . When I tell you, Mrs. Wright' he wrote to Miss Eden's elderly maid 'I am in my glory, you will excuse further details'.[1] Boatloads of plants were sent to Wallich, some for despatch to England and some for the Calcutta garden. He found between eighty and ninety species of orchid which he thought were not cultivated in Britain, besides rhododendrons and many other beautiful plants. 'He seems to be in a continual trance of rapture and admiration,' wrote Wallich to the Duke, 'and I can readily enter into his feelings when he says it is with pain that he is forced to leave a few orchideae etc. behind. He would be glad to send their forests and all if he could.' Gibson got back to Calcutta on 1 February 1837 with a final magnificent boatload of plants, and sailed for home on 4 March, with a large cargo including two carefnlly-packed amherstias. One was for the Duke, and the other was a present from Wallich to the Court of Directors of the Honourable East India Company. It was the one addressed to the Duke that died on the homeward voyage – almost the only casualty among Gibson's plants; but his Grace made such earnest representations to the Directors that he was allowed to keep the survivor. Among Gibson's orchids were *Coelogyne gardneriana*, *Dendrobium devonianum*, *D. gibsonii*, and *Thunia alba*; he also introduced *Rhododendron formosum*.

In 1842 Wallich's health again broke down, and he was obliged to go for a prolonged holiday at the Cape of Good Hope. During his absence, from August 1842 to June 1844, the Calcutta garden was temporarily put in charge of Griffith, who had meantime been making extensive and dangerous journeys, often in unexplored country, in Upper Burma, Bhutan and Afghanistan. Griffith has been described as a man of genius, and perhaps the most acute botanist ever to visit India; but his interest in plants was entirely morphological and the possibility that they might have aesthetic or other values seems never to have occurred to him. His first act at Calcutta was to present a report, openly or overtly critical of his predecessor's management; he then embarked on a well-meant but ill-judged attempt to remodel the garden on strictly botanical-textbook lines. The avenue of sago-palms, the groves of teak, mahogany, cinnamon and cloves were ruthlessly felled to make way for 'open clay beds, disposed in concentric circles, and baking into brick under the fervent heat of a Bengal sun',[2] and the fine amherstia was nearly killed by the exposure of its roots to the heat. Griffith's intention was to put the place on what he

[1] Markham *Paxton and the Bachelor Duke* 1935
[2] Hooker *Himalayan Journals* 1855

thought 'a proper footing', but the havoc in the garden when he returned in 1844 must have broken Wallich's heart. The veteran botanist stayed in India for three more years, and retired, finally defeated by the climate, in 1847; he died in London seven years later.

During his long reign in Calcutta, Wallich was associated with many botanists, British and foreign, besides those already mentioned. One of his earliest visitors was the French **Jean-Baptiste Leschenault de la Tour** (1773–1826), well-known as the naturalist who accompanied Captain Baudin's expedition to Australia in 1800. He was sent to India at the close of the Napoleonic Wars to collect economic plants and to establish a botanic garden at Pondicherry. He sailed in May 1816, and with his usual liberality in the cause of science, Sir Joseph Banks furnished him with recommendations and letters of introduction, and asked permission for him to have plants from the Calcutta garden. Leschenault was well received by the British authorities and permitted to travel in Madras, Bengal and Ceylon; he returned to France in 1822. Another French naturalist, **Victor Jacquemont** (1801–1832), came to India in 1829 when Wallich was on leave in England. He had started his travels three years before to distract his mind from an unfortunate love-affair, and was staying with a brother in Haiti when he received an invitation from the Jardin des Plantes to go on a collecting expedition in a country of his own choice. He chose British India and the Himalayas, and before setting out paid a visit to England to secure the necessary authorizations.

After several months in Calcutta Jacquemont left on 20 November 1829 and made a slow progress across northern India, visiting Royle at Saharanpur the following April. From there he went by an indirect mountain route to Simla, and from Simla explored the valley of the upper Sutlej and penetrated over the border as far as Tashigong in Tibet. He returned in October to Simla, where he received an invitation to visit Kashmir, with a letter of introduction to its independent monarch, Rundjet Singh. The season of 1831 was spent, therefore, in various excursions in unexplored Kashmir, where he made himself so popular that Rundjet Singh is said to have offered him the vice-regency of the kingdom. Jacquemont's travels were made largely at his own expense, for the allowance he received from France was quite inadequate; but the British everywhere gave him help and kind, if sometimes uncomprehending, hospitality. The exhausting hardships of his travels in the hills were, however, less deadly than the miasmas of the plains; on his way back to the south he contracted cholera, and though he made a partial recovery, he fell ill again and died at Bombay in December 1832. The Governor-General, 'praiseworthy to the last', consigned his collections to the French consul, and accorded the young victim of science a handsome funeral. Many Himalayan plants commemorate his name, among them the beautiful birch, *Betula jacquemontii*, introduced by Hooker.

It is odd that the British authorities should have been so kind to these two aliens, Leschenault and Jacquemont, and so unhelpful to Hove and to Hooker. Even the great Humboldt, with a reputation as high as Everest, for years hoped in vain to travel in the Himalayas and was frustrated, it is said, largely by the 'obstructive attitude' of the East India Company Hooker's complaint against the Company and its successor, the India Board, was that it 'snubbed him before he set out, refusing him assistance and official letters of introduction to India, and even a passage out', while not hesitating afterwards to exploit his hard-won knowledge and to annex the rewards of his trials and labours. Fortunately he was little hindered by this lack of official support, as he had so many good friends, all of whom he gracefully commemorated in the names of new-found rhododendrons.[1]

It was five years since Dr **Joseph Dalton Hooker** (1817–1911) had returned from Captain Ross's scientific expedition to the Antarctic, but at the time of his departure to India in 1847 he was still comparatively little known. He was commissioned to collect plants for Kew; but he also wished to gain some experience of tropical vegetation, and in particular to compare the Antarctic flora with that of very high altitudes in the tropics. It was recommended that he should go to Sikkim, nominally under British authority and hitherto entirely unexplored. During his three years in India Hooker made four journeys, but two were in tropical regions and one, being a winter excursion, was of greater importance to geography than to horticulture. The vital trip was the momentous third.

When Hooker arrived at Calcutta in January 1848, it was too early to go direct to the mountains, so he joined the party of a Mr Williams of the Geological Survey, who was going to Bijaigarh, near Sulkun, south of the Ganges, to inspect a recently-discovered coalfield. He left Calcutta on 28 January to join Williams at Burdwan, and from there they took the Great Trunk Road to Barun on the Son river. Then they followed the river upstream, along the base of the Kaimur Hills, with no roads worthy of the name, till it was time to cross north over the hills to Sulkun. There Hooker left Williams and his party and pushed on north-west to Shanganj and thence to Mirzapur on the Ganges. Transport on this journey was by elephant – intelligent animals who could be taught to toss up geological specimens from the roads or reach down flowers from high branches. From Mirzapur Hooker descended the Ganges by boat as far as Bhagalpur, and then struck north-east by slow stages to Darjeeling, arriving on 16 April. This was to be his base for the next two years.

On this occasion Hooker stayed in Darjeeling for more than six months – partly due to the advent of the rainy season, when the place was wrapped in what he called a 'dear, delightful, double-distilled Greenock fog'[2] and partly to the unexpected obstacles that hindered his further progress.

Some thirty years before, the Rajah of Sikkim had been replaced on his throne by the British, after having been driven from it by the warlike Nepalese; and Britain had guaranteed to maintain the integrity of his kingdom in return for certain privileges and concessions. In 1848, the now aged Rajah was willing enough to honour his treaty obligations, but was completely dominated by a powerful, unscrupulous and violently

[1] *R. R. aucklandii, dalhousiae, falconeri, hodgsonii, campbellae* and *thomsonii*
[2] Situated near Glasgow, where Hooker spent his youth, Greenock on the Clyde was notorious for bad weather

anti-European Prime Minister or Dewan[1], 'unsurpassed' says Hooker, 'for insolence and avarice', who knew that the British would quickly put a stop to certain nefarious but highly profitable practices in which he indulged, and was determined to keep them out of the kingdom. The application made by the Governor-General of India for a safe-conduct for Hooker was at first decidedly refused; and though it was subsequently granted, every difficulty was put in his way. Meantime, a similar request was made to Jung Bahadur of Nepal, who not only gave permission, but offered a guard of Gurkhas to escort Hooker to any part of his kingdom where he wished to go. It was eventually arranged that Hooker should explore the north-east frontier of Nepal, and crossing the mountains, return by way of Sikkim. While these negotiations were taking place, he made local excursions in the hills about Darjeeling, and assembled and trained a team of local collectors, 'very liberally' paid from 8s to 16s a month. Many of his rhododendron finds date from this period.

On 27 October he at last took the road, with a motley train of fifty-six – personal servant, interpreter, Sirdar, military escort with officer, plant-collectors, zoological collector, and all their attendants and coolies – amiable Lepchas and surly Bhutanese. A series of up-hill-and-down-dale marches brought him westward to the valley of the Tamur river, a tributary of the Arun; he turned upstream and went north to Chingtam and Mywa Guola. So far, the vegetation was tropical; yet only six marches farther on, at the village of Wallanchoon (reached on 23 November) the vegetation was sub-alpine and the inhabitants 'good-natured, intolerably dirty Tibetans'. They cared very little for the authority of the Maharajah of Nepal, and both the local headman and his own men were reluctant for him to proceed to the Wallanchoon Pass; it was only by the exercise of great firmness and tact, together with ministrations to the local sick, that Hooker was enabled to climb to the 16,750 ft pass with a small party and take the barometric and other measurements he wanted. At this season, of course, no plant was in flower; but he had no difficulty in recognizing the grass *Poa annua* and Shepherd's Purse,[2] – seeds of which, he thought, must have been accidentally transported over Central Asia by men and yaks. 'I could not but regard these little wanderers from the north with the deepest interest.'

Returning to Wallanchoon, Hooker sent the greater part of his train back to Darjeeling with the collections so far made, and set off with nineteen picked men and seven days' food on a journey that he knew would be one of great hardship. This time he followed the valley of the Yangma, a tributary of the Tamur, and approached as near as possible to the Kanglachem Pass at its head, which had been closed by snow since October. Then retracing his steps for a short way downstream he branched off by a lateral valley to the south-east, and edging round the great west flank of Kinchinjunga, crossed in turn the passes of Nango, Kambachan and Chunjerma. Finding the pass of Kanglanama closed, he turned southwards for a time through warmer and more cultivated country to the

[1] He was a Tibetan, and related to one of the Rajah's wives
[2] *Capsella bursa-pastoris*

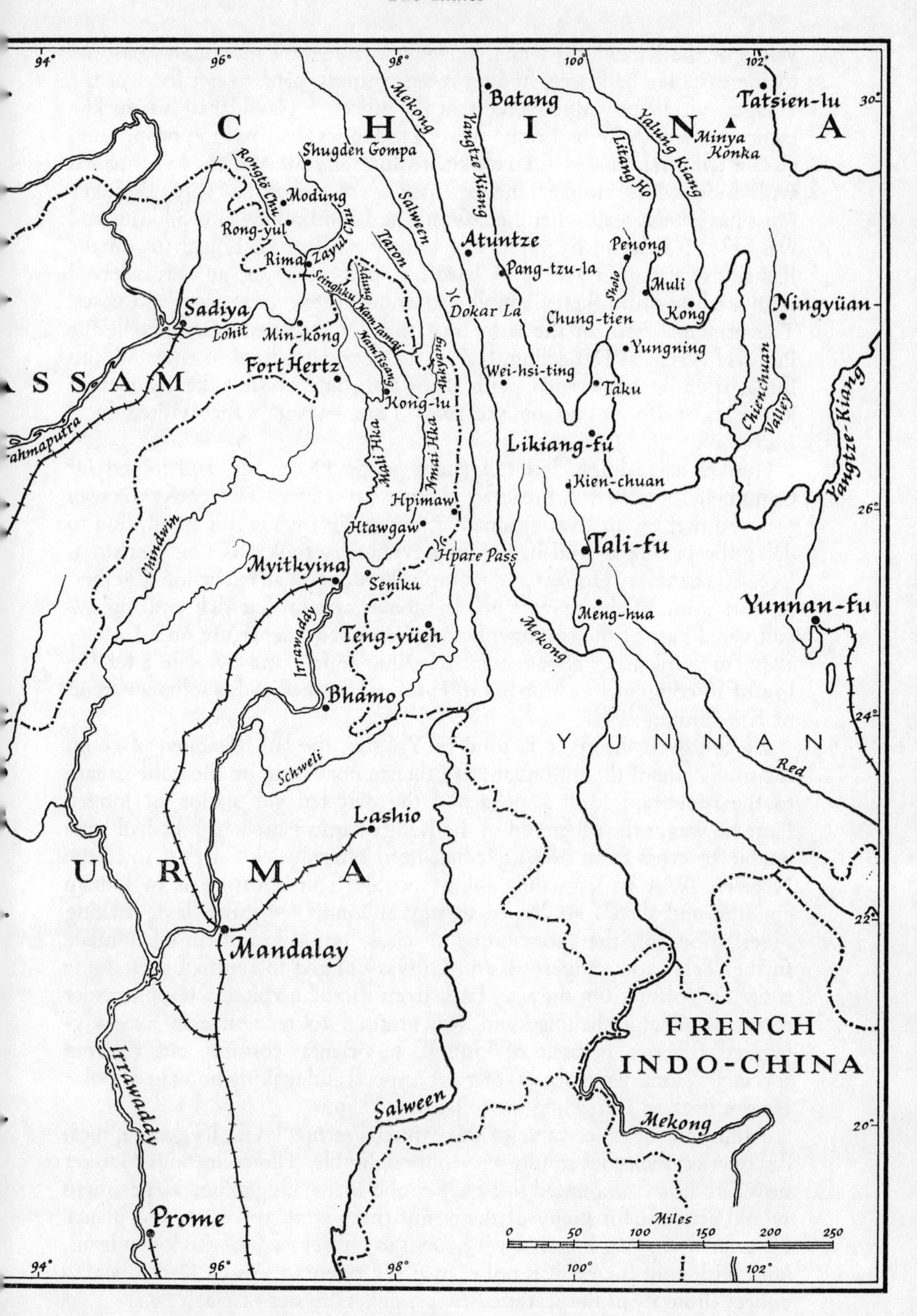

The Forrest and Kingdon-Ward country
Prepared by M. A. Verity

valley of the Khabil, up which he again assaulted the mountain-fortresses to the east. He had been finding it surprisingly hard to get food in the villages, and he now discovered that the officer or Havildar of his Gurkha escort, to whom he had done many kindnesses, had been appropriating all the tribute intended for Hooker to himself; his soldiers were loaded with food while Hooker and his men were almost starving. Hooker must have been glad when he crossed the Islumbo Pass into Sikkim, and was able to descend by rapid stages to the village of Lingcham on the Kuhait, where after weeks of hardship and privation, he was received with a salute of musketry, ample provisions, money, messages and news. The principal item of the latter was that Dr Archibald Campbell,[1] the Political Agent at Darjeeling, had at last been permitted to enter Sikkim for a 'friendly conference' with the Rajah, and would like Hooker to join him at Bhomsong on the river Tista – never before visited by a European.

Hooker accordingly left Lingcham on 20 December, and joined Dr Campbell in time for breakfast a few days later. The crafty Dewan received them with an appearance of hospitality, while still attempting to delay the promised audience, which eventually took place on Christmas Eve. The next day Hooker and Campbell set out on an excursion together, to visit some of the many notable monasteries in that rich and highly-cultivated part of the country, before Campbell's departure on 2 January 1849, to fulfil other engagements. Feeling lonely and desolate after the loss of so congenial a companion, Hooker returned to his solitary circuit of Kinchinjunga.

He followed the river Ratong to Yoksun, the last inhabited place on the south side of the mountain, and thence northwest up the same stream to the mountain Mon Lepcha and the deserted yak-station of Jongri. Here he was at the other end of the Kanglanamo Pass, which he had been unable to cross from Nepal; it had been officially closed ever since the Nepalese War to stop the Sikkim people from crossing it to kidnap children and slaves. He hoped to stay at Jongri for some days, making observations on the surrounding glaciers, but a heavy snowfall made further delay too dangerous, and he was obliged to return by the same route to Yoksun. On the way back from there he visited a temple under repair at Changachelling, and was amused to recognize in a newly-painted fresco a portrait of himself, in oriental costume but wearing spectacles (considered objects of great respect) and making notes in a book.[2] He got back to Darjeeling on 19 January 1849.

Although the chief value of this winter journey lay in its geographical discoveries, the other results were not negligible. The collections Hooker now sent home amounted to eighty coolie-loads; his geological specimens would account for many of these, but there were also seeds and plants. He refers to having lingered at 13,000 ft to collect seeds of rhododendrons, 'and with cold fingers it is not easy at the ripening season, December, to collect those from the scattered twigs, generally out of reach'.

[1] Not the same as Major-General Sir Archibald Campbell of the Burmese War
[2] Unfortunately this portrait had disappeared by 1909

Most of March was passed on an excursion in the Plains, where he found the atmosphere oppressive after the pure mountain air; April was spent in organizing his next journey, and he was ready to set out on 3 May – this time for the north-eastern frontier between Sikkim and Tibet. Permission had been obtained from the Rajah, but the Dewan was more than ever determined to obstruct his progress. He took the very simple means of forbidding the villagers to sell the foreigner any food. Foreseeing difficulties, Hooker had arranged for supplies of rice to be sent to him periodically from Darjeeling; but the Dewan had given orders that the roads were not to be repaired, which (in the rainy season especially) caused difficulties and delays. Bridges were destroyed or made unsafe, stepping-stones removed and paths blocked by branches of trees. One consignment completely failed to get through, and another was so delayed that the coolies carrying it were obliged to eat most of it before they arrived. No wonder that on passing through Bhomsong Hooker remembered with nostalgia his time there with Dr Campbell, and faced the future with loneliness and apprehension.

He was following the tropical valley of the Tista, which during this wet season (the rains began on 10 May) was full of leeches and insect pests. At Chakung the temperate and tropical floras met, with orchids growing on oaks and tree-ferns mingling with bracken. Higher up, at Chungtam, the river divided into two branches, confusingly called the Lachen and the Lachung; the former rose in the Cholamo Lake, beyond the Kongra-Lama in Tibet, and the latter near the Donkia Pass, both of which Hooker wanted to explore. He was delayed for some time at Chungtam, however, by the difficulty of getting provisions. He divided his forces, and taking with him only fifteen men, set off up the Lachen on 25 May. At the large village of Lamteng, three marches upstream, the Tibetan inhabitants were friendly, and regarded a doctor who could also paint, draw and write with extreme veneration; but they dared not defy the Dewan's stringent orders, though some of them surreptitiously brought food to his tent by night. Moreover, he was implored not to fire his gun, as it would bring down rain that might injure the crops – another device of the Dewan's, to prevent him supplementing his larder with game. He could get no information about the distance and position of the Tibetan frontier, nor even, where the valley forked, which of the two rivers was the Lachen; and in that wild country of tortuous valleys and half-obliterated paths it was almost impossible to find the way unaided. But the flora was magnificent, and Hooker went so far as to call the scenery 'very pretty'.

Undeterred by any difficulty, Hooker took a tent and eight followers, and with provisions reduced to a kid, a few handfuls of flour and some potatoes, pressed on up the Lachen, which shortly afterwards was joined by a stream called the Zemu, which he was permitted to explore. For a month he struggled to reach the sources of this river and its tributary the Thlonok, hoping to find a route to the north; but the way was barred by impassable cliffs and thickets. Meanwhile he and his uncomplaining men lived on wild herbs – leeks, nettles, a Polygonum which made a good spinach and was called Pollup-bi, bamboo shoots, mushrooms, and a species of Smilacina ('Chokli-bi') with a few pounds of secretly-purchased meal and a little rice when their provisions got through. The weather was wet, Hooker had rheumatism and some of his people were ill. Having taken such observations as he could, and collected many fine plants (including *Rheum nobile*) he returned to the Zemu-Lachen junction at the beginning of July.

Here he received letters announcing that Dr Campbell had repeatedly made strong representations to the Rajah about the treatment Hooker was receiving, and that the Rajah had sent peremptory orders to the local authorities that he was to be guided to the (unspecified) pass, together with a handsome, but in the circumstances singularly useless, present of sweetmeats, fine cloth and an embroidered silk gown. Even so, Hooker had to wait till 11 July for his next consignment of provisions to arrive, before he could set out, and for a long time could get no further than Tallum, which the unwilling local authorities tried to persuade him was in Tibet. His unwearying patience and persistence at last had its effect, and on 25 July he stood on the frontier, on the summit of the Kongra Lama pass, feeling that he had attained the object of many years' ambition.

Once he had got so far, no objection was made to a longer stay, and he spent a pleasant week at Tungu just below the pass, among kind and hospitable Tibetans. He left on 30 July to descend to Chungtam, which he reached on 5 August, and then prepared to explore the more easterly river, the Lachung. From the village of the same name, two days up the valley, he made a side-trip to the pass of Tunkra-La leading to the Machu valley, where some magnificent views of the surrounding mountains were obtained. On 29 August he continued his journey up the river from Lachung through Yeumtong to Momay, the highest yak-grazing station in Sikkim and a few miles south-west of the Donkia pass, which he visited next day. Here he decided to stay for a month, to the disgust of his official guide, who retired to Yeumtong, leaving Hooker 'truly happy' at the prospect of 'uninterruptedly following my pursuits at an elevation little below that of Mont Blanc'. One of the plants found here was *Myosotis hookeri*.

On 30 September Hooker left Momay to meet Dr Campbell, who was expected at Chungtam. The latter arrived on 2 October, and though he was wearied by travel, they decided to make an expedition together, and crossing the Kongra Lama pass into Tibet, circle round by the Cholamu Lake and re-enter Sikkim by the Donkia pass. The country beyond the passes was uninhabited, so the Tibetan authorities were unlikely to make any objection. They reached the Kongra Lama pass on 16 October, and were met on the summit by a party of Tibetan frontier guards with an officer, who demanded a parley. Fearing that he would not be allowed to complete his cherished survey, Hooker broke away on his pony and galloped ahead, determined at least to trace the river to its source in the Chulamo lake – a distance of some fifteen miles. Giddy from altitude and exertion, and leading the exhausted horse by a plaid (the bridle having broken), he reached the lake about one o'clock, happy to have achieved his aim, though faced with the possibility of having to sleep out, at 17,000 ft, with only the pony to keep him warm. Two hours later, however, having partly retraced his steps, he found his party encamped; the Tibetan officer – who is said to have 'admired' rum and water – had not only allowed them to cross the pass, but had given permission for them to stay for a day on the Tibetan side. They were able therefore to climb the peak of Bhomtso, and to return to Chungtam by the Donkia pass and the Lachung river, as they had planned.

The next move was south to the capital, Tumlong; but attempts to obtained an interview with the Rajah were barred by his officials – all creatures of the Dewan. They then turned east to explore the Cho-la and Yak-la passes – Hooker gathering seeds of twenty-four species of rhododendron in two days on the way. They reached the top of the Cho-la, another Tibetan frontier, but were not allowed to cross, so they retired to camp a little below the summit. There seemed to be an unusual number of Sikkim soldiers about, and their bearing was disrespectful.

That evening, 7 November, when Dr Campbell stepped outside his tent, he was savagely attacked, beaten and taken prisoner; Hooker was forcibly prevented from going to his aid, but was not otherwise molested. This was done by accomplices of the Dewan (who did not himself appear) and without the knowledge of the Rajah. The object was to hold Campbell as a hostage, to obtain certain concessions from Britain – such as a relaxation of the laws against slavery; and Hooker, having no official status, would have been allowed to continue his journey unhindered. The assailants were surprised when he not only elected to stay with his friend, but insisted on seeing him, if only for a moment, to assure himself that although bruised and bound, he was not seriously injured. The taking of political hostages was not unusual in Nepalese and Tibetan warfare, and the Sikkim officials had no conception of the seriousness with which this outrage to their representative would be taken by the British Raj.

For a considerable time, however, the British Raj was unaware of the insult that had been offered to it. Hooker and Campbell were not allowed to write, and a letter sent by the Dewan to Darjeeling was couched in Tibetan. It was concerned chiefly with the terms he wished to impose, and only in the last line, which an interpreter failed to read out, was it revealed that the plenipotentiary was being held till an answer was received; so the letter was put aside for Campbell to deal with on his return. It was only when the Sikkimese became worried at receiving no reply that Hooker was permitted to write; and he sent his letter direct to the Governor-General, Lord Dalhousie, who happened to be absent from Calcutta on a visit to Bombay. When at last the facts became known, the reaction was prompt; a peremptory demand for their release was despatched to the Rajah, and a military force was sent to Darjeeling, with a threat of invasion if he did not comply.

Meantime Campbell was ignominiously carried to Tumlong, and Hooker 'kept as near as I was allowed, quietly gathering rhododendron seeds by the way'. After about a week the stringency of Campbell's captivity was slightly relaxed, and Hooker was allowed to share it. Their men were imprisoned for a time and then sent home, chiefly because the impoverished Sikkimese could not afford to feed them. Hooker and Campbell supplemented their scanty rations by purchases as long as their money lasted, and were constrained by pity to feed their guards – the poor wretches were so starved. When at last the order came for their release, the Dewan was sent to escort them to Darjeeling; but they were still treated as prisoners along the route. As the boundary was approached, however, the Dewan got more and more nervous, and when the newly-built barracks at Darjeeling came in sight he was stricken with such terror that Hooker began to think he would murder his captives from sheer fright. A little timely persuasion on the advisability of releasing them in time for the great British festival of Christmas induced him to let them go, and they galloped the few remaining miles to Darjeeling on 24 December, where they were received as though returned from the dead.

Much was threatened as a reprisal for this outrage, but little was done. The poor old Rajah was deprived of the most profitable part of his estates; the Dewan and his accomplices were disgraced and shorn of their offices and emoluments, but were not otherwise punished; and the lower part of Sikkim was annexed – to its own content – by the British. But Hooker received no compensation for the loss of all his instruments and part of his notes and collections; and the presents afterwards made to him by the repentant Rajah were actually sold by the India Board on behalf of the Government.[1]

When Hooker got back to Darjeeling after this eventful journey, he found his friend Thomson waiting for him. A fellow-student of Hooker's in Glasgow, **Dr John Thomson** (1817–1878) had himself had an adventurous time since his appointment as an Assistant-Surgeon to the East India Company in 1839. He was taken prisoner in the Afghan war, and was destined for the slave-markets of Bokhara, but escaped and returned to India. In 1847 he was appointed to the Kashmir-Tibet boundary commission – abortive because the Chinese commissioners failed to arrive – and made a journey to the Himalayas which covered much of Jacquemont's ground, but extended much farther north to Leh and Iskardo in Ladahk, and eventually to the Karakoram Pass, getting back with some difficulty at the end of 1848 owing to the outbreak of the second Sikh war. He was 'devotedly fond of botany', and was ready to sacrifice a year's furlough to accompany Hooker on his next excursion.

Bhutan was closed to Europeans, and they could not go to Nepal because the friendly Jung Bahadur had just left for a visit to England, and felt unable to guarantee their safety while he was away. They decided therefore to go to the Khasia mountains, and set off on 1 May 1850, by more-than-leisurely waterways, reaching their base at Cherrapunji about the middle of June. Here they were in country already extensively botanized by Wallich, Griffith and Gibson; and although they stayed longer than Wallich, they did not travel nearly so far. In July they went to

[1] No official of the Honourable East India Company was allowed to receive presents, and gifts from native potentates were always sold for the Company's benefit. But Hooker was not in the Company's service

Nunklao on the northern slopes of the hills, and were back in Cherrapunji by 7 August, living in a bungalow and keeping three large coal fires constantly burning to dry themselves and their plants – necessitated by a climate in which thirty inches of rain sometimes fell in twenty-four hours, and fifty feet in a year. On 13 September, they started a longer trip north-east to Jowai and Nartiang in the Jaintia Hills; on the way back (4 October) they were so fortunate as to find near Pomrang an abundance of the rare and beautiful blue orchid, *Vanda caerulea*. (They collected seven mens' loads of it to send to Kew, but unfortunately only a few of the plants survived the journey.) Before leaving the district they made another excursion south and east to Sylhet and Silchar and a little beyond, to the frontier of Manipur. Then they went back to Chattuc to retrieve collections deposited there, and by boat to Chittagong on the Bay of Bengal where they made excursions in the neighbourhood of the town, but were not able to botanize in the mountains of the interior owing to the activities of head-hunters. They returned to Calcutta on 28 January 1851, and Hooker sailed for England on 7 February.

One cannot but admire Hooker's courage and determination in the face of difficulties; those of the Sikkim climate and terrain would have been enough to daunt most people, without the semi-starvation and other impediments imposed on him by hostile authority. His chief assets were his medical skill and the attachment of his men; he tells some touching stories of the devotion of his Lepcha lads to his incomprehensible interests. The map of Sikkim that he made under such trying circumstances was so accurate that the officers of the Sikkim-Tibet Boundary Commission of 1903 sent the eighty-six-year-old botanist a telegram of congratulation on his work. Many fine Himalayan plants bear his name; his numerous introductions included *Primula sikkimensis*, *P. capitata* and *Meconopsis napaulensis* (= *wallichii*). He will chiefly be remembered, however, for his rhododendrons; he found forty-three species and collected seeds of a large proportion of them, which were successfully raised at Kew. As soon as he arrived at Darjeeling he began to send home drawings and descriptions of his finds, and these were published during his absence by his father, Sir William Hooker, as *Rhododendrons of the Sikkim Himalaya*. The first part of the book reached Darjeeling in the spring of 1850, when Hooker was resting there after his release from captivity – less than two years after his first arrival; and it created a sensation. 'All the Indian world,' he wrote, 'is in love with my Rhododendron book.' It was the first revelation of the glories of the genus. Among the species that he introduced were *R.R. campylocarpum*, *ciliatum*, *cinnabarinum*, *falconeri*, *griffithianum*, *maddenii* and *thomsonii*, and after his return home he was urgent with his friends in the south and west of England to plant Indian rhododendrons, which were found to thrive in Cornwall even better than in Sikkim.

Such was the influence of the rhododendron book that it led to the despatch of another collector to India, when only the first few sections had been published and when Hooker himself was still in Sikkim. Thomas Nuttall, reluctantly retired from the USA and living in Lancashire, was fired with the desire for discovery and possession; and he had a nephew, unemployed at twenty, who was eager to embark on a plant-hunting adventure. **Jonas Thomas Booth** (b. 1829) seems to have had a character similar to that of William Bartram – always starting on some new career with an enthusiasm that quickly waned in the face of harsh reality. Nuttall sent him to Sir William Hooker at Kew for a preliminary briefing on Indian plant-collecting; he was also to collect birds for Lord Derby, but neither Sir William nor Lord Derby was very hopeful about the probable results. Booth's inexperience, his ignorance of science, and his optimism about the ease with which the necessary knowledge could be 'picked up', were not promising; but his uncle thought he had sufficient resolution, and that the experience might do him good. He was furnished with letters of introduction to Dr Joseph Hooker and to Dr Falconer (then superintendent of the Calcutta garden) and sailed from Liverpool on 30 June 1849.

Booth arrived at Calcutta in the autumn, and soon began to discover that there were difficulties in the way. 'The Rajah of Sikkim is very mutch opposed to Europeans', he wrote to some cousins on 15 November, 'and as jelous as the Chinese, his immediate neighbours.' Soon afterwards the news of Hooker and Campbell's imprisonment put an end to his hopes of travel in Sikkim. On Falconer's advice Booth went east to Gauhati in Assam, and by mid-December was among the mountains of the Balipari Frontier Tract, north of Tezpur and Bishnath – the area drained by the river Bhareli and its tributaries. North-west lay the Bhutan boundary, north-east the hills inhabited by the Akas and the Duphlas; the temperature ranged from 94°F. in the valleys to 7°F. on the hills, with snow and ice at about 10,000 ft. Here he found a wealth of rhododendrons and other plants, and stayed so late into the summer of 1850 that he was caught by the monsoon rains while still far in the interior, and suffered great privations and hardships before he got back to Calcutta. When the dry weather returned he went back to the same area for another season's work. Despite repeated statements[1] that his plants came from Bhutan, there is no evidence that he ever crossed the boundary; he wrote on one occasion that he had been to the frontier of Assam, beyond which he was not allowed to travel by the Rajah of the district.[2] He sent home two shipments of living plants, specimens and seeds, and brought a third consignment on his return in June 1851. All were solicitously cherished by his uncle at Nutgrove, and many new plants were successfully grown and distributed. The most beautiful, but unfortunately the least hardy, of his new rhododendrons was named *R. nuttallii*; he also introduced *R. boothii* and *R. hookeri*, besides several dendrobiums, *Agapetes buxifolia, Begonia xanthina* and *Primula mollis*. He did, in fact, a great deal better than might have been expected; but he never returned to plant-collecting.

When Joseph Hooker expressed the view that plants (especially orchids) sent to Kew were less likely to be successfully cultivated than those sent

[1] In the *Botanical Magazine* and elsewhere

[2] Nevertheless, two of his plants, both rare (*Primula boothii* and *Rhododendron batemannii*) were subsequently rediscovered in Bhutan

to a professional nurseryman, he may have been thinking of **Thomas Lobb**, who was in India at the same time, collecting for Messrs. Veitch. This was his second journey; he had already spent three years in the East Indies, before leaving England for Calcutta on Christmas Day 1848. Hooker heard of his arrival in March 1849, when he was at Darjeeling preparing for his second venture into Sikkim. Plant-hunters were often jealous of the possible encroachments of a rival, but not Hooker; on 23 June, when he was far up the Lachen valley and in the midst of his own difficulties, he wrote to his father with a most generous offer. 'Tell Veitch by all means to send Lobb to Darj. before October if possible, he shall have every opportunity, facility and information I can afford both as to living and collecting. May use my collection as much as he pleases in instructing himself on his own – I hope to return to these parts in October, for seeds, and I will (with Campbell's sanction) let Lobb accompany me, when he shall be shewn everything I can shew and have every facility I can afford, subject to whatever advantage to Kew you may think fair . . . it is a chance Lobb may never get again, certainly never so cheaply. . . . You must tell V. that I travel as a poor man and Lobb must not expect great tents and serv'ts.'

As we know, Hooker's plans were disrupted by the arrest of Campbell, and he did not get back to Darjeeling till Christmas. In any case, it seems doubtful whether Lobb would have accepted his offer. He was in Darjeeling in March 1850, and Hooker being absent on a visit to Calcutta, Dr Thomson made friendly overtures; but Lobb did not respond, and indeed seemed to avoid them both. They found him 'a most steady respectable man' and 'very modest and well-behaved in his deportment', but dreadfully conceited; 'He pooh-poohed Sikkim and has a very poor opinion of Lindley and Wallich!!!' At Cherrapunji in June they found that Lobb had been there shortly before them; and they all met at Myrung in October, when Hooker and Thomson were on their way back from the Jaintia Hills. 'We spent the evening and following morning most pleasantly,' wrote Hooker, 'but he would not stay even a day with us, though in no hurry – it appears odd to me – he talks very slightingly of the plants and seeds as usual and to judge by what he says he cannot be worth 6d to Veitch and Co. His plants he says die en route to Calcutta and that it is almost useless sending roots, bulbs or cuttings straight home from this.' All the same, two of Hooker's namesake plants, *Berberis hookeri* and *Hypericum hookerianum*, were successfully introduced by Lobb, together with Wallich's *Lilium giganteum* and a number of Khasia orchids – *Aerides fieldingii*, *Cypripedium villosum* and *Pleione lagenaria*. He also sent home plants of *Vanda caerulea*, and it seems likely that Hooker told him where it grew; he had found it just before they met.

The greatness of mind displayed in Hooker's generosity towards Lobb is also apparent in his reaction when at last he reached the summit of the Kongra Pass. 'I am very pleased to think that *any one* may now go,' he wrote to his father; 'the egg-shell is broken: the intricate route once known and the nature of the impediments, it is easy to forestall the one and follow the other.' Actually no one seemed in a hurry to face the difficulties and hardships that he had undergone; and it was twenty years before his example was followed (in 1870) by Elwes and Blanford. Hooker's *Himalayan Journal* was the inspiration and guidebook of their journey.

Henry John Elwes was not at this time the great plantsman and dendrologist he afterwards became; his interest in gardening only developed after his marriage the following year. He had come to India to pursue his hobby of ornithology, and he joined forces at Darjeeling with a naturalist of similar tastes, W. S. Blanford of the Geological Survey, who wished to investigate the zoology of upper Sikkim. It was not a good time for travel as the rainy season was not yet over, but Blanford had only three months' leave, and had to go when he could. It is from his journals that we learn the details of the expedition.

Little had changed since Hooker's day. The weak Rajah had been succeeded by an equally ineffective son, and the government was in the hands of his forceful but not unfriendly brother. Hooker's malicious Dewan was not allowed to enter Sikkim, but he was governor of the town of Chumbi, just over the border in Tibet, and from there still exercised a baneful influence. Elwes and Blanford wanted to cross the border at one of the south-eastern passes and take a short cut through Tibet to the Tunkra-La, thus avoiding a long tedious trek in the rains through the steamy Tista valley. But the Tibetans were obdurate in prohibiting them from crossing the frontier; partly, Blanford thought, because of orders from the Dewan, and partly because such a rush of Europeans – three in one year,[1] after twenty years' quiescence – had 'alarmed the Celestials'. They had to retreat, and follow Hooker's route up the Tista and Lachung rivers to the Donkia Pass. Here Blanford had another argument with the Tibetan frontier guards; and in the meantime Elwes, who had been left behind in camp because he was lame from leech-bites, had roused himself, explored and finally crossed a little-known pass two or three miles to the west, and astonished the indignant Tibetans by appearing at the Donkia from the northern end. They thought they had been played a trick; but Elwes' lameness was genuine, and so much aggravated by crossing two 18,000 ft passes in one afternoon, that he was laid up for several days on the way back down the Lachung, and it was thought at one time that he would have to be sent back to Darjeeling in a litter. However, he recovered sufficiently to continue down to Chungtam and up the Lachen to Kongra Lama pass – again in Hooker's tracks. They then returned via Tumlong to Darjeeling 'and the nineteenth century', arriving in mid-October just when the fine weather was beginning.

This was not a botanical excursion; the discoveries made were ornithological and geographical, though Elwes also collected specimens of 200 different ferns. But during this visit he acquired a financial stake in Sikkim which brought him back on two subsequent occasions. It had been found that the area of lower Sikkim that was annexed by Britain as a reprisal for the imprisonment of Hooker and Campbell, was particularly

[1] A sportsman, Captain Chamer, had obtained a permit a month or two before

suitable for the cultivation of the Assam tea-plant investigated by Wallich and Griffith; the manager of a flourishing tea-plantation suggested to Elwes that they should go into partnership, and it became one of the most profitable investments that he ever made. The affairs of this plantation gave Elwes an excuse to revisit Sikkim in 1876 and 1879; and on the first occasion in particular, he took advantage of the opportunity to make a fortnight's collecting-journey into Western Sikkim (Sanga Choling, Tonglu, Simana and back) from which he introduced *Pleione hookeriana*, *Satyrium nepalense*, and three arisaemas, *A. nepenthoides*, *A. utile* and *A. griffithii*. The 1879–80 visit, being in the winter, yielded birds and insects rather than plants.

In 1886, on Hooker's recommendation, Elwes was appointed naturalist to a Government mission to Lhasa. While waiting at Darjeeling for the dilatory and over-numerous officials to assemble, he made a preliminary trip to Pashiteng and the Rishi-La to survey a possible route for the party. On his return to Darjeeling he found that the mission had been abandoned; so he made a profitable short trip (though in the rains) north-west to the Singalila Range; and another to the much-botanized Khasia mountains, where he nevertheless found a new species of Hedychium, *H. elwesii*. His final Indian visit was a short journey to Nepal in 1914, at the invitation of the Maharajah, to whom he took the entirely characteristic present of a pedigree Aberdeen-Angus bull. Elwes was a big, powerful, bearded figure, loud-voiced, dominant, generous in material ways but without the sensitiveness and sweetness of Hooker's character. His interests ranged from big-game shooting to butterflies, and he was so constant a traveller that he hardly spent an entire year in England after the age of seventeen.

When Elwes was in Nepal, Cooper was in Bhutan. In 1913 that insatiable patron, A. K. Bulley, had again applied for a collector to Dr Bayley Balfour of the Royal Botanic Gardens, Edinburgh – the source that had already supplied Forrest (1904) and Kingdon-Ward (1911); and this time **Roland Edgar Cooper** (1890–1967) was recommended. He had been working at the Edinburgh garden since 1910, and was particularly well-qualified for the post as he had lived for three years in India and had studied at the botanic gardens of Calcutta and Darjeeling. Bulley sent him to Sikkim, but nothing is recorded of this (1913) expedition, which according to Cooper himself, yielded little of interest. He seems then to have broken with Bulley, for in 1914 and 1915 he travelled in Bhutan 'solely for collecting botanical material' – probably on behalf of the Botanical Survey of India, to which his notes and journals were consigned.

Bhutan was the most difficult of access of all the native states, and had not been botanically explored since Griffith went there in the capacity of surgeon to Major Pemberton's Embassy in 1937–8. Permission was obtained for Cooper's researches, but his plans for the first season were disrupted by finding that he was expected to arrange visits to the Dzongpens or governors of all the valleys he crossed, as well as to pay a formal call on the Maharajah. In two seasons he covered a great deal of ground, often in country little visited even by the native Bhutanese, and seldom stayed two nights in one place; his collections were 'little more than an occasional mouthful snatched on the move' – a survey to guide future travellers. But none followed until Ludlow and Sherriff began their series of expeditions in 1933; and except for their last journey in 1949, they worked for the most part farther to the east, overlapping Cooper's country only to a small extent.

His first season in Bhutan was a short one, for he did not leave Buxa Duars on the Bengal boundary till July, and his time was further curtailed by the necessary visits to the Rajah and lesser authorities. He went almost directly north up the Rydak river and its eastern branch the Thimbu Chu, to Lingshi Dzong[1] and the pass of Philey La on the border of Tibet. Later in the season he went east via Punakha for his interview with the Maharajah and returned to Buxa in November. Plants found in 1914 included *Ceratostigma griffithii*, *Cotoneaster cooperi* and *Viburnum grandiflorum.*

Most of the rivers of Bhutan run from north to south, from the boundaries of Tibet to those of Bengal and Assam; on his second tour Cooper traversed almost the whole country from west to east, crossing the divides from valley to valley, and ascending several of the rivers as far as possible to the north, before going on to the next, and then returning by a more southerly route with detours to some of the isolated high lands at the southern end of the ranges. He left Buxa on 18 April 1915, and went north up the Rydak river and its north-western branch to Paro, and explored the Pa-chu valley to Ghassa, the Temo-la and the peak of Chumolhari. Then he returned to Paro and embarked on the first stage of his eastern journey, crossing as quickly as possible over the Thimbu valley and the Dokey-La to Punakha on the Mo-chu (23 May) where he had been before. From there an eight-day tour was made up an eastern tributary, the Po-chu, and a longer one north up the Mo-chu; but in both cases further progress was barred by flooded streams and broken bridges. The way up the Po-chu was particularly bad; cliffs had to be scaled by means of notched poles, or skirted on platforms of branches which had rotted in places and let the unsuspecting traveller through; but in this valley Cooper found *Rhododendron rhabdotum* (unfortunately not hardy) – 'a marvel with large red stripes down the four-inch-wide corolla'.

The switchback eastern journey was then resumed – over the Pe-Le-La to Tongoa on the Mati-chu and over the Yato La to Biaka[2] (= Pumthang?) on the Pumthang-chu, the capital of the country and the seat of the Maharajah (12 June). Here the weary traveller took a few days' rest, to recover from leech and *pipsee*-fly bites which had 'become rather badly septic' – meantime, however, sending native collectors about the district. Then he moved upstream to Champa, where he explored two lateral valleys, and north again to the base of the Melakatchu Pass. He also made an excursion south-west from Biaka to encircle two peaks of the lower Yato-la range, the Tibdeh La and Pegnay La, where he made a find of great botanical interest, *Lobelia nubigens*, the only Asiatic representative of the giant tree-lobelias of East Africa.

[1] Dzong in Tibetan means a fort, and La a pass

[2] Also spelt Bya Gha, Byagur and Beyaka

Eastward again, on 22 July; the next pass to be crossed was the Rudong La, whose name means the Pass of Horns, because some of the rock-passages were formerly so narrow that the horns of the yak got stuck as they went through. Here Cooper slipped from a precipitous track, but his fall was providentially stopped by a bush which turned out to be a new species of buddleia – *B. cooperi.* This pass brought him to Lingtse Dzong in the valley of the Kuru-chu; and again he explored north up the river and its eastern branch the Komo-chu, and camped at a monastery on a flat called Narim Thang, (14,000 ft) below the Gong La pass to Tibet, where he found many primulas, including 'the gem of the whole of the Eastern Himalaya' – *P. eburnea.* Then he returned to Lingtse.

The next ridge to the east was crossed by the Dongo La, and the next river, the Dangma-chu, was reached at Tashiyang-tse. Although he made a further excursion to the north-east, this was really Cooper's turning-point; he went south instead of north, down the river to Tashigong Dzong,[1] and then circled west and north-west, crossing the Kuru-Chu at a more southerly point than on his outward journey, and over two more passes to the Pumthang valley and Biaka again. Here collections were sorted, packed and despatched, and seed-collectors sent to places previously visited, with instructions to meet again at Angduphorong on the Mo-chu. Cooper meanwhile made a detour to the south between Tongsa and Angduphorong, to visit the Black Mountain (16,130 ft) of Central Bhutan.

Four days before Cooper reached Angduphorong, a landslide higher up the stream had blocked the Mo-chu, and the inhabitants were hourly expecting the dam to burst and flood the valley. All his collectors but one had arrived, and at the urgent advice of the Dzongpen they crossed the river and camped on the farther side. At 3.30 a.m. on 23 September, the waters came roaring down, and swept away the bridge; and communication with eastern Bhutan could only be maintained by arrows bearing messages which were shot across the river once a day. Cooper waited a week for news of the missing collector – the one who had been sent right back to Narim Thang for seed of *Primula eburnea*; then, leaving a message of instructions, he continued his westward journey, making another detour to the south to explore the high peaks of the Taka La. At Chapcha in the Rydak valley he tried to get a permit to revisit the Tremo La for seed, but this being refused, he returned earlier than he had planned to Buxa, on 10 October. A month later, when he was about to leave, the missing Lepcha collector arrived, bearing with bashful triumph a small and grubby packet of seed. Although unwell with fever, he had reached the remote monastery where they had camped, and had gathered some seed of the primula the same afternoon; but the next morning the plants were buried in snow, and he had to beat a hasty retreat before the passes were closed. When he got back to lower levels, his fever had returned with such severity that he was delirious for eight days, and he resumed his journey when he was still too weak to travel more than half a stage at a time. Nevertheless, he volunteered to go with Cooper the following year,

[1] Visited by Ludlow and Sherriff in 1934

(1916), when the botanist made a journey to the western end of the Himalayas, in Kulu and Lahaul; but the country investigated, between the Rohtang and Bara Larcha passes, was too high and arid to support much vegetation. After a period of military service and two botanical appointments in Burma, Cooper returned to Scotland, to become in due course the Curator of the Edinburgh Botanic Garden.

Although his survey of Bhutan was so rapid and its object primarily botanical, Cooper introduced (by seeds to Edinburgh) a number of plants, which would be more familiar if they had proved more hardy. Like Forrest and Sherriff, he was so dedicated a primula-addict that he had little interest in any region where plants of this genus were not to be found.

For excellent reasons, the eastern end of the Himalayan chain, where Bhutan, Assam and Burma impinge upon Tibet, was the last to be explored. Difficult of access, peopled by hostile tribes, with the most execrable climate, the most terrifying mountains, the most numerous and turbulent rivers, and the most magnificent flora in the world, the area was for forty years the chosen hunting-ground of that redoubtable explorer and collector, **Frank Kingdon-Ward.** It may have been the presence of other collectors in China (notably Forrest) that drove him, after his first two journeys, farther west; and also the fact that, unlike Farrer, he did not take to the Chinese people, preferring the Tibetans and Burmese. But undoubtedly for him the most powerful attraction was that of the unmapped and unknown, a country where even the courses of the rivers – the chief and often the only channels of communication – were only tentatively indicated by dotted lines. (See map on page 159).

Above the railhead of Myitkina in Burma the great Irrawaddi divides into two branches, the Mali Kha to the west and the Nmai Kha to the east. High on the Mali stands Fort Herz (Putao) then the most northerly European station in Burma and the starting or finishing-point of many an expedition. At about the same latitude, the Nmai divides into two branches, the east and principal being the Taron, down which Kingdon-Ward travelled in 1922 when he crossed the mountains from China, and the Tamai or Nam Tamai, all flowing approximately north to south. The Tamai again forks into the Seinghku and the Adung, both of which he explored to their hitherto-undiscovered sources. Altogether he made six journeys in Burma before our date-limit of 1939, three in Tibet and two in Assam; but he did not necessarily confine himself to one country on one trip, crossing frontiers with more nonchalance than most of us cross roads.

A new frontier, indeed, seems an irresistible lure to a collector; no sooner is a mud fort built, with a handful of native soldiers and two or three British officers in residence, than the botanist is there. In 1914 the British had recently occupied Hkamti Long on the Nmai Kha, and the valleys east and north-east of Myitkina; and the frontier with China had been defined for some way to the north. Kingdon-Ward hoped to follow this frontier northwards through ever-higher mountains, and then possibly strike over to Assam. He had only a month at Rangoon to prepare for the journey, after a year's hard work and hard travel in China; and he arrived at the border-fort of Hpimaw, on the Nmai-Salween divide north-east of Htawgaw, on 18 May 1914.

This was the start of one of his most trying seasons. It seldom or never stopped raining; he was never dry, and rarely free of fever. The bamboo forests of the lower levels were exhausting and dismal, destitute of wild life except for the over-abundant insects. All travellers in the lower Himalayas speak feelingly of the leeches, which at some seasons infest the soaking vegetation; they are most difficult to repel (they will get inside boots by the holes for the laces) and their bites cause sores that often take long to heal. At higher levels the sandflies made life a misery.

Kingdon-Ward stayed at Hpimaw Fort till the middle of August. He visited two local passes and looked over into China; but his first attempt to climb the mountain of Imaw Bum, across the Ngawchang to the north-west, had to be abandoned, as he several times collapsed with fever and had to be carried into camp. After a week at the fort he tried again and this time reached the summit, and shortly afterwards he also conquered the neighbouring peak of Lacksang Bum. He then had to lie up for some time with a bad foot and a recurrence of fever, and to depend on the efforts of two hired Lashio collectors who were 'lazy and unenterprising to a degree'.

His health being so precarious, Kingdon-Ward felt obliged to return to civilization; but he chose to do so by another route. On 18 August he set off to the north up the Ngawchang River, encouraged by the thought of seeing some new country; but found it 'vastly the same . . . for many weary miles'. On the 26th he crossed the Wu-law Pass, tormented by rain, cold and sandflies; but he never saw a more wonderful place for flowers, or one more difficult of access – 'I hardly think a white man could spend a season there, and live.' Thence he descended into the warm Laking valley with its tropical flora, and turned northwards up the river Nmai Kha. His food was now running short; he had prepared for a four-week journey, but in that time only half the distance had been accomplished. It was impossible to hurry his carriers; if they did more one day, they only did less the next. Faced with a shortage of porters and a stretch of country where there were no more villages at which food could be obtained, he decided to press on with a small party and leave his servant to bring on the greater part of the baggage when transport and supplies could be had; and left the valley of the Nmai to climb westward over the pass of Shingrup Kyet to the valley of the Mali Kha. It was still raining, his feet and ankles were covered with dreadful sores from the leeches and constant wet, and he could not sleep. At last he reached the fort of Konglu (an outpost of civilization, where there were two British officers) and got a bath, a shave, clean clothes, food – and news of the outbreak of the first World War. He left on 23 September, and four days later was ferried across the Mali river to the settlement of Fort Herz (four Europeans) arriving in a state of great exhaustion. Next day he again collapsed with fever, and was ill for six weeks. He abandoned all thought of going on to Assam, and as soon as he recovered travelled gently down to Myitkina, which he reached in time for Christmas Day.

In some autobiographical notes written late in life, Kingdon-Ward refers to an occasion when he was offered a permanent botanical post in England after a 'disastrous journey' in his early years, but refused it because he was ashamed to admit defeat. This would seem to be the journey in question; in no other of his recorded travels did he suffer so much, nor was obliged to leave the field before the season of the full seed-harvest. But the journey cannot be regarded as a failure; to Kingdon-Ward himself, as a geographer, it was of great interest, and it was valuable for its revelation, to himself and others, of the great wealth of flowers in the region, especially primulas and rhododendrons. He later remarked that owing to his discoveries 'all the plant-collectors in Asia . . . came buzzing round the Htawgaw hills like flies after jam', and that Hpimaw became 'probably the most collected spot on the frontier'. This was perhaps a slight exaggeration, but Farrer and Cox made it their base in 1919, and it was further combed by Forrest in 1924. Kingdon-Ward himself returned to the area in 1919, and when Farrer was stationed at Hpimaw, Kingdon-Ward was working on the westward side of Imaw Bum.

Never was a momentous expedition undertaken with such light-hearted irresponsibility as this of **Farrer** and **Cox**. It was all settled within half an hour, when Euan Cox went to visit Farrer in the nursing-home where he was recuperating from an operation, shortly after the Armistice; and when the two arrived in Rangoon in March 1919, their ultimate destination was still uncertain. They wanted to go to the Tibet-Assam frontier at the source of the Mali Kha, but in this uninhabited region neither food nor transport was to be had, so they decided to make Hpimaw their centre, and travelled by what was now a fairly well-engineered road from the railhead at Myitkina north-east to Seniku, north up the Nmai Kha, due east to Htawgaw, and then north-east again to the fort.

Hpimaw lies at the foot of one of the passes into China, the frontier running almost due north and south along the range of mountains that walls in the Salween river. Cox estimated that they traversed the short distance between the fort and the pass at least eighteen times during their stay, and every time found something new. It was a monsoon area, and rich in dwarf woody plants – rhododendron, berberis, cassiope, bamboo and willow, but with rather a disappointing absence of herbs; on the drier Chinese side of the range the flora was quite different. Farrer was much annoyed by the arrival at the fort of a party of Forrest's native collectors, which he regarded as an encroachment on his territory; but actually they were working on the Chinese side of the boundary, where Europeans could not go, and only visited Hpimaw out of curiosity.

Farrer and Cox went several local excursions, chiefly to a camp they established in the flowery Chimili valley farther to the north, where they found a more alpine and less tropical vegetation. They also visited the passes of Fung Schweling and Hpare to the south, but found them unproductive, and returned to the comparative comfort of Hpimaw for the worst of the monsoon rains. Towards the end of the season they had a four-day visit from Kingdon-Ward, who 'strolled over from the far side of Imaw Bum' to see how they were progressing; in this case there was no jealousy – 'never was there such talk and such opening of bottles and tins'. Then they settled down to the seed-harvest, revisiting the Chimili camp early in October, and leaving Hpimaw for good on 18 November. At Myitkina they met Kingdon-Ward, very weary after his season on Imaw Bum, and all travelled together down to Rangoon.

It had been a good year, with much pleasure and little hardship, and many beautiful plants had been found. Unfortunately very few of them remain in cultivation. Even in Britain it is impossible to provide the constantly-saturated soil and atmosphere to which the plants of this area are accustomed, and only a few species have proved sufficiently adaptable to survive in our conditions. They include *Primula sonchifolia*, *Rhododendron aperantum* and *R. calostratum*, *Jasminum farreri* and *Nomocharis farreri* – but even these are difficult. One of the most interesting introductions of the season was the Chinese coffin-tree, *Juniperus coxii*. Its fragrant wood was so much valued for coffins by the wealthy Chinese that the tree had been almost exterminated in all but the most inaccessible localities, and planks sold in Tengyueh for £70 apiece. It is now safely established in cultivation.

At Rangoon the friends separated; Cox had to return to England, but Farrer decided to remain for another season. He spent the winter months at Maymo, east of Mandalay, before starting again in the spring of 1920, to continue his exploration of the Chinese frontier-ranges from a point farther north than had been reached the previous year. At Myitkina he found all his former staff waiting for him, including his loyal Gurkha orderly, Jange Bhaju. This time he went up the Mali Kha to Fort Hertz, and then struck west to the Nmai Kha by way of Konglu and the Shingrup Kyet pass – the route which Kingdon-Ward had followed from the other direction in 1914. He forded the Nmai Kha with difficulty, the bridge having been swept away; followed the river downstream for a time, then turned eastwards up its tributary the Ah-Kyang to his base for the season at Nyitadi. This 'city' of four huts was situated at the junction of two streams, within convenient reach of three passes, each of which Farrer hoped to visit three times in the course of the season.

Conditions were very different from those he had experienced the previous year. At Hpimaw he had had a dry bungalow, a congenial companion, a telephone and a weekly postal delivery; at Nyitadi, in an exceptionally wet season, he had only a leaky bamboo hut, and his retreat was cut off by the flooded Nmai Kha. Yet he was strangely content. The solitude he had dreaded turned out to be congenial; his men behaved admirably and gave him no trouble; his health was good, and he was so happy, he wrote, 'that I go to bed at 8 and just lie there being it . . . though I should not *positively* object if the rain would sometimes leave off for five minutes'. In camp on the heights, however, it was less agreeable, for here there was mist as well as rain, and for three weeks at a time he was unable to see more than a few yards ahead. This became so oppressive that when he descended again to Nyitadi, where it was merely wet, not foggy, 'to see daylight and visible objects again was like having a load of lead lifted off'. The whole countryside, as he wrote in another letter, wept incessantly at having been torn from its immemorial mistress, China; 'never have I known a country to cry so constantly'.

In September, when the period of exploration and discovery was over, Farrer was able to rest for a little before the final flurry of the seed-harvest, when better weather might also be expected. But the strain had told. On 1 October he fell ill, and a fortnight later became very much worse. His devoted staff did all they could, and Bhaju made a fantastic non-stop four-day journey to Konglu and back to try to get medical aid; but 'without giving any pain or trouble to us' as Bhaju put it, 'he breathe his last on the morning of 17th October, 1920 at about 11.30 a.m.'[1] His men made a coffin and carried his body to Konglu, where he was buried on the hill above the fort. They were not able to bring away all his possessions, so salvaged what seemed to them most valuable – tent, stores and equipment; his papers and the seeds he had begun to collect were left behind.

It seems particularly tragic that Farrer's death should have occurred on what, in many ways, was a thoroughly unsatisfactory expedition. Burma had only been chosen because Forrest was monopolizing Yunnan, and because Nepal and Szechuan were politically inaccessible. The chief riches of the region were in rhododendrons, which had never been Farrer's plants (his garden at Ingleborough being on limestone); and as Cox wrote, 'the sad truth is that these Burmese hills do not breed species of Alpine that give any return for care or kindness at the hands of the gardener'. Other collectors have died in the field, but Farrer's end seems particularly pointless, like an air broken off in the middle.

After his third and last trip to China in 1921–2, Kingdon-Ward made one of his most important and interesting journeys – his first visit to Tibet. In 1848 Hooker had closely questioned the treacherous Dewan about the unknown course of the river Yaru Tsang-po, which the Dewan assured him was known to all Tibetans to be the same as the Brahmaputra. Flowing eastward over the high Tibetan plateau, it is a broad, almost sluggish, navigable river; then it turns south, vanishes into the Himalayas and reappears in Assam, having bored its way through the mightiest mountain range in the world and descended 11,000 ft in the process. There were persistent rumours of big waterfalls in the Tsang-po gorges, but these had never been discovered. In 1913 Lieut.-Col. F. M. Bailey had got part of the way up the river, but fifty miles of impossible country remained to be explored, and the only way that Kingdon-Ward could reach it was by crossing the Sikkim mountains into Tibet, travelling eastward along the back of the Himalayan ranges and approaching the gorges from the northern end. The river itself was almost impossible to follow, so deep and narrow were the valleys, so violent the current, and so unclimbable the knife-edge ridges in between.

It was largely through Bailey, now Political Officer for Sikkim, that Kingdon-Ward obtained permits from the Indian and Tibetan govern-

[1] His death was attributed to diphtheria, but was more probably caused by pneumonia

Pierre Belon
By courtesy of the Royal Botanic Gardens, Kew
Photo by Kenneth Collier

Philip Barker Webb in Turkish dress.
Painted by Lalagero di Bernardis 1820
By courtesy of the National Portrait Gallery

IOANNES GEORGIVS GMELINVS
Medicinæ Doctor, ejusdemque ut et Botanicæ et Chemiæ
Prof. P. Ord. in Acad. Tubing.
nat. Tubingæ d. 12. Aug. A° 1709.

3 *Opposite*
Johann Georg Gmelin
By courtesy of the Linnean Society
Photo by John R. Freeman & Co

4 John Tradescant
Painting attributed to Emmanuel de Critz
By courtesy of the Ashmolean Museum, Oxford

5 Carl Peter Thunberg
Painting by P. Krafft 1808
By courtesy of the University of Uppsala
Photo by Svenska Porträttarkivet, Stockholm

Clarke Abel. M.D. F.R.S.

Opposite
The reception of Lord Macartney's Embassy by the Emperor of China. The page, young George Staunton, is just visible on the right. Engraving after W. Alexander
By courtesy of the British Museum

Opposite bottom
Clark Abel
By courtesy of the Royal College of Physicians

Chinese drawing of a white Camellia. From the Reeves Collection
By courtesy of the Royal Horticultural Society

Abbé Armand David
Photo by Mountrey, Condé Nast Publications.

Augustine Henry. The foliage is that of *Rhododendron augustinii*. Painted by Celia Harrison 1929
By courtesy of the Henry Herbarium, Glasnevin
Photo by J. E. Downward

11 Frontispiece from Fortune's *A Journey to the Tea-Countries of China* 1852
Photo by Kenneth Collier

12 A Fern Valley, Tasmania. From Backhouse *Narrative of a Visit to the Australian Colonies* 1843

By courtesy of the Royal Botanic Gardens, Kew
Photo by Kenneth Collier

13 Map of Sikkim and Eastern Nepal (detail) by J. D. Hooker

The N. boundary of Sikkim is from Donkiah by Kongra Lama to Chomiomo thence S.S.W. to Kangchan-Junga
Donkiah Mt.
M
I
Mt. Tunkra
THIBET
Phari
Guareum Mt.
17,556
THIBET
Choombi R.
THIBET
Choombi
Chola Mt.
Tungiao River
Machoo River
TUMLOONG
THIBET
THIBET
BHOTAN
BHOTAN
1. The ranges E. of Pundim and W. of the Lachen river have not been visited, nor the E. heads of the Gt. Rungeet.
2. Bhomtso (Thibet) was well determined
3. The position of Tukcham also requires confirmation, to 2 miles.
The following positions and their elevations are taken from Col. Waugh's map (As. Journ. Novr. 1848.)
Kinchin-junga
Donkiah
Guareum
Chola
Darjiling
Gipmochi
All other positions, all rivers, villages, mts. passes &c. from a rough survey by
Jos. D. Hooker M.D. R.N. F.R.S. &c.
Darjiling } March 6th 1850

14 Sir Joseph Hooker in the Himalaya. Engraved by W. Walker after Frank Stone
Copyright photograph, British Museum

William Roxburgh
Engraving by Charles Warren

16 Nathaniel Wallich
Lithograph by Thomas Herbert Maguire, 1849
By courtesy of the Royal Botanic Gardens, Kew

Opposite
Dionaea muscipula, Venus' Fly-Trap or Tippitiwitchet. Drawing by Redouté from Ventenat *Le Jardin de la Malmaison* 1803

By courtesy of the Royal Botanic Gardens, Kew
Photo by Kenneth Collier

Amherstia nobilis. From Wallich *Plantae Asiaticae Rariores* 1830–2

By courtesy of the Royal Botanic Gardens, Kew
Photo by Kenneth Collier

Allan Cunningham. From a water-colour by J. E. H. Robinson

By courtesy of the Linnean Society
Photo by John R. Freeman & Co

Francis Masson. Painted by George Garrard

By courtesy of the Linnean Society
Photo by John R. Freeman & Co.

John Fraser. The flower is *Zenobia pulverulenta*, one of Fraser's introductions. From the *Companion to the Botanical Magazine* II, 1836

Photo by Logan, Birmingham

Thomas Drummond. From a crayon drawing by Sir Daniel MacNee PRSA, before 1825

By courtesy of the Royal Botanic Gardens, Kew

Sir Hans Sloane. The drawing of *Lagetta lintearia*, the Lacebark Tree, was reproduced in *A Voyage to . . . Jamaica* 1707. Painted by S. Slaughter 1736

By courtesy of the National Portrait Gallery

24 Map of North America (detail) showing Douglas's travels and the Hudson Bay route to the Columbia. From Hooker *Flora Boreali Americana* 1840

By courtesy of the Linnean Society

Photo by John R. Freeman & Co

GREENLAND
BAFFINS BAY
DAVIS STRAIT
HUDSON STRAIT
HUDSON'S BAY
LABRADOR
LOWER CANADA
NEWFOUNDLAND
NOVA SCOTIA
ICELAND
Smith Sound
Jones Sd.
Lancaster Sound
Cockburn Id.
Melville Peninsula
Southampton Id.
Cumberland Strait
Frobisher Strait
Resolution Id.
C. Chudleigh
James Bay
Hannah Bay House
Mistissiny L.
Anticosti
Quebec
Three Rivers
Montreal
Kingston
L. Ontario
Lake Erie
Lake Huron
Lake Michigan
Sacketts Harb.
Falls of Niagara
Champlain
Portland
Boston
C. Cod
Halifax
C. Sable
Bay of Fundy
Fredericton
St. John
C. Breton Id.
Pr. Edward
Gulf of St. Lawrence
Sandwich Bay
C. Charles
Belle Isle
C. Farewell
Clear Water L.
Richmond Gulf
East Main River
Albany
Moose
Nipissing
Ottawa
York
Longitude West of Greenwich
80°
70°
60°
20°
30°
40°
50°

25 Humboldt and Bonpland in the Andes. From a painting by F. G. Weitsch
By courtesy of Verwaltung der Staatlichen Schlosser und Garten *Photo by Elsa Poste*

26 Clarence Elliott on the way home. From a drawing by John Nash
By courtesy of Mr Joe Elliott *Photo by P. Siviter Smith*

ments for himself and one European companion. The Rt Hon. the Earl of Cawdor, interested in ethnology and natural history, volunteered to go with him, and the two left England in February 1924. Darjeeling could now be reached from Calcutta by train; but from there onwards was still a matter of coolie and pony transport, as it had been for centuries before. They went first to stay for a day or two with Bailey, who was stationed at Gangtok in Sikkim, and gave them much helpful information. On 23 March they crossed the Nathu La pass and entered Tibet, finding the still-wintry plateau 'all dust and ice and a raving wind'.

They travelled north to Gyantse, where there were six Europeans – the last they were to see for ten months – and then took the Lhasa road as far as Lake Yamdok. For three days they skirted the south shore of the lake; then, crossing the Shamda La pass, they reached Tse-tang on the Tsang-po on 21 April. From there they followed the placid river downstream – not always happily; Cawdor was unwell, a servant was bitten in the hand by a dog, it rained every day and their quarters were filthy, though the people were hospitable and kind. After a stay of a few weeks at Tsela Dzong, at the junction of the Gyamda and the Tsang-po, they left the river, crossed the divide north-eastward between the Gyamda and the Rong-chu by the Temo-la pass, and descended to Tumbatse, a village on the Rong-chu, which was to be their base for the next five months.

Here they were in a paradise of flowers; the blue poppy (*Meconopsis betonicifolia* var. *baileyi*), hitherto regarded as a rarity, was vigorous and abundant; a new primula was found which Kingdon-Ward named *P. florindae* in honour of his wife; and plants of Hooker's *Rheum nobile* 'stood up out of the Rhododendron sea like light-houses'. They went several short local excursions and a longer one in August and September, north over unexplored country to the Kingdom of Pome. Having followed the River Tongkyuk to its source in a hitherto-unmapped lake ten miles long, they crossed to a tributary of the Gyamda river and went north up-river to the Pasum Kye La on the Salween divide, finally reaching Lake Atsa Tso, where they were again on known territory. (A heavy snowfall occurred on 29 August.) They then turned south to Gyamda, on the river of the same name – a metropolis with a post-office and shops. At the next halt, Napo Dzong, they saw a courtyard gay with hollyhocks, asters, sunflowers, dahlias, pansies, geraniums, poppies, stocks and nasturtiums; the Dzong-pen (local governor) had been to Calcutta, and brought back a tin of Sutton's seeds. (Both Kingdon-Ward and Farrer remarked on the cultivation by Tibetans of European garden-flowers, especially at the monasteries; and later in the same journey Kingdon-Ward received a letter from the Tibetan government asking him to send them some of the seeds he collected, 'as the Dalai Lama is very fond of flowers, and at his private residence on the outskirts of Lhasa grows a great many, which he tends with loving care'.) They descended the Gyamda river to its junction with the Tsang-po, and from there returned by their original route to Tumbatse, after an absence of five weeks.

From mid-September to mid-November they engaged on a carefully-worked-out programme of seed-collection; rhododendron-seed can be gathered right up till Christmas, if one is prepared, as Kingdon-Ward was, to dig for the bushes under three feet of snow. On 16 November, when fine and dry weather might be expected and the water in the rivers was low, they set out for their exploration of the Tsang-po gorges. They seem to have made a circuit to a point on the river below that reached by Bailey, and then worked slowly upstream for three or four weeks, till they came to some falls, beyond which it was impossible to pass. They were obliged to detour, strike the river higher up, and work downstream towards their point of departure, finding two more small falls of a mere thirty or forty feet; but there was a gap of about five miles that they were unable to cover. They formed the opinion, however, (confirmed by local tradition) that there was a series of small falls rather than one big one. Then they retreated northwards up the river, and late in December got back to Tumbatse. In the gorges Kingdon-Ward collected seed of every rhododendron he met, though he had not seen their flowers; the result was ten new and worth-while species including *R. leucaspis* and *R. pemakoense*, the latter never since seen wild.

The journey home was started on 28 December, but by an indirect route – up the Tong Kyuk again and over to Gyamda, where they stayed from 18–24 January 1925, making local excursions; then down to Tse-tang (which they had visited on the outward way) south to Pang-chen and over the border into Bhutan. This was a miserable journey, owing to the altitude and the bitter dust-laden wind that sprang up every day at noon and blew with increasing violence till nightfall, by which time, Kingdon-Ward said, 'we felt stony all over, like a saxifrage'. (Because of this wind, the Tibetans themselves preferred to travel by night, when there was moonlight.) They reached the railway and civilization at Rangiya on 23 February.

The seeds that Kingdon-Ward had gathered so carefully crossed Tibet in 40° of frost, and were subjected in India to a temperature of 90°F.; they were shipped to England in cold storage, the first consignment arriving on 20 March and the last on 20 April. Before the end of the latter month all were sown, at Kew, Edinburgh, Wisley and 'a hundred other gardens', and consignments sent to New Zealand, South Africa, North and South America and elsewhere. So successful was his harvesting and packing that of more than 250 species only two per cent failed to germinate. The glorious blue poppy (*Meconopsis betonicifolia*) had previously been found by Delavay, Bailey and Forrest but had not been successfully introduced; it was Kingdon-Ward's seed from Tumbatse that established it safely in cultivation. To introduce from one journey two plants of such garden-value as this and *Primula florindae* would suffice most collectors for a lifetime; and there were other treasures – *Berberis hookeri*, *Lilium wardii*, *Primula cawdoriana* and *P. baileyana* – the primula genus, more than most others, being a dictionary of collectors and their friends.

Space does not permit of more than a brief epitome of the majority of Kingdon-Ward's journeys. There is no lack of documentation, for he wrote a book after almost every expedition, partly to make money but partly because he enjoyed writing and found it a resource in lonely places to be able to 'talk in his own language to paper'. In the best of his books he rivals Farrer in his plant-descriptions, and has a less self-conscious sense of humour; who but Kingdon-Ward would compare himself, after a flea-ravaged night, to the flowers of a certain rhododendron – white, with pink spots? Of his twenty-three books (fourteen of which are travel books) there are few that are not worth reading.

After his year in Tibet, Kingdon-Ward's next journey (1926) was to the Seingkhu river in Burma; in 1928 he went to the unfriendly Mishmi Hills in north-eastern Assam, with a companion H. M. Clutterbuck (also commemorated in a primula) who is endearingly referred to as 'Buttercup'. Back in Burma in 1930–1 he explored the Adung valley, accompanied again by Lord Cawdor; and in 1933 he made his second visit to Tibet.

It had hitherto been believed that the Himalayan mountains swung southwards into Burma in which was known as the Malay Arc; but a new theory postulated that the range actually continued eastwards to China, and that the Burmese mountains were formed only by the rivers cutting their way through a high plateau. Kingdon-Ward thought that in that jumble of mountains and valleys too much attention had been paid to the courses of the rivers and not enough to the alignment of the principal snowy peaks; and on this journey he hoped to contribute to the solution of this geographical problem. Since much of the route was to lie in unknown country he asked the Royal Geographical Society for an assistant, and they introduced him to the young, likeable and enthusiastic Ronald Kaulbeck; the party was completed by an amiable but accident-prone photographer, R. B. Brooks-Carrington. Unfortunately these two were recalled by the Indian authorities (when they had already been two months in Tibet) on the grounds that they were not named in Kingdon-Ward's passport, which had been made out for himself 'and party' long before he had known who were to be his companions. The itinerary was from Sadiya in Assam up the Lohit river and its continuation the Zayul to Rima; up the Rong-to valley north to Rongyul and Modung; north-east across the Ata-Kang and Cheti passes to Shugden Gompa and an excursion from there east to the Salween river, returning more or less the same way. Funds for the journey came from the Royal Horticultural Society and from private gardeners, so the collecting of plants was still a main object; the principal prize of the expedition being the Carmine Cherry, *Prunus cerasoides* var. *rubra*.

On his next venture, also to Tibet, it was Kingdon-Ward himself who had an inadequate passport. This was one of his most audacious exploits, when the passion for discovery lured him far beyond his original intention. He started, modestly enough, in April 1935 from Tezpur on the Brahmaputra, northwards into that lost province which seems to belong neither to Assam, Bhutan nor Tibet. At Shergoan in Monyul he applied to the High Priest for permission to enter Tibet, and was told to write to the Dzong-pen or governor of Tsona, the first place of consequence over the border, and that a favourable reply might be expected in about a fortnight. Meantime, Kingdon-Ward continued his journey northward, through Dirang Dzong and Senge; he received the answer to his request when he was in the small and primitive village of Lugathang, but neither he nor his servant could read Tibetan, and he decided to press on to the larger village of Karta, where an interpreter might be found. But before he had got so far, the letter had got lost among the baggage, and did not turn up again till he was back in India. It was then read, and proved to be a refusal; and the whole of his subsequent journey in Tibet had been without permission and without passport. This seems afterwards to have caused some diplomatic repercussions; but Kingdon-Ward did not receive the slightest hindrance from the local Tibetan authorities while on the journey, except when they forbade him to enter regions inhabited by savage tribes not under their control.

Once over the border and into Tibet, Kingdon-Ward was in country where Europeans were practically unknown, and where crowds gathered to see him; at Karta the last white man had passed through about twelve years before, 'and although it had caused comment at the time, the excitement had since died down'. The Lamas granted him transport to go north to Chayul Dzong on the Loro Chu, the southern branch of the river Subansiri. It was now mid-June, and although Chayul at 12,000 ft lay at a height where eggs would not boil and hens were not kept, the common white jasmine (*Jasminum officinale*) was blooming on the hills and on gravel-pits in the river. The Chayul Dzongpen would not allow him to go eastwards down the Loro Chu into the dangerous Lopa country, but said he might go north to Sanga Choling on the other branch of the Subansiri, the Char Chu. Here he found two monasteries which he compared to St Mark's, Venice, where he was greeted by the monks as a friend. He arranged to go on to the holy mountain of Tsa-ri; and conceived the daring project of continuing north from there to the Tsang-po and linking up with his journey of 1924.

Accordingly he left the 'Arabian Nights palace' of Sanga Choling on 30 June, and finding on the way *Dracocephalum hemsleyanum* and the beautiful *Thalictrum diffusiflorum*, crossed the Cha La to Chosam in the Tsari valley and the pilgrim village of Chickchar. This was in a sacred region where nothing might be killed, but there was no rule, Kingdon-Ward found, against taking live creatures away – 'Several fleas which had become attached to me left Chickchar under my protection.' He made a side trip of a few days to see the sacred lake of Tsogar, then resumed his march on 11 July, north-east to Kyimdong on the river of the same name, where he was only five miles south of the Tsang-po; but he turned east instead and crossed the divide to Molo on the next tributary, the Lilung. One last forced march brought him on 18 July to the village of Lilung at the junction with the Tsang-po, where he and Lord Cawdor had stayed ten years before.

On that previous visit they had glimpsed a great range of snowy peaks fifty or sixty miles to the north; but although they had skirted the base of the range when they ascended the river Tongkyuk, they never again saw the peaks, owing to thick weather. The indefatigable explorer now decided to go to look for this lost range. He descended the Tsang-po to Tsela

Dzong and crossed the familiar pass to his old base at Tumbatse; but although he received a great welcome he stayed only for a day before pushing on to Tongkyuk Dzong, where he was held up by a broken bridge. Here he met Colonel Yuri, Chief of the Lhasa Police, and had some difficulty in convincing him that he was not Urush Marpo, a Bolshevik agent for whom Yuri was hunting. Kingdon-Ward had thought of going south-east to Showa on the Po-Tsangpo, but Yuri discouraged travel in that direction and offered his escort north in a few days' time to the river Po-Ygrong. Actually the botanist and his party went in advance; and approaching the steep and difficult pass of Sobhe La (open for only three months of the year) the sky cleared in the evening, and he found himself in the midst of the lost range, with magnificent snow-peaks all around. On arriving at the Po-Ygrong he was mistaken for the expected Colonel Yuri, and given an embarrassingly royal reception.

He next decided to explore the river to its source, and followed it westward for eighteen days, collecting in that period a hundred species of plants and noting others; he thought it a good possible area for future plant-hunting. On 20 August he reached the glacier at the head of the valley, and had a wonderful view of the source of the Po-Ygrong – 'it epitomised a life's ambition; a worth-while discovery in Asia, truly finished'. Then he crossed the Lochen La to the Gyala river, and descended southwards to Gyamda, where he had been before. After four days' rest he followed the Gyamda river westward to De, and turned south – again in unmapped country – to where the 17,000 ft pass of Ashang Kang La led from a southern tributary of the Gyamda to a northern tributary of the Tsang-po. His face was now turned homeward, and though many weary marches still lay ahead, he made south like a homing pigeon – across the Tsang-po and the pass of Kongmo La to Chosam again; by a slightly different route to Sanga Choling; by another loop westward and an unmapped pass to Chayul Dzong; once more to Karta and Mago, then, with a detour to the east, to what he considered his starting-point, Dirang Dzong, which he reached in the second week of October, rather footsore, after nearly 1000 miles during which he rarely retraced his steps. He had also had a good deal of illness to contend with, and in crossing the Po-Ygrong by a rope-bridge in August he had lost his irreplaceable spectacles, which resulted in severe headaches from eyestrain. Moreover, he was now fifty years of age; but nothing seemed to damp his ardour.

Indeed, it seems pointless to mention age in connection with Kingdon-Ward. In the four years before the outbreak of war in 1939 he made three more journeys; two in Burma (mostly over ground already covered) and one in Assam. During the war he taught jungle-survival in India, and was afterwards employed by the American Government to search for wrecked planes. In 1947 he married again (his first marriage having been dissolved ten years before) and his intrepid second wife, deservedly commemorated in *Lilium mackliniae*, accompanied him on five subsequent journeys, including that of 1950, when they were nearly annihilated in the Lohit valley by one of the greatest earthquakes ever recorded, whose epicentre was only some twenty-five miles away. His last trip (1956–7) was to Ceylon, and he was planning a new journey when he died in London in 1958, aged seventy-three.

Kingdon-Ward's collecting methods were largely determined by the nature of the country in which he worked. In Burma particularly, he could not employ large numbers of native helpers, as Forrest did in China – the country would not support them, for there was neither cultivation nor game. In those thinly-populated areas coolies for transport were hard to recruit, and baggage had often to be brought up in relays; with all food to be carried, other requirements had to be cut to a minimum, and almost his only luxury was the homely hot-water-bottle. For these reasons, he usually travelled alone, and did all his seed-collecting in person, often marking a particularly fine plant in flower to be returned to later on. He had a phenomenal memory for such localities, and rarely failed to collect seed of every plant he had noted, even when he had only found a single specimen. These methods led to quality and accuracy, rather than quantity, in his collections. By his own estimate he found and introduced nearly a hundred species of rhododendron, but only a few have survived in cultivation.

Some of Kingdon-Ward's Tibetan country was also explored by Ludlow and Sherriff – another pair of heavenly botanical twins as seemingly inseparable as Lewis and Clark or Humboldt and Bonpland. Their trails crossed in several places, and in 1935 Kingdon-Ward's servant picked up on a high Tibetan pass a blackened rupee which they thought must have been dropped by Ludlow or Sherriff, who had camped on the same spot the year before.

Frank Ludlow (b. 1885) and Major **George Sherriff** (1898–1967) met for the first time at Kashgar in Chinese Turkestan, in September 1929. Ludlow had come to India originally as a teacher; at this time he was on his way to the Tien Shan mountains, and had stopped to visit his friend Mr F. Williamson, now British Consul at Kashgar; and Sherriff had been appointed Williamson's Vice-Consul the year before. Both Ludlow and Sherriff were keen ornithologists and were already experienced travellers, and all three men were fascinated by Tibet. Ludlow had met Williamson at Gyantse, where he had lived from 1923–6; they were probably among the six European residents Kingdon-Ward found there in 1924. The plans of Ludlow and Sherriff for natural history exploration were made soon after their first meeting, but did not materialize till four years later, when Williamson had been appointed Political Officer for Sikkim, Bhutan and Tibet, and was able to procure for them the necessary passports. In the spring of 1933 they all set out together – Ludlow, Sherriff, Williamson and his new wife.

They travelled eastward from Gangtok in Sikkim to Bumthang the capital of Bhutan, where the two naturalists were introduced to the Maharajah, whose friendship became of great importance in making possible their later travels. There they parted from the Williamsons, and went north-east to the Me-la, whose Tibetan name means 'The Pass of Flowers'. Having crossed this and another pass, the Kang-la, they proceeded north-west across Tibet by the lakes of Phome Tso and Yamdrok

Tso to Nangkartse Dzong on the main Lhasa road, and thence west to Gyantse; from where they returned to India. This was, in a way, an experimental journey, and the plant-collections they made were small compared to those of later years.

Ludlow and Sherriff had found that they could work harmoniously together, and they now made more ambitious plans – for nothing less than a botanical and ornithological survey of the Himalayan chain from Monyul eastward to the Tsangpo gorges, in a series of journeys each of which was to overlap slightly with the one before. The next year, accordingly, they started late in June from the railhead of Rangiya and went north-east to Chungkar, Trashigang Dzong, Sakden and over the Tibet border to Tsona Dzong, all to the east of their previous journey. From Tsona they made an excursion eastward to the district of Mago and back, then continued north-west to Dongkar Dzong and made a backward loop from there south-west to the Me-La pass of the previous year.

This of 1934 was an unlucky expedition; they were delayed at the start by the non-arrival of passports, and then by the breaking of heavy monsoon rains, and at Sakden they were held up by malaria, which attacked every member of the party except Sherriff and two Lepchas, and almost caused the project to be abandoned. The side-trip to Mago had been mainly in search of birds, but the results were disappointing. They did, however, find some interesting plants. Ludlow has described how at Chungkar Sherriff stood on a branch of a large shrub to reach a lovely mauve primula on a cliff-face, and saw at the same time another primula 'a little scrubby thing', growing in a clump of moss. All three proved to be new – *Primula sherriffae*,[1] *P. ludlowii*, and the shrub *Luculia grandifolia*.

There was no expedition in 1935; Sherriff returned for a visit to England, solicitously nursing on the voyage a single live plant of his new primula, which got safely to Edinburgh in June and bloomed in July. The exploration of the Himalayas was resumed in 1936, and this, their third attempt, was an outstanding success. Accompanied this time by Dr K. Lumsden, they went first to Tibetan Tsona, and then north-east over the high pass Nyala La to the valley of the Chayul river, an eastward-flowing branch of the Subansiri. They followed the river downstream to Lung, then crossed north-east again to the valleys of the Char Chu and the Tsari-chu – the latter the sacred region which Kingdon-Ward had visited only the year before. They found it, as he did, a paradise of flowers – all cultivation and grazing being forbidden. Sherriff remained in the Tsari neighbourhood while Ludlow and Lumsden went still farther north-east to Molo on the Lilung river, and then south over the Lo La pass into the province of Pachak Shiri.

From this journey Ludlow and Sherriff brought back some 2000 specimens, two crates of living plants and large quantities of seeds; their finds had included sixty-nine rhododendrons, of which thirteen were new, and fifty-nine primulas, fourteen of them new. Among plants

[1] Named in honour of Sherriff's mother; he did not marry till later

introduced were the lovely rose-pink *Meconopsis sherriffii*, the delicate twining *Codonopsis vinciflora*; *Rhododendron sherriffii*, shy-blooming and refined, *Cyananthus sherriffii* and primulas *kingii* and *muscarioides*.

In 1937 there was again no combined expedition, though Sherriff made a solo visit to the Black Mountain in central Bhutan; but in 1938 the partners resumed the trail where they had left off, travelling this time from Kalimpong in Sikkim and eastward across Tibet by the well-known route along the Tsang-po river. They started in February, and crossed the Lo-la pass into Pachakshiri (prospected by Ludlow two years before) in late April. After twelve days at Lhalung on the Siyom (a southward-flowing tributary of the Dihang, or lower Tsang-po) they returned north-west on 17 May to Molo, where they had arranged to be joined by Dr (later Sir George) Taylor; he arrived from England before they had finished pitching camp. A few days later, the party divided, Ludlow and Taylor working the main ranges eastward towards the Tsang-po gorges while Sherriff went south and west to Tsari Sama and the Kuchu La. They met again at Tsela Dzong (at the junction of the Tsang-po and the Gyamda) and again divided, Ludlow going alone to the Pasum La,[1] far to the north, while Sherriff and Taylor worked the lower Gyamda ranges. In August, Taylor alarmed his friends by developing a severe attack of suspected appendicitis, and the future Director of Kew was still weak when they embarked on the long laborious journey back by Tsari, Tsona and East Bhutan – taking with them more than 5000 gatherings of plants, including many novelties. The season's introductions included primulas *ioessa* and *sandemaniana*, and *Rhododendron viscidifolium*; but perhaps the most valuable contribution to horticulture was *Paeonia lutea* var. *ludlowii*,[2] which holds its outsize buttercups boldly erect instead of with bending necks as in the type, and has made possible the breeding of less heavy-headed varieties of yellow and yellow-toned tree-paeonies.

After the 1939–45 war there were two more Ludlow and Sherriff expeditions, in 1946–7 and 1949 (the latter bringing us *Euphorbia griffithii*); then both men retired to Britain – Ludlow to a post at Kew and Sherriff to an estate at Kirriemuir in Scotland, where he created a garden and grew Himalayan plants with remarkable success. He was particularly interested in primulas, and boasted that he had seen every species known to science, though not all of them in the wild. Of the twenty-seven new primulas he discovered, at least seven received horticultural Awards of Merit.

Besides the novelties, Ludlow and Sherriff introduced a number of superior forms of plants already known, and also some which, like *Corydalis cashmiriana*, had not been securely established in cultivation. They were most discriminating collectors; theirs were plants for the connoisseur – exquisite, but not always easy to grow. Sherriff was one of the first to send plants home by air, which enabled him to introduce species whose seeds did not long retain their viability; though even this presented difficulties when it took twelve or more days of hard travel to

[1] Visited by Cawdor and Kingdon-Ward in 1934
[2] Sometimes given specific rank as *P. ludlowii*

get from the mountains to the nearest railhead or airport. Many crates were sent to Kew, Wisley, Edinburgh and private gardens, at considerable personal expense; for although Sherriff received occasional grants from the British Museum and other institutions, his expeditions were mainly financed by himself. He and his companion should be remembered with gratitude, more especially as the region in which they worked has never since been accessible to the collector.

The East Indies

The products of the tropical East Indies are of little importance to the outdoor gardener; but so large and populous a portion of the earth's surface cannot be dismissed without at least a brief survey. The earliest plant-hunters to visit the islands were interested chiefly in spices, the later ones in nepenthes and orchids. There were also some notable resident botanists.

The botanists came first – each in his own island. Premier in time and importance was **Georg Eberhard Rumpf**, or Rumphius (1628–1702). He was German by birth, but entered the service of the Dutch East India Company and was sent to Amboina in 1653; he remained there for the rest of his life, eventually rising to a high rank. He was an ardent naturalist (he has been called the Indian Pliny), but suffered more than his share of the bludgeonings of fate. In 1670 he was smitten with blindness, caused, it is said, by overwork in the light of dim lamps and candles; in 1674 he lost his wife and one of his daughters in an earthquake. About 1686 he despatched a voluminous manuscript representing more than thirty years' work, to Holland to be printed; the Dutch vessel that carried it was attacked and sunk by the French, and the manuscript was lost. With the aid of his surviving daughters, Rumpf set to work to rewrite it; but a fire swept the Dutch quarters in 1687, in which all his pre-blindness drawings were destroyed. In spite of all difficulties, the first three volumes were ready for despatch in 1690, and nine more awaited their final polish. All were safely received by the Directors of the Dutch East India Company, but they were reluctant to undertake the expense of publication, and Rumpf died without knowing whether his book would ever see the light. Many years later the manuscript came into the hands of Thunberg's patron, the botanist Johannes Burmann, who edited and published the whole twelve volumes in 1741. By this time Linnaeus had reaped the credit for the first descriptions of many of Rumpf's plants; but even at so late a date, the *Herbarium Amboinense* was considered one of the most remarkable books of its time. More than seventeen hundred plants were described and 1060 illustrated; but botanists today find some of them hard to identify, especially as the flora of Amboina has greatly changed. Rumpf is said to have introduced the tree-fern *Cycas revoluta* to Holland.

While Rumpf was on Amboina, an Austrian missionary called **Georg Joseph Kamel** (1661–1706) arrived at Manila on Luzon. Born at Brunn in Moravia, Kamel became a Jesuit lay-brother in 1682, and was sent to

15 Blind Rumphius (Georg Eberhard Rumpf). From a drawing by his son in *Herbarium Amboinense* 1741 *By courtesy of the Royal Botanic Gardens, Kew Photo by Kenneth Collier*

Manila in 1688, where he established a pharmacy for the free distribution of medicines to the poor. 'Being made, as I may say, for the advancing of natural knowledge',[1] he studied the plants and also the lizards and insects of the island, and made many fine drawings. He corresponded with Ray and Petiver in England, and with Samuel Browne at Madras, but a consignment of plant-specimens and drawings that he sent to the latter was 'taken by I know not what Corsaires, on the Cost of Bulocondor'[2] (Pulo Condore). The three parts of his *Syllabus Stirpium in Insula Luzone*, however, arrived safely in England; the first two, on herbaceous and woody plants respectively, occupied a ninety-six-page appendix to the third volume of John Ray's *Historia Plantarum* in 1703, and the third section, describing 212 climbers, was published in the *Philosophical Transactions* the following year. Kamel died at Manila in 1706, and was commemorated by Linnaeus in the Camellia – a plant that seems to have been unknown to him.

The third of the trio, the Portuguese **João Loureiro** (1710–1791) was not originally a botanist, nor was he situated on any of the East Indian islands; but he lived for thirty-five years at Hue, the capital of Cochin-China on the adjacent mainland. He had already been in Goa for three years and in Macao for four, when he was sent on a special mission to Cochin-China in 1742; and missionaries, as such, being unwelcome there, he took service with the King as a mathematician and naturalist. His interest in botany was aroused by the study of local medicinal plants, to which he was obliged to resort because European drugs were unobtainable. Botany-books were equally impossible to procure, and it was not until Loureiro returned to Canton in 1777 that he got some of the works of Linnaeus, through the captain of an English ship. The same captain put him in touch with Sir Joseph Banks, and in his first letter, which accompanied a manuscript and an important collection of dried plants, Loureiro asked Banks to send him a copy of the *Systema Naturae*.

He left Canton to return to Portugal in March 1781; but his ship met with such bad weather off the Cape that she was forced to put back to Mozambique, and Loureiro botanized there and in Zanzibar before finally getting back to Lisbon early in 1782. His *Flora Cochinchinensis* appeared in 1790 – an important work but marred by its lack of plates and its author's lack of science, which led him, among other errors, to classify the hydrangea among the primulas. He complained that the work was published by the Academy of Lisbon without permission and without waiting for the index; but it was as well that the production was not further delayed, for Loureiro was already seventy-five, and died the following year.

After the scientists came the spice-hunters. During the seventeenth and eighteenth centuries the Dutch went to extraordinary lengths to preserve their lucrative monopoly of the East Indian spice trade. With the object of keeping the commodity scarce and the price high, they endeavoured to limit the production of each spice to one island, which

[1] John Ray *Correspondence* quoted Britten and Dandy
[2] Kamel to Browne, translated by Sloane

they could defend, and exterminate it everywhere else; thus nutmegs were cultivated on Amboina, cloves on Banda and cinnamon in Ceylon. If the native princelings elsewhere would not destroy their spice-trees, the Dutch offered good sums for the leaves – knowing that the natives did not realize that if stripped of their leaves for three successive years the trees would die. The export of seeds or spice-plants was punishable by death; John Evelyn in 1681 noted that it was impossible to obtain a single nutmeg in a state fit for germination.

Naturally this only made the representatives of other nations more anxious to procure the plants and break the Dutch monopoly; and one of the most intrepid and adventurous of these spice-hunters was **Pierre Poivre** (1719–1786). In 1740 he went to China and Cochin-China as a missionary; but on his homeward voyage in 1745 his ship was engaged and captured off Sumatra by a British man-of-war. Poivre was struck in the right wrist by a cannon-ball and this injury resulted in the loss of his arm, which disqualified him for the priesthood, for which he had been training. Released in Batavia,[1] he slowly made his way homeward by Pondicherry, Mauritius, the Cape and Martinique; but when in sight of France was again captured by the British and imprisoned in Guernsey till 1748. The following year he was commissioned by the *Compagnie des Indes* to open trade between France and Cochin-China, and he succeeded in establishing a 'factory' at Tai-jo on the bay of Turan, where he collected every kind of valuable economic plant, including that of the true black pepper, which was named Poivrea in his honour. In 1745 he made a dangerous trip in a leaky little vessel called the *Colombe*, to Manila, the Moluccas and Timor. How he obtained his results is not known, but he eventually got safe to Mauritius, with 3000 nutmeg seeds, a number of other spices and various kinds of tropical fruit-trees. He spent the winter in Madagascar, and on the way home from there was captured a third time by the British – but that is another story.

Thunberg spent some time in Batavia, both on his way to Japan in 1775 and on his return voyage; and being a distinguished employee of the Dutch East India Co. had little to fear from anything – except the climate; for Batavia was one of the most unhealthy spots in the East, and of twelve persons with whom he dined before his departure in June 1775, only one survived when he returned eighteen months later. The town was strategically placed near the entrance to the Sunda Straits, and was an important port of call, being the only harbour in the East Indies where major ship-repairs could be carried out; but it was situated on marshy ground in a network of stagnant and unclean waterways, and the death-rate was appalling. On his second visit (his first lasted only a month) Thunberg had a 'fever that was not very slight'; when he recovered he obtained a commission from the Governor to look for medicinal plants, and stayed on Java for seven months (January–July 1777), making several journeys into the interior. He then got an appointment as surgeon on a ship bound for Ceylon; the voyage lasted nearly two months, for he

[1] Now Jakarta

sailed on 5 July and reached Colombo on 30 August. He explored the neighbourhood, and made some longer excursions, mostly to inspect the cinnamon plantations. The coastal areas of the whole island now belonged to the Dutch, and though the wild cinnamon-tree still grew in the central provinces ruled by the native King, he was unable to sell any of the spice owing to the Dutch monopoly of the ports. Thunberg spent five months examining the products of Ceylon, and sailed for Europe on 6 February 1778, with many seeds and a number of living plants; but some of the plants perished from gales and cold before reaching Cape Town, and others were lost in a storm in the English Channel.

By the end of the century the power of the Dutch had much diminished. France having vanquished Holland in 1794, laid claim to her possessions in the East, and in 1811 the British were obliged to seize Java, to forestall its occupation by the French. During this free-for-all period it seems to have been much easier to gain access to the islands, and Smith worked in the Moluccas for eight or nine years without apparent hindrance.

Christopher Smith and **James Wiles** were the two gardeners sent by Kew to look after the plants on Captain Bligh's second bread-fruit expedition in 1791–3. This time there was no mutiny; the *Providence* carried its shipload of breadfruit and other useful trees safely from Tahiti to Jamaica, and there reloaded with West Indian plants for Kew. Wiles remained behind at Kingston, but Smith saw the consignment home, and was then sent by Banks to assist Roxburgh at Calcutta. In the autumn of 1795 Roxburgh sent him to collect plants in the Moluccas, chiefly of nutmegs and cloves; he was also to establish the various spices in a botanic garden on Prince of Wales Island in the Torres Straits. Smith was very industrious and sent large numbers of plants to Kew, Calcutta, Madras and Penang, though, in his first season at least, he was not very successful in their packing and many were dead on arrival. He is said to have distributed 127,520 plants of nutmeg and cloves, and nearly 29,000 other rare and valuable species, besides quantities of seeds. Banks attributed his success to his 'character as a botanist, added to his unwearied attention to the duties of his profession', and about 1804 he was appointed superintendant of the botanic gardens on Prince of Wales Island, to cultivate the plants he had collected. He died in Penang, at an uncertain date between 1806 and 1808.

Collectors were busy just then. **Leschenault de la Tour**, who had accompanied Captain Baudin's expedition to Australia, fell ill on the homeward voyage and was left behind on the island of Timor in 1803. When he recovered, he made his way to Batavia, but owing to the war was unable to get a passage back to France. With the consent of the Dutch governor, he spent the next three years on a thorough botanical exploration of Java, untouched except for Thunberg's explorations thirty years before. (His finds included *Hypericum leschenaultii*.) At the same time (1804–6) William Kerr, fresh from his first season in China, was working at Manila in the Philippines.

The orchid-hunters hardly began to appear until the 1840s; one of the earliest was **Thomas Lobb** (1817–1894), who signed a three-year contract to collect in the Far East for the firm of Veitch, in 1843. His instructions were to go to Singapore, and if possible to proceed from there to China; failing that, to whatever island he thought most promising, with a preference for Java. Details of his movements in 1843–4 are lacking; he seems to have spent most of 1845 on or about the Malay Peninsula – Singapore, Malacca and Penang – and part of 1846 exploring the mountains in the extreme west of Java, in the neighbourhood of Bantam and Buitenzorg, with perhaps a visit to Bali. He introduced the beautiful greenhouse rhododendrons *javanicum* and *jasminiflorum*, and the orchids *Phalaenopsis amabilis* and *Vanda tricolor* with its variety *suavis*.

The most exasperating of collectors, Lobb never seemed to stay long in one place, but hopped with flea-like agility from one end of the map to the other; indeed, if one believes all one reads, he must sometimes have been in two places at once. He left England again at the end of 1848 for a three-year spell in India; but though he was certainly in Northern India in March 1849, he is also reported to have sent plants that year from the Southern Malay Peninsula, Mt Ophir in Sumatra, and Java. Likewise, he was indubitably in Darjeeling in March 1850, and in the Khasia Mountains of Assam in June and October; but later in the same year he is supposed to have been at Moulmein in Tenasserim, in Lower Burma and in the Malay Peninsula again. In subsequent years we hear of him in North Borneo, Sarawak, Manila and the Philippines, with repeat visits to Singapore, Java, Mt Ophir, Borneo and Sumatra, but with no details until at last we find him in 1856 trying to climb Mt Kinabalu in British North Borneo. The ascent had been made a few years before by Hugh Low, who had given generous presents to the natives, and had thus established an unfortunate precedent. Lobb having only a small party and not being prepared to pay so liberally, the tribesmen refused to help him, and he was obliged to turn back at the village of Kiau. He managed, however, to ascend the adjacent peak of Saduc-saduc.

No more is heard after this till 1860, when Lobb retired to Devoran in his native Cornwall, minus a leg which was lost as the result of exposure – but where, and under what circumstances is never revealed. He lived for thirty-four years in retirement, and seems to have become something of a recluse. *Aeschynanthus lobbianus*, *Aerides multiflorum* var. *lobbii* and *Cryptomeria japonica* var. *lobbii* were among the plants he found. He also gathered large numbers of plant-specimens, which were named, sorted into collections, and sold by auction in London.

Hugh (later Sir Hugh) **Low** (1824–1905), Lobb's precursor on Mt Kinabalu, came to the East as a horticultural collector and stayed to become one of its most able administrators. In a special sense he was brought up in a nursery; for his father became the managing partner of Mackie's nursery at Clapton (where he had previously been foreman) when Hugh was seven years old. Low sailed for the East Indies on 17 July 1844, to collect plants for his father's firm, and gathered his first orchid on the islet of Pulo Chalot while his ship was becalmed off Tanjong Boulers in the Sunda Straits. He reached Singapore on 23 November, and on 6 January 1845 took ship for Borneo, arriving at Kuching on the 16th.

Here he met the celebrated Rajah Brooke, but it was not till later that he became his secretary.

Borneo has been described as so densely forested that an orang-utang could swing from branch to branch from one end of the island to the other without touching the ground. Low penetrated further into these jungles than any previous European, finding the Dyaks of the interior very honest and hospitable, and only moderately addicted to head-hunting. (He usually carried his own food-supplies, in order not to be a burden on the villages.) He made a number of excursions, one of the most important being undertaken in November 1845, when he went by boat up the River Sarawak to its junction with the Sebuloh, and then overland on foot to a Dyak village in the interior. Here he was pressed to make a long stay, but was in a hurry to return as the plants he had collected on the way up 'were too rare and valuable to be thrown away'. They included living specimens of the epiphytic *Rhododendron brookeanum*, which he named in honour of the Rajah, and of *Hoya imperialis*, both of which he managed, successfully to introduce. By 1847 he had added *Clerodendrum bethuneanum* the orchids *Arachnanthe lowii* and *Cypripedium lowii* and *Nepenthes hookeriana* to his list. The latter was one of eight new species of giant pitcher-plants which he discovered – 'the great urned Nepenthes, never before seen or imagined in dreams'[1] – but he failed in his attempts to introduce the others.

Low came home in October 1847, and his book, *Sarawak*, was published in January 1848. It was a comprehensive survey of the natural history and ethnology of the country, and was considered a remarkable feat for a young man of twenty-four. Rajah Brooke was also in England at this time, being feted and honoured by Queen Victoria; he had already invited Low to become his Colonial Secretary, and when Brooke returned to Sarawak in February 1848, Low accompanied him in that capacity, and lived for the next twenty-eight years on the island of Labuan, which had been ceded to Britain in 1846. From there he made several excursions to the adjacent mainland; he was the first European to reach the summit of Mt Kinabalu in 1851 (spending the previous night in a cave at 10,000 ft) and ascended it again in 1858. He made some further introductions, especially of orchids (*Coelogyne asperata*, *C. pandurata*, *Dendrobium lowii*) but his conscientious performance of his administrative duties made it increasingly difficult for him to pursue the occupation he loved.

At that time hothouse exotics were very popular in Britain, and the plants Low sent home were of sufficient interest to determine the great firm of Veitch to send out a collector – or rather, a whole troop of collectors, for F. W. Burbidge and Peter Veitch were closely followed by Charles Curtis and David Burke.

The plant-hunting career of **Frederick William Burbidge** (1847–1905) was confined to a single season in British North Borneo; yet he was obviously a born collector, his introductions successful and his book, *Gardens of the Sun* (1880) a typical record of a collector's triumphs and

[1] Burbidge in *The Gardener's Chronicle* 29 April 1905

trials. 'No description,' he remarks at one point, 'could possibly convey any idea of the delight which fills one when new and beautiful objects of natural history are discovered for the first time.' He had had a comprehensive horticultural training, for he had worked at the Horticultural Society's garden at Chiswick, and subsequently at Kew; had been for five years on the staff of William Robinson's paper *The Garden*, and had published at least three horticultural books before coming to Borneo at the age of thirty. When he arrived at Labuan (via Singapore) late in 1877, Low had recently left to take up the post of British Resident in Perak, and Burbidge lived for some time in what had been Low's house, appreciating to the full his garden and orchard. Soon after his arrival he was joined by **Peter Veitch**[1] of the Exeter branch of the firm, who had been wandering about the world since 1875 and was now on his way home from Australia and New Zealand.

The first trip made by Burbidge and Veitch in company was to Mt Kinabalu at the end of November (1877). It was now possible to go commodiously by steamer from Labuan to Gantisan on Gaya Bay; from there much of their route followed the Tawaren river. They spent several days on the mountain, delighted to find themselves among Low's orchids and pitcher-plants – 'vegetable treasures which Imperial Kew had longed for in vain'.[2] After their return to Labuan about a month later both had severe attacks of 'intermittent fever' (malaria), but were able to go to Brunei in January (where they were given an audience with the Sultan of Borneo) and for a short and not very productive boat journey up the rivers Limbang and Pandrowan to Bukit Sagan.

Their next considerable excursion was in April – by steamer round the northern point of Borneo to Sandakan, and from there to the Sulu Archipelago, which links north Borneo with the Philippines. They landed on the island of Meimbong and climbed two of its mountains, on one of which the remains of primitive forests still survived. In July and August they returned again to Mt Kinabalu, travelling this time by a different river, the Tampussuk, but finishing up at the same village of Kiau, from which all ascents of the mountain seem to have been made. They spent several days camping, as before, in a cold cave on the mountain. On the day of their return Burbidge's feet were scalded by the upsetting of a kettle. Rain made the steep descent slippery; his wet boots gave way like brown paper and had to be tied to his feet, which skinned in great patches. Nevertheless, he kept lagging behind to collect plants, and arrived in camp with both arms full, besides a basket on his back. On leaving the mountain he was able to continue the journey riding on a good (female) buffalo, but was nearly swept away, buffalo and all, when crossing a flooded ford.

The greatest triumph of this expedition was the successful introduction of *Nepenthes bicalcarata* and *N. rajah* – the latter the largest of the pitcher-plants, which Low had found but was unable to send home alive. The two collectors also obtained many ferns, aroids, and orchids; a new palm,

[1] Cousin of John Gould Veitch
[2] Burbidge *Gardens of the Sun*

Pinanga veitchii, and a plant of a new genus which was named *Burbidgea nitida*. After his return Burbidge was for twenty-five years the curator of the Trinity College Botanic Gardens in Dublin. In April 1905 he contributed to the *Gardener's Chronicle* an obituary notice of Sir Hugh Low; although a much younger man, he died on Christmas Eve of the same year.

Two years after Burbidge and Veitch came home in 1878, the firm of Veitch of Chelsea, still greedy for orchids, sent out another brace of collectors, **Charles Curtis** (d. 1928) and **David Burke** (1854–1897). The latter returned after a few months with the first season's harvest, mostly collected in Sarawak; the principal novelty being *Leea amabilis*. Curtis stayed on; he was anxious to find a pitcher-plant that was known only through a painting by Miss Marianne North, but the only clue which she could give to its habitat was that it had been brought to her when in Sarawak by an employee of the North Borneo Company. Sarawak is a large place and its jungles are thick; but Curtis eventually found and introduced his pitcher-plant, *Nepenthes northiana*. He then went into Dutch Borneo, but lost his first collection there, his possessions and almost his life, in a boating accident. Later he visited Sumatra and Java, and at the conclusion of his contract with Veitch became superintendent of the Botanic Gardens at Penang (from 1884 till 1903) whence he continued to send plants from time to time. Many species bear his name, perhaps the most important being *Cypripedium curtisii*, which he sent home in 1882.

David Burke made many other journeys for the firm, always to parts of the world where orchids might be found; they included South America, Upper Burma, the Philippines, New Guinea, Celebes and the Moluccas. Latterly, he is said to have become morose and eccentric; he was 'one of those natures who live more or less with the natives as a native and apparently prefer this mode of life'.[1] He was still collecting up to the time of his death in a remote part of Amboina in 1897.

[1] *Hortus Veitchii*

The Antipodes

AUSTRALIA

1791	Burton	1832	Backhouse
1800	Caley	1833	Cunningham, R.
1816	Cunningham, A.	1847	Mueller
c. 1816	Fraser	1864	Veitch, J. G.
1823	Baxter	1893	Veitch, J. H.
1829	Drummond	1929	Comber

NEW ZEALAND

1826	Cunningham, A.	1841	Sinclair
1832	Cunningham, R.	(1849	Travers)
1834	Colenso	1858	Haast
1839	Bidwill	1909	Dorrien-Smith
1839	Dieffenbach		

. . . for it is not to be imagined without experience, how in climbing crags and treading bogs and winding through narrow and obstructed passages, a little bulk will hinder, and a little weight will burden; or how often a man that has pleased himself at home with his resolution, will, in the hour of darkness and fatigue, be content to leave behind him everything but himself.' SAMUEL JOHNSON *A Tour of the Hebrides* 1774

'I have often been surprised at the great carelessness I have shown towards rare natural productions when either over-fatigued or ravenously hungry; at such times, botanical, geological and other specimens – which I had eagerly and with much pleasure collected, and carefully carried for many a weary mile – have become quite a burden, and have been one by one abandoned, to be, however, invariably regretted afterwards.' WILLIAM COLENSO *Journal* 1842

The Antipodes

Australia

It is greatly to be regretted that plants from Australia are not more hardy in Britain, for her flora is beautiful, varied, and at times as bizarre as her kangaroos and black swans. A few of the species can be grown out-of-doors in the milder counties; they do well in California and on the Riviera, and if only our islands could be pushed a little farther to the south and west, we could grow them all. The beautiful Australian eucalypts are being cultivated with increasing success, as hardier strains are found and their management becomes better understood.

The Antipodes were so late in being discovered that their botanical history is relatively short. Many of the early collectors were of the tip-and-run or maritime sort – 'one foot on sea and one on shore' – based on a survey-ship and unable to venture far from the coast. Such was the pioneer and pirate, William Dampier, in 1699–1702; such were Banks and Solander with Cook (1769–71), Leschenault with Baudin (1801–2), Brown, Bauer and Good with Flinders (1802–5). These must be dealt with elsewhere, this book being limited to land-based plant hunters. The greatest of the Australian explorer-collectors, Allan Cunningham, comes into both categories.

'New Holland' plants were already popular before the end of the eighteenth century, but it is not always possible to trace their origins. The nurserymen Lee and Kennedy had five new Australian species in 1788 – the year in which Sydney was founded – including *Banksia serrata*, claimed to be the first plant raised in this country from Botany Bay seeds; this firm was one of the first to specialize in Australian plants. In the following year (1789) Sir Joseph Banks sent out two gardeners, George Austin and James Smith, but nothing seems to be known of their subsequent history. The demand for plants from the new continent was such that in 1790 Lee and Kennedy thought it worth while to send out a collector of their own – **David Burton**.

E. J. Willson has suggested[1] that Burton may have been the son of Susannah, James Lee's eldest daughter, who married a Mr Burton, and that David was therefore the grandson of the founder of the firm. He was trained as a gardener and land-surveyor, but went to Australia as Superintendent of Convicts at Parramatta – Banks had a hand in this, too, for Burton wrote to him before his departure about his pay and kit-allowance, and Banks replied rather curtly that he had hired himself to Nepean (the Under-Secretary of State) and should apply to him. He sailed on the *Gorgon* about the end of October 1790, and arrived at Sydney on 22 September 1791; and after little more than six months in the colony, died in April 1792 of a gunshot wound accidentally contracted when duck-shooting on the banks of the Nepean. He seems to have accomplished a great deal in a short time, judging by the number of new Australian introductions made by his employers in the years 1791–3; he is also said to have collected for Kew, but though he is mentioned several times in the *Hortus Kewensis* ('Mr. Aiton's work evinces the great diligence of this unfortunate traveller'[2]) nearly all the plants were contributed indirectly, through Lee and Kennedy. 'He was a very deserving gardener' wrote R. A. Salisbury (*Paradisus Londinensis* 1806) 'sent to Port Jackson several years ago by Sir Joseph Banks, who after he had there made an ample collection, with many useful observations relative to their culture, was too soon for us called to botanize in the celestial regions.'

It has been said that once Sir Joseph Banks had bestowed his patronage, he would never admit to having been mistaken in the character of the recipient; but this hardly accounts for his steadfast loyalty to the difficult Captain Bligh or his patience with **George Caley** (1770?–1829) who was troublesome from the start. Banks doubtless made large allowances for the latter's background, for Caley's formal schooling had ended when he was set to work at the age of twelve as stable-boy to his father, a Yorkshire horse-dealer, and he had become interested in botany when looking for herbs to physic the horses. But Caley's faults were of character rather than of upbringing; he was undisciplined, obstinate and quarrelsome, though he was also brave, honest and loyal. Banks bore with great patience 'the effusion of his ill-judging spirit' for the sake of his underlying good qualities, although, as he once had occasion to remark, 'had he been born a gentleman, he would long ago have been shot in a duel'.

Their correspondence began in 1795, when Caley, writing from near Manchester, sent some specimens (including a moss believed to be new) and asked Banks if he could find him some botanical employment. Banks replied that he could arrange for him to go as a gardener's labourer at a botanic garden, which would not be remunerative but was the way in which many a famous botanist had started. Caley gladly came to

[1] In *James Lee and the Vineyard Nursery* 1961
[2] Rees *Cyclopaedia*

London on these terms, and went to work at the Chelsea Physic Garden. Early in 1798 he seems to have been given the opportunity of transferring to Kew, and to have refused because the wages were too low – 9s a week, 4s less than he was getting at Chelsea.[1] He had already become interested in the popular 'New Holland' plants, and wished to be sent out as an official collector; but Banks reasonably pointed out that he could not recommend to the Government a person who had had no scientific training and who was not familiar with the Australian plants already in cultivation at Kew. Caley chose to believe that Banks was offended by his refusal to go to Kew, and returned in a huff to Manchester. Although he had been accused by the intransigent Yorkshireman of 'not acting in a proper manner' Banks wrote in November of the same year (1798) to inform Caley that he had obtained for him a free passage to Port Jackson in the *Porpoise*, in the suite of Governor King, and that he would himself pay him a salary as a collector. He also arranged for him to have an accommodation allowance in the colony, and wrote a letter of recommendation to the Governor, John Hunter. He could hardly have done more.

The *Porpoise* might more justly have been named the *Tortoise;* although expected to leave a few weeks later, she did not sail till September of the following year, and then had to put back again owing to damage received from storms in the Bay of Biscay. Caley fretted a good deal at the delay, and Banks continued to pay his allowance. Most of the passengers were eventually transferred to the *Speedy*, and after a short stay at the Cape, they arrived at Sydney on 15 April 1800 – Caley already creating a scandal by his infatuation for a widow with a family, whose husband had died on the voyage. He was given every consideration, and offered a choice of residence; and having chosen Parramatta, a botanic garden was marked out there which was ready to receive plants by the beginning of May; he was also allowed the use of Government House in which to dry his specimens. From Parramatta he made at first short and then longer excursions, sometimes in country so desolate that a companion reported they had seen no living thing but two crows – and he believed those to have lost their way.

From 8 March to 14 May 1801 Caley accompanied Lieut Grant in the *Lady Nelson* on a survey-cruise of the southern coast to Wilson's Promontory and Western Port; here again his prickly nature became apparent, for although two other members of the party also made collections, 'Caley received everything they found, and refused to give up or part with a duplicate'. It was after this (August 1801) that Governor Paterson advised him to investigate the flora of the Hunter River – a considerable distance to the north. Between October and the following April he made various shorter excursions, to the Nepean, Tench and Hawkesbury rivers and to Mt Hunter.

On 9 May 1802 Captain Flinders in the *Investigator* arrived at Sydney in the course of his survey of the Australian coast. His scientific staff included Robert Brown, at the start of his brilliant career as a botanist;

[1] He is often said to have worked at Kew, but it seems doubtful if he was ever employed there

Ferdinand Bauer, the botanical draughtsman, and Peter Good, a gardener from Kew. Banks had authorized Caley to join the party if he wished to do so, but he refused the opportunity; it was reported that he was 'very angry at having Mr. Brown here, who he cannot help considering as a labourer in the field that ought to be wrought by himself'.[1] (He was perhaps wise not to subject himself to the naval discipline, the cramped quarters, and the presence of three of his own trade, all better qualified than himself.) The *Investigator* left on 22 July, and after another short excursion in October, Caley started on 4 December with a friend and a pack-mare, southward along the Nepean river, the principal eastern tributary of the Wollondilly. Some distance beyond Menangle they left the river, and striking westward, discovered Poppy Brook and Scirpus Mere (in the vicinity of Picton) both of which Caley named from the plants he found there. From Poppy Brook they travelled NNE to Mundogra, and thence home. Although this trip only lasted nine days, it was as much as could be managed when all provisions had to be carried, and very little transport was to be had. In the following April (1803) Banks received from Caley seeds of 170 species of plants, including *Epacris purpurascens* and *Jacksonia scoparia*.

Early in 1804 Caley was sent to discover the extent of an area near Poppy Brook which he called Vaccary Forest, though the prosaic settlers simply dubbed it the Cowpastures – the range of some cattle that had escaped and established themselves as a wild herd. But he had always been drawn to the Blue Mountains, then almost unexplored, and had made more than one excursion in that direction; and his most ambitious attempt to cross them was made in November of the same year, with four stout fellows lent to him by Governor Paterson. They went by boat up the Hawkesbury and Wollondilly rivers as far as Richmond, and then toiled westward up the pathless valley of the Grose. The nature of the terrain can be seen by Caley's names – Swamp Valley, Dark Valley, Devil's Wilderness and Skeleton Rocks, where he had to talk his men into a good humour – they were 'not so [much] overcome by fatigue as overawed by the dangers through which they had passed'.[2] After twelve gruelling days they reached Mt Banks, named by Caley in honour of his patron; 'Being the first European in visiting these parts, I claim the privilege of giving names to a few places which appear to me the most obvious.'[3] Here the men rested for a day, while the indefatigable Caley botanized, surveyed, and got his first view of central Australia – looking deceptively near at hand, but barred, he surmised, by a series of impassable valleys; 'one comes on them all at once, like a ha-ha'.[4] His men were exhausted and his supplies low, so he returned by the same route. In subsequent years he made further attempts to cross the mountains, but never succeeded in discovering a pass. A cairn on the ridge between Linden and Woodford was subsequently named Caley's Repulse, and was thought to mark the furthest spot that he attained; but earlier travellers attributed the cairn

[1] Governor King to Banks, quoted by Maiden
[2] and [4] Ida Lee *Early Explorers in Australia* 1925
[3] Caley to Banks, quoted by Maiden

to Bass,[1] and there is no evidence to associate it with either explorer.

By April 1805 Caley was anxious to return home, feeling that he had achieved all that could be done by a single individual at small expense; but it was to be another five years before he quitted the colony. He acquired a small farm, but continued to collect, and in 1805–6 made an excursion to Norfolk Island and Tasmania. In 1808 Banks wrote to say that owing to his (Banks') age and failing eyesight he no longer required the services of a collector; he offered Caley a pension of £50 a year for life, whether he returned to England or remained in Australia. Caley stayed in Australia; he was deeply involved in an unsuccessful courtship of Margaret Catchpole, a notorious young woman who had been transported for stealing a horse.

Meantime stirring events had been taking place. In 1805 Captain Bligh, of the ill-fated *Bounty*, was made Governor General of New South Wales, in succession to the easy-going Colonel Paterson. Bligh had excellent ideas about the administration of the colony, but enforced them in so harsh and arbitrary a manner that on 26 January 1808 a section of the New South Wales Corps under Major George Johnstone rebelled, deposed and imprisoned him, and set up a government of their own. Caley, who had hitherto been highly critical of Bligh, now became his champion, visited him in prison, and did his best to sway public opinion in his favour; and it is said that when he eventually left the colony in May 1810 to return to England, it was to give evidence at Major Johnstone's court-martial on Bligh's behalf. Some years later Caley was appointed director of the Botanic Garden at St Vincent in the West Indies, where he again behaved so strangely and gave so much trouble that one suspects some mental instability.

Two first-rank botanists with experience of local conditions gave Caley's work high praise. Robert Brown called him *Botanicus peritus et accuratus*, and Cunningham 'a most accurate, intelligent and diligent botanist'. Among the plants named after him were *Banksia caleyi*, *Grevillea caleyi* and *Acacia caleyi* (= *podalyriaefolia*); also *Caleana*, a small genus of botanically-interesting orchids, remarkable for their irritability. His introductions included the Stag's Horn Fern, *Platycerium bifurcatum*.

No excuses are required for **Allan Cunningham** (1791–1839) whose behaviour seems always to have been exemplary. He was the elder of two brothers, born in Wimbledon but of Scottish extraction, and was destined for the law; but he preferred botany and gardening, and went to Kew, where he was employed by the younger Aiton on the preparation of the second edition of *Hortus Kewensis*. In 1814, Aiton was anxious to resume the sending out of collectors, which had been interrupted by the Napoleonic Wars, and with the assistance of Banks arrangements were made for the despatch of Cunningham and Bowie. They worked together at Rio de Janeiro for nearly two years, and then Bowie was instructed to proceed to the Cape and Cunningham to New South Wales. He arrived at Sydney on 20 December 1816.

[1] Dr George Bass of the Bass Straits, the first European to explore the Blue Mountains

The young colony, after a slow start, had been making phenomenally rapid progress. Caley, as we know, never crossed the Blue Mountains, but in 1813 a pass had been found, and by 1817 a well-marked road with bridges and inns ran past Caley's Repulse to the promising settlement of Bathurst in the plains beyond. Cunningham visited Bathurst at least nine times, and it was from there rather than from Parramatta that many of his excursions began. Oxley and Cunningham discovered the Wellington Valley in 1817; a settlement was founded in 1823 which was a town by 1825. Similarly, in September 1824 he went with Oxley to investigate the possibility of establishing a settlement at Moreton Bay; and by the time he left Australia in 1831, Brisbane was a thriving port. In this rapid expansion Cunningham's explorations played an important part.

A little more than three months after his arrival, he was directed by the Governor (there was no time to consult Banks or Aiton) to join an expedition that was about to set out for the west under Surveyor-General John Oxley and his assistant G. W. Evans. He left on 3 April 1817, and had six days to botanize round about Bathurst before moving on to the next stage, a new depot on the Lachlan river, which was then the last human habitation to the west. Here there was another halt for a few days, but botanizing was only possible with an armed guard, for fear of native attack. The party was now joined by **Charles Fraser**, (d. 1831) a private in the 46th Regiment, who had been at the depot for about a month and had already been on the hills 'in his pursuits of the Flora (to which he is very much attached)'.[1] He was to go with the expedition in the capacity of collector for Lord Bathurst.

It was then believed that the Lachlan river was a tributary of the Macquarie,[2] and when the expedition finally launched into the unknown on 28 April, it was to trace the river downstream to its supposed junction with the Macquarie – from which they were actually moving ever farther and farther away. The party numbered thirteen in all, with fourteen packhorses, and two boats loaded with some of the stores. After three weeks of travel they came to a place where the river divided into two branches, each of which diminished and lost itself in labyrinthine marshes; here they abandoned the boats and struck south-west to the hills that they christened Peel's Range. From 25 May till 23 June they skirted the base of the range westward, halting for a day from time to time to rest the horses, but often having extreme difficulty in finding water. They then rejoined the river, now reunited and clear of the swamps, and again followed its course, but in a series of arcs as the actual banks were often boggy and impassable. They would strike hopefully towards the hills in the morning, and be driven back to the river by lack of water at night.

On 10 July they turned homewards, having reached approximately the spot where the town of Oxley now stands. They followed a similar route until they approached the swamps again; then they made a raft, crossed the river, and struck north-east into the hills, almost at once meeting with water difficulties, until they found the Bell River, which led

[1] Cunningham, quoted by Lee
[2] Hence, probably, the name – Lachlan Macquarie being the patronym of the then Governor

them at last to the Macquarie, in beautiful, fertile country which they christened Wellington Vale. (Their rations of pork and flour were now extremely low, but fortunately game was abundant.) A journey up the Macquarie of about a week brought them back to Bathurst on 29 August – weary, tattered, half-starved but triumphant, after nineteen weeks and 1200 miles of hard travel. The botanical finds had included *Acacia pendula* and *A. spectabilis*, *Grevillea acanthifolia*, *Hovea celsii* and *Pandorea pandorana* (= *Tecoma oxleyi*).

One is apt to picture the early Australian explorer as a keen-eyed figure on horseback gazing into the sunset from under a broad-brimmed hat. Nothing could be further from the truth. Horses were at a premium. All livestock had to be brought, by sail, from India or the Cape, and although brood-mares and stallions had been imported at an early stage, there were many losses and the supply was nothing like equal to the demand. Most of the precious beasts were under the control of the governor, and were hired or lent where he considered them most essential; they were used almost exclusively as pack-animals, and even when Cunningham was weak after a bad attack of fever he was unable to ride – 'as the whole of us walk, our horses being very heavily laden, I had no resource or alternative but to walk likewise'. Many a day's march was through swamps or through shoulder-high scrub, and many a march resolved itself into a desperate hunt for water; natives were eagerly looked for, as possible guides to water-holes, but they were timid and scarce.

The day after his return to Parramatta on 8 September, Cunningham received instructions from Banks that he was to accompany Lieutenant Philip Parker King's[1] surveying expedition in the *Mermaid*, now fitting out. The voyage started on 21 December and lasted seven months, and was followed at short intervals by four others, in all of which Cunningham took part; but their story must be told elsewhere. After his first voyage Banks commended him for having volunteered for a second – knowing well the hardships he must have undergone; he added that he entirely approved the whole of Cunningham's conduct, as did Aiton. Cunningham received this treasured letter on his return from his fourth voyage, and with it the news of Banks' death; he 'immediately put on that outward garb of sorrow which at best is but a poor indication of that heartfelt grief I even now feel for the loss we have sustained. . . .'[2] While he was absent on his fifth and last voyage, Governor Macquarie was replaced by Sir Thomas Brisbane. Macquarie had been unhelpful, to say the least – perhaps his obstructive attitude influenced Cunningham's decision to spend so many of the previous years at sea; but Brisbane encouraged and supported his explorations, as far as the scanty resources of the colony would permit.

It is impossible to detail more than a few of Cunningham's travels during his fifteen years in Australia; as he said himself, he was always on his legs – a wanderer. At an early stage he decided that he could 'blend discovery with botanical research tolerably well', and this decision, while

[1] Son of the former Governor, Philip Gidley King
[2] Quoted by Woolls

putting him into the front rank of Australian explorers, deprived him of the reputation he might have gained as a botanist, as he never had time to work up his collections before he was off on some new adventure. No longer a follower, he was now a leader of expeditions, and some of his geographical discoveries were of great importance; yet his portrait shows a fragile clerkly person, who looks as though he had never travelled farther than to a desk in the city.

On 31 March 1823, almost a year after his return from sea, Cunningham set out with five men and five strong pack-horses, first to Bathurst and then north-east to Dabee on the Cudegong river. This was not altogether new country, as he had explored parts of the Turon and Cudegong rivers the year before; but this time he went much farther north, and reaching the main Liverpool Range, began to search for a pass across the mountains. Having worked his way eastward for some time without success, he circled back to the Goulburn river and tried again to the west. At last he came to an opening which he called Pandora Pass, because 'a Hope . . . at the close of their journey led them to persevere westerly'. He could go no farther as his men were fatigued and his provisions exhausted, so he buried a message in a bottle, and, without crossing the pass, returned to Bathurst towards the end of June. In March 1825 another expedition was undertaken to complete the discovery. The mountains were approached by a more easterly route, by Wollombi and up the Hunter river to Mt Dangar, where the party struck NNW to Smith's Rivulet, and eventually gained Pandora Pass. This time it was crossed, and a road opened to the Liverpool Plains – fifty miles of level treeless pasture-land. They returned by a different route to the new settlement of Mudgee on the Cudegong, and thence to Bathurst. The last three months of the year Cunningham spent botanizing in the neighbourhood of Wellington, and in Croker's Range, where he made a large collection of seeds and tuberous roots, including those of twenty-five species of orchid.

An even more important pass was discovered by Cunningham, which also took two expeditions to confirm. Sir Thomas Brisbane had been replaced by General Darling, and in 1827 the latter planned an expedition into Queensland, which Cunningham (who had spent most of the intervening year in New Zealand) volunteered to lead. Seven men and eleven horses assembled at a rendezvous on the Hunter River (Segenhoe, about eleven miles south of Scone) and set off northwards on 30 April. They crossed the Liverpool Range by a new but very steep pass, and skirted the eastern border of the Liverpool Plains. They found the Namoi and Gwydir rivers, and followed the latter till it turned west, about the neighbourhood of Warralda; then skirted the Masterton Range and crossed the Mackintyre, Severn and Dumaresque rivers – all low after a season of drought. At last, on 5 June, they reached the Condamine River, and Cunningham got his first view of that wonderful grazing country, the Darling Downs. They continued to explore for a few more days, looking for a pass over the mountains to Moreton Bay; but the weather turned rainy, and though he could see a promising opening two or three miles to the north-east and sent two men forward to investigate it, Cunningham turned back on the 11th without himself visiting Cunningham's Gap. He may have been unwell; but tells us only that he 'felt unable' to continue his journey further. The party returned at first by a higher and more easterly route, then turned south-west, crossed the Gwydir to the west of their previous crossing-point and the Liverpool Range by a pass also farther to the west. They were back at their starting-point on the Hunter river on 28 July – having travelled 800 miles in thirteen weeks; and after a rest, Cunningham returned to Parramatta on 31 August.

In 1818, when Cunningham was absent on his first voyage with Lieut. King, Governor Macquarie had appointed young Charles Fraser as first Colonial Botanist (Cunningham ranked as a King's Botanist), and had given him charge of a new Botanic Garden at Parramatta. Fraser was hardly more of a stay-at-home than Cunningham; the same year he accompanied Oxley's second inland expedition, in 1825 he made a brief visit to New Zealand and in 1826–7 an important journey to the Swan river. He was now charged to lay out a botanic garden for the four-year-old settlement of Brisbane, and Cunningham accompanied him to make some explorations in the neighbourhood. They sailed on 7 June 1828, and arrived at Brisbane on the 30th. Three weeks were spent in work on the garden and short local excursions; but on 24 July a more ambitious convoy got under way. It consisted of the Commandant of the settlement, Captain Logan; Cunningham and Fraser; one soldier, five convicts, and several bullocks, carrying provisions for four weeks for eight persons (someone, therefore, must have been expected to go hungry). Their direction was mainly south-west, up the valley of the River Logan; by 3 August they had penetrated as far as Mt Lindesay, which they attempted to climb. Logan reached the summit, Fraser nearly did; but Cunningham, whether delayed by botany or already suffering from the effect of years of hard travel, only got half-way up. On 6 August they turned north-west and attempted to make their way to the eastern end of Cunningham's Gap; but the rocky ridges densely clothed with araucaria forest were impossible for the bullocks, already much fatigued. Provisions were getting low, and Logan was anxious to return to his duties, so they bore ENE to regain the river Logan, which they reached on 10 August. Fraser was in a great hurry to transport the plants he had collected back to Sydney, so he accompanied Logan to Brisbane, but Cunningham remained behind; with three men and two bullocks he made for Ipswich on the Bremen, where he had arranged for a new lot of provisions to be sent. On 18 August he made a fresh start up the valley of the Bremen; and by the 24th he had found the eastern end of Cunningham's Gap. Starting at seven the next morning he crossed the pass, and recognized on the other side familiar landmarks of his journey the year before; he had opened a way from the coast to the rich grazing-grounds of the interior. After climbing Mt Mitchell he got back to Ipswich by another route, made a few more local excursions and returned to Sydney on 29 October.

In 1830 Cunningham paid a visit to remote Norfolk Island,[1] where a

[1] 10 May–28 September

penal settlement had been established since the previous century. On 21 May a Government whaleboat set him ashore on Phillip Island, about six miles away, where he wished to botanize, and returned to base. That night eleven convicts broke loose, seized the boat and some provisions, and then visited Cunningham's island camp and robbed him of chronometer, pistols, provisions, water and tent, before sailing off to a series of nefarious but ultimately profitable adventures. It is not recorded how long it was before Cunningham was rescued from his awkward predicament – probably not long, as it was known that the convicts had made off in his direction; but the colonial government refused to refund his losses, on the grounds that he was an Imperial officer and not under their jurisdiction, and he had to replace his instruments out of his own scanty salary.

Before the Brisbane expedition of 1828 Cunningham had renewed an earlier request to be allowed to go home; but permission was not received until November 1830, after his return from Norfolk Island, and was then granted rather because the Kew authorities were retrenching their expenses than to comply with his wishes. After a great packing-up and a final rush of collecting, he left on 25 February 1831, and arrived in England in mid-July, where he settled down in a 'pretty cottage', near Kew. It must have been a strange world to him; he had been away for seventeen years, fifteen of them in Australia, in surroundings as different as possible from Strand-on-the-Green. Banks was dead, and so was Cunningham's father; but his brother Richard was there to welcome him, and his old friends and correspondents W. T. Aiton and Robert Brown. He was not left long undisturbed. Charles Fraser, whom he had left flourishing at Sydney, fell ill at Bathurst on the way back from a collecting expedition, and died at Parramatta on the last day of the same year; and his post of Colonial Botanist was offered to Allan Cunningham. He refused it, in favour of his younger brother.

Richard Cunningham (1793–1835) had gone to Kew at the same time as Allan, when he was only fifteen; and had remained there ever since as Aiton's amanuensis and clerk. In this capacity he had handled all his brother's collections, and was therefore familiar with Australian plants; but he had been 'almost wholly immured from the world, and cut off from personal intimacy with men of science'.[1] Allan Cunningham was anxious that his brother should be given more scope for his abilities, and this was backed by a strong recommendation from Robert Brown. Richard sailed for Sydney with a consignment of plants for the colony and some letters of introduction, and arrived on 5 January 1833.

Though well-qualified as a botanist, Richard Cunningham had not his brother's stability of character; he was full of zeal, but was one of those whose irresponsible behaviour causes others a great deal of needless trouble and anxiety. Towards the end of the year he went to New Zealand and remained there (without authority) till May 1834.[2] Early in the following April he accompanied a party sent out under the Surveyor-General, Major Mitchell, to explore the Darling River and its tributaries.

[1] Hooker *Companion to the Botanical Magazine* II, 1836

[2] See p. 232

Richard Cunningham was strong and hardy, but he lacked that essential part of a traveller's equipment, a sense of direction. He was always straying off to the right or left of the route to botanize, and then having difficulty in regaining the encampment at nightfall; 'he was not so often with his companions as away from them'. This must have been exasperating when conditions were such that any aberration by one member of the party might endanger the safety of them all, and Major Mitchell had several times to warn him that unless he took more care, he would never return to Sydney. On the night of 16 April, they were obliged to camp without water. When Cunningham overtook the main body of the party next morning after having lagged behind as usual, he learned that Mitchell had gone ahead to prospect for water; and he spurred on, saying he would overtake him. On reaching the dried bed of the river Bogan, Mitchell turned along it to the north, leaving signs on the bank to show he had done so; Cunningham failed to observe the signs, and continued to the west. The company camped for twelve days by the water that Mitchell had found, while they vainly scoured the country for the errant botanist; but his fate was not certainly known until a special search-party was sent out from Bathurst. They found that after following the Bogan for some twenty miles, he had wandered about aimlessly in all directions, leading his exhausted horse by the bridle, till the animal died. He then fell in with a party of aborigines, who took him in and gave him food and water; but during the night his behaviour became so strange and violent (he was probably delirious) that the frightened natives killed him.

After this tragic event, the post of Colonial Botanist was again offered to Allan Cunningham, and this time he accepted. It took nearly two years for the death of Richard to be confirmed, the appointment to be offered, and the new director to arrive; meantime, the botanic garden at Parramatta was temporarily put in charge of **James Anderson** (1797–1842). From 1826 to 1830 this Scottish botanist had accompanied Captain P. P. King in his survey of the coasts of South America, collecting meanwhile for a patron, Francis Henchman, and for the firm of Low and Mackie of Clapton. This was the Captain (then Lieutenant) with whom Cunningham had so often sailed, and it may have been King's influence that decided Anderson to come to New South Wales in 1832 as an independent botanical agent, sending collections of seeds and specimens to London to be sold. His former employer, Henchman, had previously sent **William Baxter** to collect in the new continent (1823–5 and again in 1829–32) but of his movements little is known. He seems to have worked chiefly on the south coast – Wilson's Promontory, Cape Arid and King George's Sound; and is said to have added sixty new species to the Proteaceae. About 1832 the nurseryman Joseph Knight purchased a collection of Baxter's seeds for £1500: they included *Epacris compressa*, *Pimelia hypericina* and (probably) *Sollya fusiformis*.

Allan Cunningham returned to Sydney in February 1837, but he soon found that the post of Colonial Botanist was not to his liking. His duties at the botanic garden – which included landscape gardening and the production of vegetables for the Governor and his staff – did not give sufficient freedom for the botanical exploration which he felt should be his work. A new Governor, Sir George Gipps, who arrived in February 1838, was sympathetic to his views, and suggested retaining his services as botanist, with Anderson to superintend the garden. He agreed that the salary of £450 that Cunningham asked was a reasonable one, but delayed so long in granting it, that in April the botanist washed his hands of the whole transaction and went off on a visit to New Zealand. Unfortunately the symptoms of tuberculosis had already begun to appear; the winter climate of New Zealand and the hardships he underwent there, aggravated the complaint, and by the time he returned to Sydney on 13 October he was seriously ill. 'I am past further great exertion,' he wrote; 'I can neither undertake any more expeditions nor walk about in search of any more plants . . .' On 24 June 1839 he was moved from Sydney to Parramatta; it was hoped that he would benefit from the change of air, but he died three days later, in the arms of his 'faithful friend' James Anderson. He had a kindly and affectionate nature and was mourned by many friends.

As a result of Cunningham's extensive explorations, the map of Australia is positively littered with the names of botanists. He christened Mt Aiton, Aiton's Bay and Aiton's Plains; Mounts Brown, Caley, Fraser and Greville; Buddle's River, Dryander's Head, Good's Peak, Hove's Rock, Lambert's Head, Sims Island and Smith's Plains. He and Fraser together named Mt Hooker 'in honour of the mutual friend of Mr. Fraser and myself, the . . . Regius Professor of Botany in the University of Glasgow' and he considered that Ferdinand Bauer deserved a much higher honour than the naming of a mountain 'that may never be seen by European eyes again, and doubtless will never be visited by any'. He is himself commemorated in Mts Cunningham and Allan, in Cunningham County, and in many of the plants he found – *Araucaria* and *Castanospermum cunninghamii* among them.

Mention should perhaps be made of two amateurs who were active during what might be called the Cunningham era – **James Backhouse** (1794–1869) and **Ronald Gunn** (1808–1881) both of whom collected with enthusiasm whenever their more pressing duties would permit. Backhouse was in Australia from 1831 to 1838, and visited Tasmania, Norfolk Island, New South Wales and other parts of the continent in the capacity of Quaker missionary and temperance crusader. His book, *Narrative of a Visit to the Australian Colonies* (1843), is full of references to plants, but gives no clue to which, if any, he contrived to send home to his partner and brother, with whom he had recently founded the famous nursery firm of Backhouse of York. Hooker, when naming the Backhousia in his honour, paid tribute to his work in the collection and description of Australian plants. While in Tasmania he made several botanical excursions in company with Ronald Campbell Gunn, who went there as an assistant superintendent of convicts in 1829, and remained in various official capacities till his death in 1881. Gunn was an assiduous collector of all objects of natural history, but his interest in plants was botanical rather than horticultural. He is commemorated in *Nothofagus gunnii*, *Eucalyptus gunnii* and *Olearia gunniana*, the last of which he introduced. He and Backhouse between them compiled a paper on the (negligible) esculent plants of Tasmania, the best of which, according to Backhouse, was a fungus that tasted like cold cow-heel.

All the collectors hitherto mentioned worked in eastern or south-eastern Australia, but meantime development had started in the west. The Swan River area had been visited by the French expedition under Captain Baudin in 1801, but this expedition was ill-organized and unlucky, and the boating-party under M. Heirisson got no further up the river than Heirisson's Islands. Their naturalist Leschenault, however, had time to find the beautiful shrub named after him, and some living plants and seeds were collected for the Empress Josephine's garden at Malmaison. Twenty-five years later a party was sent from Sydney to explore the area, accompanied by the botanist Charles Fraser, and this time the river was followed to its source. There was water everywhere in the subsoil, and Fraser found many beautiful new plants; the dark green of the landscape being very striking to one 'long accustomed to the monotonous brown of the vegetation of Port Jackson'.[1] The black swans that gave the river its name were extraordinarily abundant; Fraser saw about 500 of them rise at once, 'exhibiting a spectacle which, if the size and colour of the bird be taken into account, and the noise and rustling occasioned by the flapping of their wings previous to their rising, is quite unique in its kind'.[2] The sailors thought nothing of eating eight roasted swans a day.

It is said to have been largely due to Fraser's favourable report in 1826, that the Government sent out a party of colonists under Admiral Sir John Stirling, in the spring of 1829. They arrived in August of that year, and set about the founding of Perth and Freemantle, which within four months had a population of 1300. The party was either accompanied or immediately followed by **James Drummond** (1784–1863), who was to lay out a botanic garden for the new colony. The Horticultural Society supplied a collection of seeds, mostly of vegetables and fruit, but also including chrysanthemums and dahlias.

Drummond was a middle-aged man when he embarked with his family on this adventure, and had been for twenty years the curator of the botanic garden at Cork. He was the younger brother of the collector Thomas Drummond, whose untimely death in Cuba occurred six years later. On his departure J. C. Loudon prophesied[3] that within a year or two he would become a Justice of the Peace, and eventually rise to high office in a United States of Australia; but these rosy prognostications were not fulfilled. About a year after his arrival Drummond's post of Superintendent of Gardens was abolished, and he was instead appointed Government Naturalist at the usual salary of £100 a year; but in 1832, when the infant colony was having a hard struggle for survival, this too was withdrawn. Drummond is reported to have been shattered by the blow – he had a number of children to support, and times were hard; it may

[1] and [2] Fraser, in *The Botanical Miscellany* I, 1829   [3] In the *Gardener's Magazine* V, 1830

have been in consequence of this experience that he was afterwards stigmatized as grasping and money-grubbing. By 1835, however, he was farming a 3000-acre grant of land in the Toodjay Valley, and by 1837, when he was visited by James Backhouse, he had begun to work as an independent collector.

Fate had established him in the midst of one of the most interesting botanical areas in the world, where a primitive and ornamental flora had survived from an early geological era. It was chiefly on plants from this region that gardeners cast covetous eyes, and Drummond was not the only one to exploit it. From the earliest days of the colony enthusiastic amateurs had sent seeds of the local wild-flowers to correspondents at home. The wife of the Governor, Lady Stirling,[1] had a cousin, Captain James Mangles, who was a keen horticulturist and amateur botanist; he came to Swan River for a visit in April 1831, and made contact with several persons who afterwards sent him specimens and seeds. In 1836 Lady Stirling put him in touch with an ardent gardener, Mrs Georgiana Molloy, then living at Augusta and afterwards at the Vasse River; and until her early death (exhausted by hardships and childbearing) in 1843, she was one of his most helpful correspondents. In 1838, for example, she sent him a magnificent consignment of seeds, which he divided into fifteen lots and shared with his brother Robert Mangles, Joseph Paxton at Chatsworth, the nurseryman George Loddiges and others, and from these a great number of Australian plants were successfully raised. They included the Kangaroo Paw, *Anigozanthus manglesii*, *Helipterum manglesii* and *Trichinum manglesii*; but a genus named after Mrs Molloy was afterwards merged in Grevillea, and no plant now in cultivation bears her name.

It is not known whether Mangles met Drummond during his short visit, but they were in contact a few years later, and in November 1838 Drummond sent the Captain a large consignment (which Lady Stirling thought exceeded his instructions) of more than 220 different sorts of seeds, a large bale of unclassified plant-specimens, and 44 lbs of seeds which he expected Mangles to sell for him. He was later accused of ingratitude to Mangles, who had rendered him many services, and of sending seeds under 'fine names' which turned out to be 'mere fiction'; whether or not this was the case, Mangles found in Mrs Molloy a more satisfactory source of supply.

We have no account of Drummond's travels until he entered into correspondence with W. J. Hooker, who began to print extracts from his letters about 1840. In 1839 he made a number of short trips, within a radius of fifty or sixty miles from his farm; on two of them he was accompanied by his youngest son, John, who was collecting birds and insects, but who was also 'very fond of flowers'. Drummond explored the valley of the Avon and eastward to the sand-plains called by the natives Guangan (Dangin?) – and these interested him so much that he visited them twice more in the same year. In between, he joined a party which included a German botanist, J. A. L. Preiss, who had come to Perth the

[1] Née Ellen Mangles

year before, and John Gilbert, who was collecting birds for the ornithologist John Gould, on an excursion to the coast and Rottnest Island.

Early in September 1840, Drummond started on his first expedition to King George's Sound, a two-month journey covering some 300 miles. On his way south he crossed six considerable rivers, the courses of which were still largely unknown, and had an arrangement by which he received supplies from military posts along the way, in return for a report on the country through which he had travelled. On reaching the Sound he found Preiss already there,[1] and they botanized together in the neighbourhood, until Drummond returned by much the same route as he came. He was disappointed not to be able to climb Mt William and Mt Saddleback on the way home, but provisions at the nearest military posts were scanty and did not allow of any diversions or delays. Shortly after his return he cut his foot badly when bathing in the river and for a time was unable to walk – a serious hindrance, for he had seed-orders to fulfil to the value of about £100. He had entered into a contract with the firm of Low of Clapton (the employers of Baxter and Anderson) to supply them with seeds and bulbs every year.

Perhaps owing to this accident, Drummond seems to have undertaken no major expedition the following year, though he and his son accompanied the Government Resident, Captain Scully, on a prospecting-journey to the north-east, during which they discovered and named the Victoria Plains. But in 1842 he was very active. He had been shown specimens of a new plant gathered near Augusta, and of course he had to go and find it. A Mr Harris was going south to take up a post as surgeon to the Australind Company, so they travelled together – Harris in a 'cart upon springs' and Drummond on a favourite grey pony appropriately called 'Cabbine' – a native name whose approximate meaning was 'Perhaps'. Their route lay down the coast to Mandura; they were then to cross the estuary to Pinjarra (there being no bridge over the River Serpentine) but overshot the mark, landed too far south, and had to retrace their steps, reaching their destination after four hours of wading waist-deep. After a day's rest they continued their journey, reaching Australind on the third day; then Drummond went on alone to the Vasse Inlet near Busselton, where he stayed with Captain and Mrs Molloy. There was no road from there to Augusta, at the mouth of the Blackwood River – only a footpath; Drummond's journey there and back took nearly a fortnight, in continual heavy rain. His clothes were never dry, his lucifer matches would not ignite and his hands were blistered with trying (not always with success) to make a fire by friction in the native fashion; but he got his plant – *Dasypogon hookerii*, of botanical but not of horticultural interest.

He was home by the end of June, and shortly afterwards set off with John Gilbert NNE to the Wangan Hills, in the wake of his sons and Captain Scully, who were prospecting for a new sheep-station. About October he and Gilbert started again to the south, and this time climbed

[1] He had probably gone by sea

both Mt Saddleback and Mt William; but they found the latter disappointing. Drummond had planned to explore the mountains behind Cape Leeuwin and Cape Naturaliste, to follow the Blackwood River up to his previous crossing-point, and then strike south to King George's Sound; but there is no record of this part of the journey. Late in the following year (1843) he made a more prolonged and ambitious tour in the south-east, with his son and a servant. His itinerary included the Beaufort and Gordon rivers; the Stirling Range, where he climbed Hume's Peak; Toolbranup, inland from Albany, and the hills adjacent; Mt Mary Peak and Cape Riche, farther east along the coast, and the Salt River.

This was one of Drummond's most productive trips. He found the fragrant black-flowered *Boronia megastigma* 'in the swamps behind Cheyne's beach', and gathered a great harvest of seeds, the banksias and dryandras alone amounting to forty or fifty species. The plants he afterwards sent home included *Anigozanthus pulcherrimus*, *Backhousia myrtifolia* and *Leschenaultia laricina*; yet in February 1844 he wrote to Hooker in a despondent vein. It would be a pity, he thought, to give up collecting 'now that I have brought myself to live in the Bush as comfortably as in a house'; but the sale of his products hardly covered his expenses. The natives on the whole were well-disposed, partly because of their extraordinary belief that the white people were the spirits of their deceased ancestors; but it was unwise to go among unfamiliar tribes with a party of less than three well-armed men, and this was costly. He did not mention another fact which might have prompted thoughts of retirement – he was now sixty years of age.

For many more years, nevertheless, Drummond continued to collect. A journey to the south had to be abandoned half-way, because of a severe attack of opthalmia, which blinded him for a fortnight; for some time afterwards he could travel only by night, as his eyes could not bear the light of day. Undeterred, this forgotten Kingdon-Ward of Australia set off at the age of sixty-six on the longest and most dangerous of his journeys, north to the Murchison River, a trip lasting eighteen months – half of 1850 and almost the whole of 1851. (He may have accompanied an official mission, as he seems to have had an escort of mounted police.) The natives hereabouts, unlike those of the south, were what Drummond mildly called 'troublesome', and a strongly-armed party was necessary; both he and his son John, who was now at the head of the local police, had some narrow escapes from tribes that were not only murderous, but reputedly cannibalistic. We have no journal of this expedition, but its route can be traced from the localities Drummond gives for his plants. Moore River, Smith River, Dunderagan (Dandaraga?) where they stayed for some time, making local excursions; Gairdner Range, Cockleshell Plains, Irwin River, Greenough River and Champion Bay, Bowes River, Hutt River and finally Murchison River, more than 300 miles from home. Botanizing was often difficult, but the harvest was rich.

For some time Drummond had received reports from the natives of a scarlet-flowered banksia which grew in the eastern deserts, and which was more handsome than any species to be found in the coastlands. He thought the only way to procure it would be to go to the Wangan Hills in the rainy season and thence strike east on foot – there being no fodder for horses. Such a journey was made some time in 1852, under the guidance of a native called Mangerroot; they reached the head of an unmapped river called by the natives Wallemara, but ventured no farther, as the next source of water to the eastward was four days' journey away. Here Drummond found a number of new plants, but he does not say whether the coveted banksia was among them. Later in the same year he sustained a grievous blow; his son John, who for twenty-two years had been the companion of so many of his excursions, was murdered by a native spear as he lay sleeping at a sheep-station by the Moore River. Gamely his father struggled on, and made plans for yet another journey; he was still sending seeds to England in 1861, but died at Perth in 1863, aged seventy-nine.

On one occasion Drummond apologized to Hooker for some imperfection of his dried plants on the grounds that he had been a cultivator for many years, and only took to preserving specimens owing to Hooker's encouragement, late in life – a gardener, in fact, rather than a botanist. He also remarked more than once, that he collected examples of every plant he found, but seeds only of the striking or ornamental species. His introductions included *Chorizema varium*, *Pimelia spectabilis* and the exquisite *Leschenaultia biloba*. *Acacia drummondi* was named after him, and the genus 'Drummondita' for both brothers – 'an I for James and a T for Thomas'.

The ornithologist John Gilbert, after his last journey with Drummond in 1843, transferred to Brisbane, where he persuaded the explorer Ludwig Leichhardt to take him on his astounding overland expedition to Port Essington, starting in October 1844. Leichhardt returned to Sydney by sea in March 1846, when everyone had given him up for lost; but Gilbert had been killed in a night attack by natives at the headwaters of the river that bears his name. In 1848 Leichhardt set out on another trans-continental exploring expedition (from Moreton Bay to Swan River) – and was never heard of again. Seven years later another explorer, A. C. Gregory, with two journeys already behind him, was commissioned by the Royal Geographical Society to explore the Victoria River in Arnhem's Land, and at the same time to look for traces of the lost traveller. He was accompanied by Australia's most famous botanist – Ferdinand, later Baron **von Mueller** (1825–1896) who took 'a keen and almost pathetic interest' in Leichhardt's fate.

Dr Ludwig Preiss had left Australia for London in January 1842, and shortly afterwards returned to Germany. He was a friend of the Mueller family, and knowing their susceptibility to phthisis, he urged Ferdinand and his sisters to emigrate to Australia and escape from the rigorous North German climate. Reluctantly Mueller gave up the start he had made as a botanist, reluctantly he travelled to Australia – then fell in love with the flora, and remained there for the rest of his life; for this queer, hard continent more than any other claimed and kept her botanists, whether from choice or circumstance. Mueller obtained employment as a chemist in Adelaide, but spent all his leisure and private means on botanical excursions; by 1852, when he was appointed official botanist to the State of Victoria, he had already travelled 25,000 miles. He made frequent excursions from Melbourne to the Australian Alps, where almost his only precedessor had been Dr John Lhotsky, who travelled there from Sydney in 1834, and was a German as eccentric and cranky as himself. For it must be admitted that Mueller was a peculiar character and not a wholly admirable one; he had enthusiasm and endurance, but was unscrupulous, ambitious, humourless and egotistic – a far from ideal member for a small, isolated exploring party in difficult and dangerous conditions.

Gregory and his men, with all their horses and equipment, left Sydney in July 1855 in a schooner to go to their starting-point by sea; they investigated sundry islands on the way, and reached the mouth of the Victoria River in September. They not only explored the river to its source, but continued far to the south – to Mt Muller and Mt Wilson on the edge of the Great Sandy Desert; Mueller was one of the party of four who reached Termination Lake, the turning-point of the expedition. Retracing their steps down the Victoria they then turned eastward across Arnhem's Land, and skirted the Gulf of Carpentaria about 100 miles inland from the sea, to the mouth of Albert River, which they reached on 20 August 1856. 'Not meeting there with the expected supplies',[1] they followed the coast to Gilbert River and then struck south-east by the valleys of the Burdekin, Suttor, Beylando, Mackenzie and Dawson rivers, reaching their first settler's station on 22 November and Brisbane shortly afterwards – a journey second only to that of Leichhardt, of whose fate they found no trace. 'The energies of Dr Mueller', we are told, 'were here taxed to the uttermost',[2] but he brought back nearly 2000 specimens, many of them new, besides much hard-won knowledge of every sort of plant that could be eaten in an emergency – including the orchid *Cymbidium caniculatum*.

Mueller became a botanist of international repute, and reaped a great harvest of the titles, decorations and honorary degrees of which he was so inordinately fond; but he was no gardener. Shortly after his return (1857) he was appointed Director of the Melbourne Botanic Garden, and like Griffith at Calcutta fifteen years before, attempted to make it into a living botanical textbook, without the slightest concession to ornament – not even a lawn. He was removed from his post in the gardens 'in response to popular clamour' in 1873, and installed in a specially-built library and herbarium; but although he worked for another twenty-three years just outside the gates, he never entered the garden again. He was a prolific author, and is said to have added 2000 species to the Australian flora; but many of his botanical names have since been reduced to synonyms. In the course of his work he exchanged seeds and plants with other botanical institutions, and it was in this way that Drummond's *Boronia megastigma* and the New Zealand *Veronica hulkeana* were introduced to Britain.

There were few important horticultural collectors in Australia after the

[1] and [2] Hooker, J. D. *Introductory Essay to the Flora of Tasmania* 1859

death of Drummond; the visits of **John Gould Veitch** and his son **James Harry Veitch** were relatively short. Veitch-the-first landed thankfully at Sydney on 5 November 1864 after a non-stop ninety-five-day passage with hardly a glimpse of land. He spent a month inspecting local gardens and making short excursions, and then went on to Brisbane, where the possession of an 'Admiralty letter' obtained before leaving England, enabled him to embark (with four Wardian cases) on HMS *Salamander* for a cruise up the east coast. She sailed on 8 December, and after touching at Rockingham Bay, anchored for three weeks at Somerset, on the extreme northern tip of Cape York. This was a new settlement, founded only the previous July; the European population consisted of five gentlemen, two ladies and eighteen marines – two having recently been killed by the natives. During Veitch's stay, however, the natives were peaceable enough; several of them accompanied him regularly into the bush, and brought him seeds in exchange for tobacco. He also explored the Isle of Albany, and filled his cases with palms, dendrobiums and other tropical plants; but on the whole, he found the flora disappointing. He was back at Brisbane by 8 February 1865, and sent off his collections; then he visited various places in Queensland (which he thought 'wanting in picturesque scenery'), going as far afield as Rockhampton and Port Denison to obtain certain orchids.

He was back at Sydney and about to return home when an opportunity arose to sail on HMS *Curacoa* for a four-month cruise among the Polynesian islands – a chance too good to be missed. She left on 4 June, and her first call was at Norfolk Island where the buildings of the old convict settlement were now occupied by Pitcairn Islanders; before leaving a ball was held on board, at which many of the young ladies 'were dressed in strict accordance with the latest fashions, even including crinolines'. Then, after a short visit to Savage Island, the main tour began, stopping at two of the Samoan Islands, two of the Friendly Islands, three of the Fijian group, and four of the New Hebrides – but on some of the latter the civilians[1] were not allowed to land, because of the savagery of the inhabitants. Veitch had four hours ashore on Fate Island, while a native chief was held on board as a hostage for his safety, and a day on Banks' Island, a wet, unhealthy spot where in a six-hour walk he had to wade across five streams or marshes, some of them up to the armpits. He landed for a short time on five islands in the Solomon group, stayed for a week at the French island of New Caledonia, and returned to Sydney in October. Before leaving Australia Veitch visited Melbourne, where he saw Mueller's still-unfinished botanical gardens, and made a short trip by train to Geelong and Ballarat, to see the goldfields.

Veitch brought home from the islands a fine harvest of the stove foliage-plants then so much in vogue, some of which are still grown as house-plants: *Pandanus veitchii* from the Samoans, *Acalypha wilkesiana* from the New Hebrides, *Dizygotheca veitchii* from New Caledonia, many codiaeums, cordylines and dracaenas, and from Fiji a new palm, named

[1] There were four other passengers besides Veitch

after him *Veitchia joannis*. He married shortly after his return, and his elder son James Harry was born in 1868, by which time the symptoms of his father's fatal tuberculosis had already begun to appear; a generation therefore passed between one Veitch visit and the next.

Veitch-the-second was in Australia only for six months (in 1893), and very little of that time was spent actually in the wild. He arrived at Brisbane in mid-January, and after about a month removed to Albany on King George's Sound. From there he made two excursions; the first by rail to Mt Barker and then for an eight-day camping circuit, returning to Mt Barker again, and the second to Wilson's Inlet, about thirty miles west along the coast. Veitch found travel and seed-collecting a great deal harder in Australia than in Japan; 'stooping to collect on the hard-baked earth . . . in the blazing sun is not a pastime to be chosen'. He was inclined to take a defeatist attitude – everything of value had already been introduced by others; and the rest of his stay was spent in the cities – Perth, Adelaide, Melbourne and Sydney – with their nurseries and botanic gardens. Like his father, he visited Geelong and Ballarat, before leaving in June for New Zealand, where he was as much hampered by the snows of the winter as he had been in Australia by the summer heats.

Not until 1929 do we again find a professional plant-hunter working expressly for gardeners in the classical manner. In that year a syndicate headed by Lionel de Rothschild of Exbury sent **Harold F. Comber** (1897-1969) to Tasmania, to obtain plants from high altitudes that might prove hardy in Britain. Comber had an excellent garden-background. He was the son of James Comber, head-gardener to Lieut-Col Messel at Nymans, Sussex; and at seventeen went to work for H. J. Elwes at Colebrook; the gardens in both cases being notable for the skilful cultivation of beautiful and rare species from many lands. He then went for a three-year course at the Edinburgh Botanic Gardens, where he tried without success to obtain a place as assistant on one of George Forrest's Chinese expeditions; and in 1925 had been for a collecting trip in Chile and the Argentine.[1] It was a well-qualified plant-hunter therefore who landed in Hobart, though he departed from precedent by accomplishing much of his travel by train.

He went first to the National Park of Tasmania, where he had special permission to collect, but his exploration of Mt Field was delayed by snow; the previous winter had been exceptionally severe, and he was hindered by bad weather throughout his stay. Then he went south-west from Hobart and up the Huon River (an area already well-botanized) and finally on a much longer trip to the north and east coasts. Although he was able to go by train to Launceston, Burnie and Zeehan, parts of the island were still wild enough to satisfy the most exigent botanist. Exposed situations were clothed with dense, closely-interlaced brush three to ten feet deep, interspersed by clumps of cutting-grass whose twelve-foot leaves could cut to the bone – 'making a nasty mess of one's ears if due caution is not used. . . . No wonder that so many little corners of Tasmania remain unexplored!'[2] Comber crossed the Pieman River by a

[1] See p. 375 [2] Comber *New Flora and Silva* III 1931

ferry on which he was the first passenger for six months, and borrowing a boat, rowed ten miles up the river to the deserted mining settlement of Corinna, where he found the native plants being smothered by the introduced brambles. He climbed Mt Lyall, where in places much of the vegetation had been killed by sulphur-fumes from the copper-mines; he also climbed Mt Sedgewick above Lake Margaret, and after a detour to Waratah, returned to Hobart for the seed-harvest.

Comber collected seeds of 147 Tasmanian species; the garden at Nymans had to be enlarged to contain his contributions, and a part of the new area was christened 'Tasmania'; but with a few exceptions, hopes for the hardiness of his introductions were disappointed. Many of the alpines that he found were hardly worth cultivating, and some of the better species had failed to flower or seed that season. At so late a date, few botanical novelties were to be expected, but he found a new epacris (*E. comberi*) and successfully introduced such near-hardy species as *Gaultheria hispida*, *Helichrysum ledifolium*, *Richea scoparia* and *Telopea truncata*. Ironically enough, his most valuable plant was a cultivated one – a strain of *Olearia gunniana* (var. *splendens*) raised by a farmer of Colebrook near Hobart from a solitary coloured form found on Mt Seymour years before.

New Zealand

The flora of New Zealand is even more unusual than that of Australia, though perhaps less rich in ornamental species. Flung down on the edge of nowhere like the leg of a broken doll, the islands were separated from other land-masses early in geological time; and the plants developed undisturbed by man or by grazing animals until a comparatively late date. The result is that the flora contains a very large proportion of endemics (about seventy-four per cent), including such curiosities as trees of daisies, epiphytic lilies, a carrot with stiff, bayonet-sharp leaves, the largest buttercup and the smallest conifer in the world, white gentians, yellow forget-me-nots and the 'vegetable sheep'. Unfortunately most of these marvels are difficult to grow in Britain.

As might be expected in a country so much of which is coastline, much of the botanical exploration was done from the sea; Banks and Solander with Captain Cook, Raoul with Captain Lavaud, and Hooker and Lyall with Captain Ross, come into the category of maritime surveys, which are not included here. In New Zealand the period of the travelling plant-hunter, between the sea-borne scientist and the resident botanist, was particularly short; and the fact that the natives were fierce, and, not infrequently, cannibals, also acted as a deterrent to botanical enterprise.

The first botanist to explore the interior was **Allan Cunningham**. He arrived at the Bay of Islands at the end of August 1826, well-furnished with introductions to the missionaries from his friend the Rev. Samuel Marsden at Sydney. (Caley, too, had known Marsden, but had quarrelled with him.) Cunningham roamed the still-virgin kauri forests, explored the rivers Kawa-Kawa and Keri-Keri, and made the sixty-mile crossing of the watershed to the river Hokianga and the west coast, finding *Viola cunninghamii* on the way; after his return he stayed for a month at Wangaroa before sailing for Sydney in mid-January, 1827. He was 'greatly gratified by the kindness of the missionaries as well as by the general esteem in which he was held by the natives',[1] – but this was entirely due to his own good qualities; in the space of five months he had established a reputation among the local people that was to protect his headstrong brother seven years afterwards.

The said brother, **Richard Cunningham**, arrived at Sydney in January 1833, and later in the same year was appointed to HMS *Buffalo*, which was being sent to New Zealand to obtain spars; a botanist was considered necessary, as there was no one else on board who could recognize a kauri pine when he saw one. They went first to the Bay of Islands, but in the interval since Allan's visit all the accessible timber had been felled, so the *Buffalo* went farther up the coast to Wangaroa. Here 'circumstances obliged Mr. Cunningham to quit the store ship altogether'[2] and to set up a camp ashore, purchasing a canoe for coastal exploration and hiring a crew of natives. Meanwhile the commander, finding it difficult to procure trees from the ravines, recalled his felling-gangs and sailed off to try elsewhere, leaving Cunningham alone on a coast where the crew of the ship *Boyd* had been massacred a few years before. Hooker attributes this incident entirely to Cunningham's botanical zeal; but one cannot help suspecting that he had antagonized his commander, probably by undisciplined behaviour, and that the *Buffalo* thought it could manage very well without him.

Protected, however, by the regard of the natives for the remembered 'Canni-mama', Richard botanized in safety, and worked his way down the coast to the Bay of Islands, finding *Fuchsia procumbens* in its only known locality, near Matauri. Like his brother before him, he investigated the Keri-Keri River, and in January 1834 crossed the mountains to the Hokianga; he seems to have broken little fresh ground, except in returning from Hokianga to the Bay of Islands by a different route, through what was then the Great Forest. After some local excursions, during which he collected living plants for the Sydney Botanic Garden, he obtained a passage in HMS *Alligator* and returned to Australia on 13 May.

This was during the short period when the positions the two brothers had held for seventeen years were reversed; when it was Allan who sat quietly at Kew and examined the plants that Richard sent home. Among them was *Veronica speciosa*, which Richard found near the mouth of the Hokianga River, but did not introduce; 'since the country around its locality . . . is now occupied by Europeans' wrote Allan, 'let us hope soon to receive the seeds' – but the plant did not reach England till about 1842.

Three years after Richard's unfortunate death, Allan Cunningham returned to New Zealand for his second visit. Already infected with tuberculosis, he arrived at Pai-hai on the Bay of Islands on 28 April 1838,

[1] Cunningham *Companion to the Botanical Magazine* II 1836
[2] Hooker, in the same

and was established in a newly-built cottage on the property of the principal of the mission, with direct access to the woods. The winter rains that year were 'cold, heavy, and penetrating . . . with little intermission', but he managed to establish *Dendrobium cunninghamii* and other fine plants in pots, ultimately to be despatched in the new Wardian cases from Sydney to 'those truly excellent men, the Messrs. Loddiges'. He planned to visit all the neighbouring mission-stations, including those at Waimate and Hokianga, but it is uncertain how much of this programme he was able to carry out. He did, however, penetrate to some falls on the Keri-Keri River, a few miles north of Waimate, for it was there that he found *Clianthus puniceus*. This striking shrub had been grown by the savage Maoris long before the advent of white men, and was known as a cultivated plant to Sir Joseph Banks; but it had never before been seen in the wild.

Much reduced by the weather and by meagre and monotonous food, Cunningham returned to Sydney in October in a 'most deplorable state of health', and even during the subsequent hot weather felt 'the New Zealand chill' still hanging about him, though he does not seem to have recognized the nature of his complaint and still hoped to recover in time to go on an expedition with Captain Wickham in the *Beagle* the following March, and to England in 1840. His condition deteriorated steadily and he died at Sydney eight months later. His visit to New Zealand, though disastrous to himself, had far-reaching consequences; for he had handed on the flag of his enthusiasm and knowledge to a young missionary, **William Colenso** (1811–1899) whom he met at the Bay of Islands, and who afterwards became New Zealand's most notable botanist.

Colenso, a young printer from Cornwall, had been sent to the mission of Pai-hai by the British and Foreign Bible Society in 1833–4, about the time of Richard Cunningham's visit; but his first four years were very fully occupied in learning Maori, translating the New Testament into that tongue, and printing it on the press he had brought with him for the purpose. This work was completed in 1837, shortly before Allan Cunningham's second visit; and in 1840 Colenso began mission-work, taking his texts and testaments to the different stations established in the island. In all his travels he kept an interested eye on the flora; but before he was fully launched on botanical exploration two other collectors had visited the country – Dieffenbach and Bidwill, both of whom arrived in 1839.

John Carne Bidwill (1815–1853) did not meet Colenso, for though he disembarked on 5 February 1839 at the Bay of Islands, he made no stay there, but boarded on the same day a coastal schooner sailing for the south. The son of an Exeter merchant, Bidwill had emigrated to Australia a few months before, with the intention of taking up a grant of land; but having selected his plot he found the preliminary negotiations so long-drawn-out that he decided on a voyage to New Zealand to fill in the time until he could take possession. He had an agreement with a mercantile firm, but they were in no pressing hurry for his services.

Bidwill landed first at Mercury Bay, but learning from a resident that it 'would not be the proper place' for his purpose, he re-embarked and went on to Tauranga. Here the surrounding country was distracted by tribal warfare; a marauding party not long before had captured, killed and eaten about twenty of the relatively peaceable locals, within sight of their villages. Owing to this and a severe epidemic of influenza, Bidwill found it hard to recruit enough carriers for a journey into the interior; it was necessary to take one or two men belonging to the tribes among whom he was going, or his own party might meet with a similar culinary end. The missionaries helped by lending him some of their converts, and he set off at last (17 February 1839) with seven natives and a white interpreter. In three days they reached Rotorua, and after a short stay to inspect the celebrated hot springs, Bidwill resumed his journey south-westward over open grassy country to the Waikato River. It was summer, but the nights at this height were intensely cold; he describes the extraordinary sensation of wading through an icy stream, yet finding the sand underneath too hot for the feet – owing to a hot spring in the river-bed. For a few miles they ascended the Waikato in a leaky native canoe, then left the river and continued overland to Lake Taupo, finding on the way a new acaena (*A. microphylla?*) bearing decorative balls of scarlet spines – 'a very beautiful and curious plant which I hope to be able to introduce into English gardens'. They celebrated their arrival at Lake Taupo by feasting on a pig, given by the friendly chief of Pirata, the last village; Bidwill distributed some medicines (rhubarb, aloes and peppermint) 'as the people were all sickly from the influenza which had been violent among them', and a good time seems to have been had by all.

Bidwill was the second European to set eyes on the lake – and the first, a missionary, had been there only three weeks before; it was all new country, which he describes in some detail. He crossed the water in a large local canoe and camped at a village at the south-east end. His objective was the mountain of Tongariro; but after a four-hour climb of the first hill he came to, he found he had another valley and lake (Rotuite) to pass before reaching his destination. At Rotuite there was a large native encampment; he left most of his men there, as they were tired and unwell, and taking on others, proceeded as near to the mountain as the reluctant Maoris would conduct him. For Tongariro was 'tabu', and nobody might approach it; there was only one spot, and that on a road now disused, from which the peak was visible, and they were supposed to pass this point with their mats over their averted heads, and not to violate even by a look the sanctity of the summit. To this place, at the junction of two water-courses, Bidwill was conducted; although he was on the mountain all day, he never saw it till the clouds lifted at the sunset. Beneath his feet, however, he found a two-inch-high plant which he mistook at first for a moss, but which turned out to be a conifer – *Dacrydium laxifolium*.

The Maoris had a sound basis for their tabu, for the mountain he climbed next day – some say it was not Tongariro but the adjacent and still higher Ngauruhoe – was an active volcano. It made ominous noises during the night, and there was a small outburst on the far side in the morning. No native would go near it, but Bidwill pressed on, and after a hard climb reached the edge of the crater. The view was hidden by cloud and another eruption seemed imminent, so he did not stay long, and was glad to get safely back to camp.

He returned by the same route, botanizing by the way. The local chief at Lake Taupo was very angry with him for having climbed Tongariro; Bidwill pacified him with presents and with the assurance that the tabu only applied to Maoris and not to white men. He was held up by storms when crossing the lake, said a reluctant farewell to the friendly chief at Pirata, crossed the Waikato at a higher point than on the outward journey and on 12 March could smell the hot springs of Rotorua, still ten miles away. Here again the country was disturbed by tribal warfare, and he stayed for some time with a hospitable missionary before venturing on the further three-day journey back to Tauranga on the coast.

After waiting impatiently for a ship Bidwill grew restive, and against local advice launched out on a short exploring trip – north-west up the coast, then inland over the hills and the rich but marshy plains to Matamata, an abandoned mission-station on the upper Thames, and back by another route. He met with one war-party on the march; they were impertinent, but not dangerously aggressive. He was back early in April, in time to embark on the schooner *Columbine*; his firm in Sydney had sent for him, and his time was up. In August, however, he returned to New Zealand on business and lived for a time at Port Nicholson; and in March 1848 he visited Nelson on South Island, and was the first to investigate New Zealand's alpine flora.

Bidwill had a fine garden at Sydney (where he was visited by J. D. Hooker in 1841) and before his appointment as Director of the Sydney Botanic Garden in 1847 was known in England for his work on hybridization. His rapid midsummer transit in New Zealand did not afford much opportunity for seed-collecting, but during his longer sojourn in Australia he made some fine, though unfortunately not hardy, introductions, including *Sterculia bidwillii* and the great blue Australian water-lily, *Nymphaea gigantea*. He made a special journey from Sydney to a place about seventy miles north-west of Moreton Bay to see the Australian monkey-puzzle, *Araucaria bidwillii*, growing in the wild, and was able to bring back seeds and a living plant when he came to England on a visit in 1843. In 1848 he moved to Wide Bay, where he held the offices of Commissioner for Crown Lands and Chairman of the Bench of Magistrates. He died in 1853 from the after-effects of a dreadful eight days when he and a companion were lost in the bush, with no compass and hardly any food, while prospecting for a more direct route between Maryborough and Brisbane than the roundabout road then in use. This, at least, was the avowed purpose of his voluntary journey; but his friends believed 'that botanical investigation was the master-motive that sent him forth'.[1]

Most collectors of this period took for granted the heaven-decreed superiority of the white man to the coloured – without this confidence,

[1] *The Gardener's Chronicle* 12 January 1856

they could not have accomplished so much; but few showed Bidwill's total insensibility to the feelings of the natives. Kicks and blows were to him a matter of course; and his ruthless violation of Maori tabus – in some cases, just to demonstrate the folly of their superstitions – is not endearing, and caused trouble for those who came after him. The principal sufferer from the effects of this attitude was Dr **Ernest Dieffenbach**, who had the misfortune to follow very closely in Bidwill's wake.

Dieffenbach came to New Zealand as naturalist to the New Zealand Company; he stayed longer (1839–41) and travelled farther than Bidwill, but his principal object was to find land suitable for settlement and cultivation, and his principal interest was in economic plants; Hooker thought his collections disappointingly small. He arrived at Ship Bay on Queen Charlotte's Sound, and stayed for a time at Port Nicholson; explored the valley of Heretaunga, and was the first European to ascend Mt Egmont. After visits to Chatham Island and to Australia, he returned to the Bay of Islands and explored the northern tip of New Zealand with a Frenchman, Captain Bernard, 'a most agreeable companion, and various were our adventures in the little schooner of sixteen tons, belonging to him in which we coasted along the island . . . but having a great disinclination to describe personal incidents,' continues the misguided author, 'I will omit them altogether'; and he substitutes 'what will, I conceive, be more useful, a topographical description of the different parts of the country'. In consequence, the chronology of his visit is not made clear; but in the course of his travels he followed the river Waikato to its source in Lake Taupo. He got on well with the Maoris, whom he found friendly, laughter-loving and intelligent; but they were still upset by Bidwill's visit, and would not allow him to go near Mt Tongariro. In a general survey of the flora of New Zealand, Dieffenbach cautiously estimated that there might be 'as many as nine' species of shrubby veronica; there are now known to be over fifty. He made some discoveries, but no introductions.

In 1841, New Zealand was buzzing with botanists. Before Dieffenbach left in October, Captain Charles Ross had arrived, in the course of his Antarctic expedition; and his stay of more than three months at the Bay of Islands (14 August–23 November) enabled his young surgeon-naturalist Joseph Hooker to make excursions ashore in the company of the local enthusiast, William Colenso. Colenso, at thirty, was six years older than his visitor, but Hooker was the more expert botanist; and a friendship was founded that continued by correspondence for fifty years. On 4 November the two were joined by a third, of similar age – Dr **Andrew Sinclair**, then serving as surgeon in HMS *Sulphur*, who had already made collections in Mexico and Central America in the course of his voyages. Colenso could hardly avoid becoming a botanist. Already interested in natural history before leaving Cornwall, he had encountered Charles Darwin in 1835, Cunningham in 1838, and Hooker and Sinclair in 1841.

Stimulated by these contacts, Colenso set out on 19 November 1841 (four days before Ross sailed) on the first of his long journeys; he had already been for several short excursions. Nominally, the object was to visit the various missions and distribute his testaments and texts; but in his journals he makes few references to his mission work and many to his natural-history observations, and he penetrated to places where not the most ardent missionary had as yet set foot.

He began his southward journey by sea, and was set ashore at Hicks' Bay, north-west of East Point, exhausted by three days of sea sickness. The captain was in a hurry to proceed, the surf was high, and Colenso had to scramble on shore as quickly as he could – and found, too late, that his botanical paper had been left behind. Naturally he at once saw plants that he wished to collect; but time was pressing and he had far to go, 'so with a sigh that only Botanists know', he was obliged to proceed. He worked southward down the coast – to Rangitukia, where he was kept so busy with mission work that he had hardly time to eat and sleep, still less to collect; to Tokomarua Bay and Tolaga Bay, and by a slippery beach traverse to Pakarae, where he was welcomed in spite of the fact that his hosts had no food to give him except one 'raw, dead crayfish' among six, and for breakfast some sweet potatoes that were fetched from a distance during the night. (It was at the season just before harvest, when food-supplies were at their lowest.) His next stop was at Kaupapa on Poverty Bay, and he was delighted to find that his ship had called there and delivered his packet of botanical paper unharmed. After a few days' rest, he set off into the almost unknown interior.

He followed the Turanga valley and was carried by a native across the only ford – a slippery shelf of rock between a deep, dark pool and a twenty-foot waterfall, which he dared not attempt himself; as it was, he nearly fell from his mount 'through nervous excitation, into the gloomy depth below';[1] then to the river Hangaroe and by canoe to the village of Te Reinga. He had difficulty in obtaining guides and carriers for the next stage, when he 'gained nothing new in botany in the whole of this melting day's horrid march – fern, fern, nothing but dry dusty fern all around'.[2] The high hills that they crossed grew progressively more barren, and it was long after sunset before they found water at which to camp; they had no food until a three-hour march next day brought them to the village of Whataroa. A few more marches, and they arrived on Christmas Eve at Lake Waikare, which Colenso was the first European to see.

Strong and bitterly cold winds made the crossing of the lake impossible for several days, and he spent a 'very unpleasant Christmas' under canvas; the natives were hospitable and kind, but their food-supplies were low. On 29 December the party crossed the lake in a flotilla of fragile canoes, and next day resumed their march, through forests of beech, where it rained and rained and rained. Colenso astonished his men by pausing even then to botanize – snatching up some ferns and a white, fragrant violet; 'I hastily removed this interesting plant from its mossy bed to the bosom of my cloak, now nearly as wet as the bank where the flower originally grew'.[3] The night was stormy; they camped by a deserted hut at a stream-

[1] and [2] Quoted by Taylor *Early Travellers in New Zealand* 1959
[3] Colenso *London Journal of Botany* III 1844

side, and Colenso, cold, wet, lonely and starving, in light summer clothing, sat shivering under the scanty shelter, 'holding an old, worn umbrella over my head'.[1] His reluctant guide had disappeared, none of his other men knew the country, and they had no more food. They decided to go on by compass, and hope to find a village; but next day the rain cleared, and at noon they were overtaken by the defaulting guide, with a load of sweet potatoes that he had gone to fetch. Four days later they were crossing hot, uninhabited, barren hills where the tracks were ankle-deep in pumice dust. The views were fine, but Colenso found the loneliness oppressive – 'The grass grew, the flowers blossomed, the river rolled, but not for man – solitude all! . . . I could not but exclaim "Oh, Solitude, where are thy charms?" '[2] They crossed the river Rangitaiki, and near Rerewhaikatu found a colony of a very beautiful leptospermum in bloom, in an isolated situation so sterile that nothing else grew but a moss; only old trees were found, no seedlings or young plants. Here they bivouacked once more without food, being too exhausted to go the remaining three miles to the comparative civilization of Talawera, which, however, they reached next day. They had now only to paddle up a series of lakes to reach the mission-station of the Te Ngae on the east side of Lake Rotorua, and from there, after a few days' rest, to follow the now well-trodden route to Tauranga on the coast.

Most men would have regarded this as a sufficient journey, and been content to return to the Bay of Islands by sea; but Colenso got there by a zigzag overland route involving another five weeks of travel almost as hard as the eight weeks he had just accomplished. He began by going north up the coast and then striking inland, climbing Te Wairera and fording the breast-high river Waihou to Matamata, where Bidwill had been four years before. Then south-west through fern-country to the Waikato River and across it to Horahora (above Cambridge) which he reached at sunset, 'unwell, in pain and much fatigued'. (Colenso hated the fern-brakes, which yielded no new plants and whose 'dreaded, subtile yellow dust' gave him hay-fever.) He then went down the Waipo and Waikato rivers by canoe to the west coast – a comparatively restful way of travel, but affording no opportunity for botanizing. Two days of rough going along the beaches took him north to Otahutu on Manukau Bay, where he obtained a supply of rice, the only portable food available, from near-by Auckland; then over uninhabited country, with no guide, to Kaipara, and by canoe up fifty miles of harbour to the north-east end of Otimatea inlet, at that time a pathless desert. Thence he made for the east coast by compass, forcing through 'the horrid interwoven mass of shrubs and prickly creepers, fern and cutting-grass, and prostrate trees, and swamps, and mud'[3] at the rate of a mile an hour. He reached the coast at Mangawhai on the morning of 14 February, and turning north quickly crossed an extensive inlet – the tide was coming in and the water was already breast-high; when he attained the other side he found that his men had not dared to follow, and he was cut off from his supplies. He went on alone, living

[1] and [3] Taylor [2] Colenso *London Journal of Botany* III 1844

on wild fruits and shellfish, for two painful days of beach-slogging and cliff-climbing – and even then, found and recorded a new leptospermum. His train caught up with him on the Tuesday at sunset; they had not touched the food they carried but had lived on fruits and shellfish likewise, saying 'What! shall we eat while our father is fasting?'[1] The remainder of the journey was difficult, but less gruelling. Collectors in those days were commonly reported to have 'detected' this and that plant (like so many Sherlocks) and on this part of his journey Colenso cleverly detected some fine tree-ferns (*Cyathea dealbata* and *C. medullaris*) between thirty and forty feet high. He reached Waikare on the 23rd, and from there it was only a three-hour row across the Bay of Islands to his base at Paihai. He brought home more than a thousand natural history specimens, and sent to Hooker more than 600 dried plants that he thought might be new; but he makes no reference to seed-collecting, for which indeed he had few opportunities.

This will perhaps serve as a sample of Colenso's many subsequent journeys. He made another the next year, starting from a point farther south (Ahurire, on Hawkes Bay) and covering some of the same ground, but finishing at Tauranga, without the final figure-of-eight up the north part of the island. In 1844 he was ordained a deacon, and moved to Waitangi near Napier; from there he made his first trip to the Ruahine Range in 1845. Here he found a new and exciting alpine flora. 'Never did I behold at one time in New Zealand such a profusion of Flora's stores . . . I was overwhelmed with astonishment, and stood looking with all my eyes, greedily devouring and drinking in the enchanting scene before me. . . . We had left our encampment that morning taking nothing with us, so we were all empty-handed. However, as I had no time to lose, I first pulled off my jacket, a light travelling coat, and made a bag of that, and then, driven by necessity, I added thereto my shirt, and by tying the neck etc., got an excellent bag, whilst some specimens I also stowed in the crown of my hat.'[2] In 1847 he made another extended tour, again in conditions of great hardship and privation – north-west by Titiokura and the Mohaka River to Lake Taupo; south-west round the flanks of Mts Tongariro and Ruapehu to Patea; then eastward and home by the Ruahine Range, which he was the first European to cross.

In 1852 Colenso was deprived of his office of deacon for having fathered an illegitimate child, and thereafter settled down on a farm, but he never ceased to botanize; he contributed a paper to the *Transactions of the New Zealand Institute* on some plants collected during a journey made to the flanks of the Ruahine Range in his eighty-fifth year. Many species bear his name – *Fuchsia colensoi*, *Phormium colensoi*, *Pittosporum colensoi* – and his discoveries included *Olearia macrodonta* and *Ourisia macrophylla*; but he was not notable as an introducer of garden-plants.

The botanical exploration of the South Island began rather later than that of the North; Bidwill on his second New Zealand visit was the pioneer, closely followed by Sinclair. After his visit to the Bay of Islands

[1] Taylor
[2] Quoted by Anderson *The Coming of the Flowers* 1950

in 1841, Sinclair went to Australia, and there became acquainted with Captain Robert Fitzroy,[1] who was on his way to take up his appointment as Governor of New Zealand. Sinclair accompanied him to Wellington in 1843 as his private secretary, and became Colonial Secretary the following year. He botanized as much as his duties would permit, and Hooker regarded him as a collector second only to Colenso. On the establishment of Parliamentary government in 1856 he resigned his secretaryship, and this left him free to go on botanical excursions with **Julian von Haast** (1824–1887), when the latter arrived in New Zealand in December 1858. It was on one of these excursions, in 1861, that he met his end; he was drowned in an imprudent attempt to ford one of the main branches of the upper Rangitata River. 'His companions buried him in a lonely grave at the foot of the glaciers, amongst the native shrubs and other natural objects which had formed the subject of his ill-fated researches'.[2] He discovered *Veronica cupressoides* in the Wairau valley, and in the neighbourhood of Lake Ohau he and Haast found in flower the magnificent *Ranunculus lyallii*, of which David Lyall had only seen the leaves.

Haast came to New Zealand on behalf of a shipping firm who wished to ascertain the suitability of the country for German immigration. In 1861 he was appointed Government geologist to the province of Canterbury, and spent the next ten years exploring the headwaters of all the great rivers on the eastern side of the Southern Alps. He was an able all-round naturalist, and made many botanical discoveries. He sent living plants of *Hoheria lyallii* to Kew, and found *Olearia haastii*, a rare plant in the wild, in the vicinity of Lake Ohau in 1863; but its seeds had already been sent to Messrs. Veitch by an unknown correspondent. Other South Island discoveries were made by the Irish W. T. L. Travers,[3] who came to New Zealand in 1849. He was primarily an ornithologist, but his name is associated with many New Zealand plants, including *Veronica traversii* from Canterbury province, and he found the invaluable *Senecio laxifolius* in the Nelson mountains.

But these were all botanists, not gardeners – discoverers rather than introducers. It seems that no plantsmen were sent to New Zealand on purely horticultural missions – perhaps because the distance was too great and the flora insufficiently attractive or too difficult in cultivation. There was never any fashion for New Zealand plants, as there was at an early date for those of 'New Holland' – until quite recently, when New Zealand alpines have become much in demand among rock-garden enthusiasts. The olearias and shrubby veronicas in general cultivation have filtered in gradually, now from one source and now from another, without the intervention of the professional collector; one of the sources was that scion of a great gardening family, Captain **A. A. Dorrien-Smith.**

Dorrien-Smith had the advantage that his father's magnificent garden

[1] Captain of the *Beagle* during Darwin's voyage
[2] Mennel *Dictionary of Australasian Biography*
[3] There was also a slightly later botanist, H. H. Travers

of Tresco Abbey was in the Scillies, and plants could be grown there that were not hardy elsewhere in Britain. He was therefore able to regard the flora of the Antipodes with a less jaundiced eye than that of the majority of gardeners; and in 1909–10 he visited not only Australia and New Zealand, but Chatham Island, which has an interesting endemic flora, including two species as desirable as they are difficult – the giant forget-me-not *Myosotidium hortensia*, and *Olearia semidentata*, the most beautiful member of its genus. His island host took him to the 'Tobacco Country' – so called because the ground was once purchased for a handful of cigars – which contained the 'highest ground, worst bogs, and greatest quantity of *Olearia semidentata*'[1] including white and pink forms. Captain Dorrien-Smith brought home many plants, but unfortunately most of those from Chatham Island perished from heat during the voyage. The olearia was among the survivors and bloomed at Tresco in 1913.

[1] Dorrien Smith *Kew Bulletin* 1910

Africa

EQUATORIAL AFRICA

1748	Adanson	1816	Smith and Lockhart
1775	Smeathman	1822	Don
1780	Brass	1854	Welwitsch
1792	Afzelius	1934	Taylor and Synge

SOUTH AFRICA

(1672	Hermann)	1798	Niven
1761	Auge	1810	Burchell
1772	Thunberg	1816	Bowie
1772	Masson	1859	Cooper
1786	Boos and Scholl	1927	Ingram

'. . . Good Lord! How numerous, how rare and how wonderful were the plants that presented themselves to Hermann's eyes!' CARL LINNAEUS *Flora Zeylanica* 1747

'. . . I met the dangers of life, I prudently eluded ferocious tribes and beasts, and for the sake of discovering the beautiful plants of this Southern Thule, I joyfully ran, sweated and chilled.' C. P. THUNBERG *Flora Capensis* 1807 (trans. HUTCHINSON)

Africa

Equatorial Africa

The earliest plant-collecting expedition of which we have any record took place in Africa in 1495 BC when Queen Hatshepsut of Egypt sent a party in charge of Prince Nehasi ('The Negro') to the Land of Punt (Somalia) in order to obtain living specimens of the trees whose fragrant resin yielded the precious frankincense.[1] The expedition went from Deir el Bahri by boat down the Nile, across by canal to the Gulf of Suez, and down the whole length of the Red Sea to the Gulf of Aden and its destination, returning the same way. Thirty-one young trees were brought back, carefully packed in wicker baskets slung on poles and carried between two, sometimes three, pairs of slaves; besides gold, ebony, ivory and other products. The trees were planted in the garden of the Temple of Amon at Thebes, and thrived – at least, for a time. The aim of this expedition, naturally enough, was economic; but it was followed not long afterwards by one that almost appears scientific, for the Queen's nephew and successor Thutmosis III brought back from his military campaigns in Syria a large number of plants which had no apparent economic value. Two hundred and seventy-five are carved on the walls of the 'Botanic Chamber' of his temple at Karnak, but only a few of them can be identified.

Very few plant-hunters in any part of the world escaped the attacks of malaria and dysentery; but except for the notorious Dutch settlement of Batavia in the East Indies, no part of the world was more certainly fatal to the European than the coasts and rivers of equatorial Africa. It was not till 1894 that the part played by the malaria-carrying mosquito was sus-

[1] A species of boswellia; but some authorities think it was the myrrh tree, *Commiphora myrrha*, that was brought

pected, and only in 1900 was it confirmed; even now it is not always possible to guard against it, and in earlier times men were pathetically helpless against a disease, the causes of which they did not understand. Again and again the trader, the explorer and the missionary was defeated by the insect. In spite of the dangers, however, Africa was particularly rich in explorer-botanists, whose primary object was geographical discovery but who collected plants by the way, to the advancement of science if not of horticulture. Mungo Park was interested in botany and sent specimens and drawings to Banks while he was waiting at Pisania in 1795 before starting on his journey up the Gambia; he could do no more, for on his first trip he was robbed of everything except his life, and on his second lost even that. Henry Salt, on a diplomatic mission to the King of Abyssinia in 1805, made a collection of plants, and was very critical of the botanical blunders of his predecessor in the area, James Bruce, who nevertheless introduced a number of plants to Kew in 1775, including the *Brucea ferruginea* or Wooginoos, and an Abyssinian pasture-grass which was grown experimentally by André Michaux before he set out on his own plant-hunting travels.

Captain Tuckey's ill-fated exploring expedition up the Congo in 1816 was accompanied by a Norwegian botanist, Dr **Christian Smith**, (1785–1816) and a gardener from Kew, **David Lockhart** (d. 1846). Both were exceedingly hard-working and enthusiastic; when the battered little party of sixteen that explored the river above the cataracts had reached the limits of its endurance, Smith was the most reluctant to turn back; he was 'so much enraptured with the improved appearance of the country, and the magnificence of the river that it was with the utmost difficulty he was prevailed on to return'. Four days later he was seized with fever, and died before reaching the coast.[1] Lockhart, though 'on his legs every day, from morning till evening, sometimes heavily loaded with the plants he had collected' went to the farthest point and returned without mishap till the very last moment when he, too, took fever and was left in hospital at Bahia; he survived, however, and later became superintendent of the Botanic Garden of Trinidad.

Captain Trotter's Niger expedition in 1841 was almost equally unfortunate, though furnished with eight physicians, three steamboats, and the blessing of Queen Victoria and the Prince Consort. He, too, carried a botanist, Dr T. Vogel, and a gardener, Ansell, and again the gardener survived, but not the botanist. Twelve years later, Dr John Kirk accompanied Livingstone to the Zambesi in the capacity of physician and naturalist, and was the one member of the party to stand by his difficult leader through every vicissitude; three precious cases of hard-won specimens which he sent home in 1860 got mislaid in transit, and came to light in Portsmouth dockyard in 1883. Later, Kirk became British Political Resident and subsequently Consul-General at Zanzibar, and from there sent many East African plants to Britain, including the popular Busy Lizzie, *Impatiens sultani*, and the striped Pyjama Lily, *Crinum kirkii*. Colonel J. A. Grant

[1] Fourteen of the sixteen died, including the captain

went with Speke to the source of the Nile (1860–3) and brought back large natural history collections and also some seeds, including those of the handsome *Mussaenda luteola.*

These part-time or travelling botanists were necessarily of less importance than the resident ones, who went out for no other purpose and worked from a settled base. 'Settled' is perhaps too flattering a term; the River Senegal was a wild place when the French botanist **Michel Adanson** (1727–1806) lived on its banks. A pupil of Bernard de Jussieu, Adanson thought that the quickest way to achieve botanical distinction would be to explore some region which, because of its dangers, was practically unknown to science; and he obtained a clerical post on the Senegal station of the Compagnie des Indes, which afforded a good deal of leisure for natural history researches. He arrived in the summer of 1749 at his base on the Ile de Sénégal at the mouth of the river; but the forests were so full of wild men and wild beasts, the river so densely populated with hippos and crocodiles, that he was unable to venture far beyond the tiny scattered trading-stations. Nevertheless he found the bewildering diversity of the tropical flora impossible to accommodate in any existing European botanical system, and before he was twenty-three had started to devise his own. The extreme isolation of his situation encouraged egotism and eccentricity; he was cut off from contact with other scientists, and he kept aloof from his few European companions who were mostly slave-traders of the roughest and most debauched sort. He returned to France in January 1754 with large and important collections, and subsequently had a long botanical career, but was one of those tragic figures who attempt the impossible and end by accomplishing little or nothing. By a 'veritable flash of genius' he diagnosed the baobab tree as a member of the mallow family, and it was named after him, *Adansonia digitata.*

The first naturalist was sent to Sierra Leone twelve years before it became a British colony – **Henry Smeathman** (1750–1787) who was despatched by the botanically-minded Duchess of Portland and others in 1775. He collected specimens of about 450 species, and brought home drawings, impressions of ferns, and living plants (including two healthy tree-ferns in tubs) when he returned to England in 1779. He wrote a paper on the habits of the white ant which excited a good deal of interest, and later took up aeronautics and invented a dirigible balloon.

Very different in character from the versatile and ingenious Smeathman was the Duke of Northumberland's gardener, the unfortunate **William Brass.** He was sent to collect at Cape Coast Castle on the Gold Coast (with a salary of £130 a year and all found) by a syndicate whose principal member seems to have been Dr John Fothergill, but which also included Sir Joseph Banks, Northumberland, the Earl of Tankerville, Dr Pitcairn, and probably James Lee. He sailed in May 1780 on the *Endymion* with an Admiralty order for his passage and a letter of introduction from Banks to Captain Carteret, who nevertheless treated him on the voyage with great harshness. Instead of continuing down the African coast after calling at Sierra Leone, Carteret bore away without notice for the West Indies, and the bewildered Brass was carried nearly to the Cape Verde Islands before they chanced to meet two Bristol ships bound for Cape Coast Castle, in one of which he obtained a passage. Before leaving the *Endymion* he was presented with a bill of £4 for his keep; he had understood that his Admiralty order would cover all expenses, and had not so much money, so was obliged to draw a bill on Sir Joseph Banks.

He landed at his destination on 1 October, completely destitute; for not only had the Bristol ship charged an exorbitant price for his passage, but he had lost all his clothes and bedding, which had rotted owing to the excessive wet and the lack of facilities to dry them. He again had to resort to drawing bills on his patrons at home in order to re-equip himself, even in the most modest manner. He had been given a small clerical post with the East India Company (at £30 a year) and a letter of introduction to Governor Roberts, but neither proved of much service to him. His magnificent salary was largely paid in goods of inferior value, and he was obliged to provide himself with a 'Euniforme' so he was little in pocket; the Governor was willing to help, but was unpopular with his colleagues and himself so poor that he was obliged to give up his 'public table', and actually applied to Brass for financial assistance – to Brass, who received only £7 between October and the arrival of the Company's store-ship the following May! Moreover, accommodation was short, and he had nowhere to dry his seeds except a room that he shared with three others – 'which puts my things in confusion'.

In spite of these very difficult circumstances Brass contrived to make 'several little Jurnies up in the country', putting his plants in a portion of the Company's garden which had been allocated to him – though without the labour to clear it. He applied to his patrons for a grant towards his expenses, at least until his finances improved; 'I would willingly Expand my own Sillery' he wrote to Fothergill, 'sooner than Return without a Collection to my Credit'.[1] But on 12 August 1781, after more than ten month's absence, he received from his wife his first letter from home – and learned that Fothergill had died the previous December. His wife reported that his employers were dissatisfied with the smallness of the consignments he had sent back, and 'seem'd very cool about the undertaking'. No one else wrote to him, and he was at a loss to know what to do, 'left in a Distress'd condition in a distant country with hardly the Bread of a common Souldier . . . to subsist on'.

Betrayed on every side by those who should have supported him, Brass had hardly sufficient force of character to conquer adverse circumstances and stand successfully on his own feet; but he made valiant efforts to do all that was required of him. He sent seeds and letters in duplicate, and kept a 'Jurnal' in which he noted 'every possible Remark' about the soil, situation and season of the plants he discovered, and any uses he could learn from the natives, with numbers corresponding to those on the seed-packets and specimens. Lee had from him more than forty-two varieties of seeds and he sent Fothergill a herbarium containing 250 specimens of plants, which afterwards fell into the hands of Banks and was examined

[1] Brass *Letters* at Kew

eventually by Robert Brown. The great taxonomist praised Brass as an intelligent collector, and named a genus of South American orchids after him; but Brass was long past appreciating this honour. He had supported the hard conditions 'tolerable well' and considered that 'the White people's Constitutions is Ruined more with their own Debauchery than the Unhilthiness of the Climate', but he died during the voyage back to England in 1783, from what cause is unknown. Brass also sent home a collection of drawings, which were lent to Afzelius in Sierra Leone and were mostly damaged or destroyed in the disaster that followed.

The colony of Sierra Leone was founded by the abolitionist William Wilberforce in 1787, chiefly to provide a home for Africans released from slavery, and was ruled for the first thirty years by a private company. In December 1791, Wilberforce asked Sir Joseph Banks if he could recommend a botanist familiar with tropical plants, who would be willing to go out and investigate the products of the new country, and Banks suggested **Adam Afzelius** (1750–1835), who had come to London from Uppsala the year before, with a letter of introduction from Thunberg. With his usual generosity Banks supplied Afzelius with his equipment (including 'a valuable collection of drawings, which the Directors refused to insure'[1] and the latter took with him as helper and servant an Italian youth, Francisco Borone, who had been trained in botany by J. E. Smith. Freetown was only in process of being built, and the Governor, though anxious to be helpful, was at first unable to provide them with a house or even a room to themselves. Fortunately both were of a cheerful and uncomplaining disposition; 'There is something in the countenance and manner of that little man', wrote the Marchioness of Rockingham of Afzelius, 'that shows a goodness that interests very much one's good wishes';[2] and everyone liked the ingenuous Italian – 'Borone behaves exceedingly well,' wrote Afzelius, 'and I am very happy to have such an amiable young man as my companion in my travels'.[3] The crowding and confusion were at first a hindrance to their collecting, but they sent home seeds and bulbs almost immediately after their arrival at the end of June 1792, and further consignments as opportunity offered, in December and the following March. By September 1793 they were back in England, bringing with them 'many fine things, though not many specimens, because the Sierra Leone climate is very damp and *insectiferous*'.[4]

Afzelius returned to Africa the following spring, landing at Freetown on 22 April 1794; this time Borone was not with him,[5] but he took a gardener from Kew and a 'plant-hutch' specially presented by the King. Conditions had improved, and he began to cultivate a piece of ground for the reception of useful or rare plants. But in September, Freetown was attacked from the sea by a French naval force, with few casualties but with almost complete destruction of property – houses, clothing and provisions; a hard blow for the infant colony, where supplies in any case were necessarily short. Afzelius lost almost everything; books, botanical paper, microscope, writing materials, most of his journal, and his collection of

[1] Banks *Letters* [2], [3] and [4] Smith *Correspondence* [5] See p. 26

shells, insects, animals and plants, which he valued at between £1500 and £1600 – 'though manuscripts, particularly when they are intended for publication, are invaluable, and may be worth double the sum you first thought of'.[1] The French threw his plant-specimens overboard, wantonly destroyed the plant-hutch and killed the man in charge of it; and the little that was left was damaged by the weather or stolen by the natives. Communications were so slow that it was November before Afzelius was able to notify Banks of the catastrophe, and the following June before he received fresh supplies. Banks advised him to come home, but it was not till April 1796 that he was able to get a passage. He worked hard in the interval to recoup his losses, and brought home 'very extraordinary plants from Africa and truly paradoxical'[2] which were described in his *Genera Plantarum Guineensis* (1804). Brown called his herbarium 'the most extensive and valuable collection ever brought from the west coast of equinoctial Africa'.[3] He also brought back seeds, some of which were raised by Messrs Loddiges; but they yielded nothing of great horticultural importance.

The territories of both Afzelius and Brass were revisited in 1822 by **George Don** (1798–1856), the first collector to be sent out by the Horticultural Society. The eldest son in a Scottish nurseryman's family of fifteen, Don came to London as a youth, and by the time of his engagement by the Horticultural Society had risen to be foreman at the Chelsea Physic Garden. He sailed on HMS *Iphegenia*, whose commander, Captain Sabine, was the brother of the Society's secretary – the same Captain Sabine who afterwards coached David Douglas in surveying. Don stayed with the *Iphegenia* for the whole of the cruise, and his itinerary was limited to that of the ship; fortunately her duties detained her for long periods in certain waters, and he was able to have nearly two months ashore in Sierra Leone, and over a week on the Portuguese Isle of St Thomas.

The start was much delayed by adverse winds in the Channel, but they left Plymouth at last on 4 January 1822, and after the usual short visits to Madeira, Teneriffe, the Cape Verde Islands and Gambia, reached Freetown on 18 February, where Don was lodged by the Governor at his farm, about a mile out of the town. Despite an attack of fever – the first of several – he despatched a consignment of plants ten days later, and another on 19 March, spending long laborious days exploring the vicinity and making local excursions. Early in April he spent a few days at York, and then returned to Freetown to pack and send home six cases of living plants, bulbs, seeds, dried specimens and insects before the *Iphegenia* sailed again on the 11th of the month. The next port of call was Cape Coast Castle, which had been Brass's headquarters; and after touching at Accra, Little Popo and Whyda they made St Thomas's Island at the beginning of June. The Portuguese authorities were obstructive, but Don managed to make a number of explorations before his ship sailed again on 11 June, but unfortunately all the tropical plants collected on this island perished in the cold weather encountered later in the voyage. After leaving the African

[1] and [2] Smith *Correspondence*
[3] Tuckey *Narrative of an Expedition* . . . (appendix)

coast the *Iphegenia* visited South America, the West Indies, and finally (in December) New York, and returned to England in February 1823. In the following year Don published a paper in the *Edinburgh Philosophical Journal* on some new species of plants found in Sierra Leone – whereupon the Horticultural Society sued him for breach of contract, although they did not themselves make any part of his notes or journals public.

In 1854 the Austrian **Friedrich M. J. Welwitsch** (1806–1872) expressed surprise that the English, so long established at Sierra Leone, knew so little about the adjoining provinces. He himself had arrived the previous October at St Paul de Loanda, and had begun an exploration of Angola that was to last for seven and a half years. He was not a young man when he came to Africa, and for the previous fourteen years he had directed various botanic gardens in his adopted country of Portugal; but he was full of energy and enthusiasm, and within seven months had investigated the coastlands from Kwansa to Ambriz. He had an allowance from the Portuguese government of £45 a month, but found this quite inadequate to finance continual expeditions where all baggage – food, water, plant-presses, paper, beds, cooking-utensils and articles for barter – had to be carried on the heads of negro porters. To supplement his income Welwitsch sent collections to London for sale – living plants, seeds, insects and herbarium specimens. After about a year in Loanda, he moved inland to Golungo Alto, and there met David Livingstone, who dissuaded him from a projected trans-continental journey to Portuguese East Africa – no undertaking for a middle-aged man who had already suffered severely from fever and scurvy. In October 1856 Welwitsch transferred to a new base at Pungo Adongo; and after a further year in Loanda working on his discoveries, he went south down the coast (in June 1859) to Benguela, and by sea to Mossamedes, where his health was much improved by the better climate. He was blockaded for two months at Lopollo in the district of Huilla by a native war, and returned to Lisbon in January 1861. Finding it necessary to compare his specimens with those in British collections he obtained permission to come to London in 1863, and lived there for the remaining nine years of his life. He will always be remembered for his discovery near Cape Negro (Mossamedes) of the extraordinary *Welwitschia bainesii*[1] (=*mirabilis*), a unique xerophytic plant of a very large size, akin to no other member of the vegetable kingdom.

Indeed, some of Flora's most aberrant products are natives of central Africa – witness the giant lobelias and senecios of the equatorial mountains. They cannot be called easy, or even very desirable, garden plants, but several of the lobelias were among plants successfully introduced by the second British Museum expedition to Uganda in 1934–5.[2] The aims of this mission were purely scientific, but its members included two notable horticulturists – Dr George Taylor, later Director of Kew, and Patrick M. Synge, now a prominent member of the Royal Horticultural Society and editor of its *Dictionary* and *Journal*. Both had previously been on plant-collecting journeys, Taylor in South Africa in 1927 and Synge in

[1] Thomas Baines was a draughtsman who sent Hooker a drawing of the plant
[2] The first was led by Dr A. F. R. Wollaston in 1906

Sarawak (where he nearly died of septicaemia from poisoned leech-bites) in 1932. The principal object of the 1934 expedition was the exploration of the Ruwenzori range, the Mountains of the Moon, which rise to 16,794 ft on almost the very equator; but Mt Elgon, the Aberdare mountains and the Birunga or Mfumbiro volcanoes were also investigated – each with its isolated, distinctive yet related flora. Plants from such altitudes might be expected to prove almost, if not quite, hardy, but their cultivation may present other difficulties, and few of the species introduced or re-introduced seem to have survived, though some of them – *Delphinium macrocentron, Hypericum becquartii*, and *Impatiens elegantissima* – would be worth every effort to preserve. Two greenhouse plants, *Canarina emenii* and *Choananthus cyrtanthiflorus*, both of which received horticultural awards of merit, are still in cultivation.

South Africa

The exploration of equatorial Africa, though exciting to the geographer and to the botanist, resulted in few permanent contributions to northern gardens; for these we must turn to the extreme south. Much of North Africa is desert; the tropics are not particularly rich in species, as tropics go; but the flora of the Cape is one of the most varied and ornamental in the world, as though the flowers that should have been dispersed over the whole continent had all filtered down to its southernmost tip. The city of Cape Town itself began with a garden – the vegetable-ground laid out by the Dutch East India Company to provide supplies for its vessels, and the fort built to protect it. This garden was founded in 1652, but remained for a hundred years a place of strict utility, with few pretensions to ornament or scientific interest. Long before this, however, plants from the Cape had been brought to Europe; Parkinson grew several before 1629, including the blue agapanthus and an ornithogalum (*O. aethiopica*) which was 'gathered by some Hollanders on the West side of the Cape of Good Hope'. Unfortunately the names of the introducers are seldom recorded.

The first professional botanist to collect at the Cape was Paul Hermann (1646–1695), subsequently Professor of Botany at Leyden, who visited the colony on his way to the East in 1672 and again on the homeward voyage in 1680. His pupil Hendrich Bernard Oldenland, a Dane, became superintendent of the Cape Town garden, and in 1689 accompanied a cattle-buying expedition to the Inqua Hottentots; he sent many specimens of South African plants to Europe, but died suddenly in 1697. His master-gardener was Jan Hartog; he, too, made journeys into the interior in 1699, 1705 and 1707, from the first of which, at least, he brought back seeds, bulbs and plants as well as specimens. After this, little seems to have been done until Auge initiated the great era of systematic collecting.

Johan Andreas Auge (1711–1805) was a German gardener who 'from an irresistible propensity to the study of plants'[1] went to Holland as a young man and became the pupil of the great Boerhaave. In 1747 he was ap-

[1] Lichtenstein *Travels in Southern Africa* 1812

pointed assistant-gardener at Cape Town, where in due course he rose to be superintendent, and was encouraged by the botanically-minded Governor Rijk Tulbagh to make the establishment into a true botanic garden. Thunberg tells us that by 1772 Auge had made eighteen plant-collecting journeys into the interior; but the particulars of his travels do not seem to have been preserved, except for a trip from July 1761 to April 1762, when he accompanied Captain Hop's expedition to Namaqualand, reaching a point near the Great Fish River, about 120 miles to the northward. He made collections for sale, and supplied Governor Tulbagh with the specimens of plants and insects that he sent to Linnaeus and others. Thunberg, with the usual lofty scorn of the botanist for the mere gardener, said that 'Mr. Auge's knowledge of botany was not very considerable, nor did his collections in general extend much further than the great and the beautiful' – but he praises the Company's 'extensive and beautiful' garden. Auge retired to a country farm in 1778 owing to failing sight; for the last twenty years of his life he was totally blind, but up to his death at the age of ninety-four he was still keenly interested in botany and gardening. Meantime, in 1772, he had acted as guide to Thunberg's first long-range expedition.

The Swedish **Carl Peter Thunberg** (1743–1828) – a pupil of Linnaeus – had left Uppsala in August 1770 with a medical degree and a travelling scholarship which amounted to about £45 over three years, intending to make further studies in Paris. On the way he stopped at Amsterdam, where he met Professor Burmann; and (by Thunberg's own account) the Professor was so impressed by his botanical skill that he offered to procure him an official appointment to go either to the Cape or to Surinam – an offer which Thunberg gladly accepted. He then resumed his delayed journey to Paris and spent the winter in a whirl of lectures and sightseeing, returning to Amsterdam in the following August. In the meantime Burmann had conceived a plan of sending him to Japan, and had been collecting subscriptions from wealthy garden-lovers for the purpose. 'And as no nation, except the Dutch, is suffered to come into Japan,' Thunberg explains, 'it was necessary for me to understand Dutch and to speak it; to obtain this I requested to be permitted previously to pass a couple of years at the Cape of Good Hope, and to be taken into the service of the Dutch East India Company.'

Thunberg sailed on 30 December 1771, nominally (and appropriately) as 'surgeon extraordinary' to the Dutch East Indiaman *Schoonzist*, and arrived at Cape Town on 16 April 1772, still shaky from the effects of having eaten on board pancakes made half with flour and half with white lead, carelessly given out by mistake by the ship's caterer. His colleague and friend **Andreas Sparrman** (1748–1820) had arrived four days before, as tutor to the Resident's children; and the two were able to do some botanizing together, so far as the winter season would permit. In August, when the spring flowers began, Thunberg made preparations for a long excursion with two companions – a youth, Immelman, and a sergeant, Leonhardi, who came for the shooting – with Auge as a 'sure and faithful guide'.

'Long' is a relative term, for though the trip occupied four laborious months, the actual distance travelled was not great[1] – the farthest point reached being the Gamtoos River, about 400 miles away as the crow flies. There were no roads, only tracks from farm to farm, clear enough near the towns but in remote areas barely visible. The only means of baggage-transport was the ox-waggon, heavy, cumbrous, and even under the best conditions incapable of a greater speed than eight miles an hour. Lions were sometimes uncomfortably plentiful, water frequently short; it was no picnic on which they embarked on 7 September 1772, with a saddle-horse apiece, a cart and three yoke of oxen, and two Hottentots to drive them.

They began by making a big detour northwards to Saldanha Bay and the Berg River, and across the mountains at Rood Zand before circling back to Zwellendam and continuing their journey to the east. On 29 October they arranged to divide forces, sending young Immelman with the waggon to a rendezvous at Lange Kloof, while the rest of the party, guided by Auge, made a detour on horseback through the wooded Outniquas mountains. On 3 November the party was attacked by a rogue buffalo in a thicket. The leader, Auge, managed to elude the animal and get behind a tree, but the sergeant's horse was charged and disembowelled. Thunberg, lagging behind to botanize, was unaware of the danger till he emerged from the thicket, but fortunately the buffalo's attention was distracted from him by the sergeant's baggage-horse, which he also attacked and killed, thus giving Thunberg time to leave his own mount and climb a tree. Satisfied, the buffalo retreated, and Thunberg was able to descend and seek his companions, whom he found 'sitting fast, like two cats, on the trunk of a tree with their guns on their backs, loaded with fine shot, and unable to utter a single word . . . the sergeant at last burst into tears, deploring the loss of his two spirited steeds; but the gardener was so strongly affected, that he could scarcely speak for some days after'. (It should be remembered that Auge was then sixty-one years old.) This incident caused some discouragement and delay, and the sergeant had to ride an ox for a day or two, till another horse could be procured.

They rejoined Immelman and the waggon on 11 November, and reached their farthest point, in Kaffir country, without further incident. At the beginning of December they began the return journey by a shorter route, and were back at Cape Town by 2 January 1773. Thunberg sent a collection of seeds, plant-specimens, insects, bulbs (including those of the Bird of Paradise flower, *Strelitzia reginae*) and trees, to the botanic gardens of Leyden and Amsterdam, and some also to friends in Sweden.

By this time the Cape had become a great resort for botanists, and officers of visiting ships purchased collections much as tourists today buy souvenirs. During Thunberg's absence a new collector had arrived from England – **Francis Masson** (1741–1806).

Nothing is known of his early life except that he was born at Aberdeen and employed as a gardener at Kew. On returning from his

[1] It could now be covered in two days by car or train, or two hours by plane

voyage with Captain Cook, Banks had suggested to the King that a collector should be sent to the Cape, and Masson was appointed, according to his own modest account, because he was the only applicant for the post; that he was eminently well-qualified for it, may be seen by the success of his subsequent career. He was the first collector officially to be sent out from Kew; his 'recompence' was to be £100 a year, payable on his return,[1] and an expense-allowance of up to £200 yearly. He was given a passage with Captain Cook, who was about to set out on his second voyage of exploration. Forster, the naturalist of the expedition, referred to Masson slightingly as a 'Scots garden-hand', but Forster was a self-important, prickly, tiresome character who antagonized everyone on board.

The *Endeavour* anchored at Cape Town on 30 October 1772, after a six-month passage, and sailed again on 22 November, taking with her young Sparrman, whom Forster had pressed into service (with Cook's consent) to help with the natural history of the voyage. On 10 December Masson set out on a two-month excursion, and was away when Thunberg returned from 'Caffraria' on 2 January. On this occasion Masson took with him a Dane called Oldenburg as interpreter and guide. This preliminary trip, a curtain-raiser as it were to his later travels, did not take him far afield: – Stellenbosch, the Hottentot Holland mountains, the warm bath at Zwart Berg, the Company's woods at the River Zonder Einde, then across the Breede River to Swellendam, and back by the same route.

It is not known how Masson and Thunberg met – possibly through the agency of Major (later Colonel) Gordon, a Dutch officer who made a number of journeys of exploration in South Africa; he was interested in natural history and the friend of several botanists. At any rate, it was in company with Gordon that Masson and Thunberg made a six-day excursion (13–19 May 1773) on foot among the mountains between Cape Town and False Bay. In what language did they converse? Thunberg may have known some English; probably by this time both were fairly fluent in Dutch, and they could eke out with botanical Latin. Evidently they got on sufficiently well to contemplate making a long and difficult journey together, yet they were very different in character. Masson was the elder by two years, burly in build, reserved, possibly a little slow, cautious, generous, modest and good-humoured; Thunberg was small, impetuous, active and enormously conceited, but with an artlessness that made even his egotism endearing. They were united, however, in enthusiasm for their work, and the economic benefits of a shared journey probably carried a great deal of weight.

Thunberg at this time was in considerable financial difficulties. He had a small medical practice in Cape Town which he did not wish to enlarge as he had to keep free for his botanical excursions; but he had received no supplies from Holland since his arrival, and the two governors to whom he had letters had both died – Tulbagh just before Thunberg reached Cape Town, and van Oudshoorn on his voyage out. Moreover, Thun-

[1] Presumably if he did not return, the Government would be the gainer, as he had no dependants

berg's ship had sailed without her muster-roll, which meant that no one on board could receive his pay, or leave to go home, for two or three years. Fortunately he was helped by M. Berg, secretary of the police, who lent him enough money to purchase a new strong waggon, and medicines to distribute in the interior. Masson, with his princely £200 a year, was 'well equipped with a large and strong waggon, tilted with sailcloth, which was driven by a European servant upon whom he could depend'. They had a saddle-horse apiece, a suitable number of oxen, and four Hottentots.

Both Masson and Thunberg have left accounts of this journey,[1] which covered much the same ground as Thunberg's previous trip, but extended somewhat further, both northward and eastward. They started on 11 September 1773, and proceeded to Saldanha Bay, which they crossed by boat; they then followed the coast to St Helen's Bay and the mouth of the Berg River. The river was in spate and they had to go four days' journey upstream to find a ferry, crossing on 7 October. Then over the hills in heavy rain to Olyfant's river, which they followed for four days; but in attempting to ascend the mountains a waggon was overturned and a shaft broken. 'After some warm debates' it was decided to send the waggons back to Rood Zand for repair, while Thunberg and Masson made a detour on horseback, first by the Coude Bokkeveld, where the country 'upon the whole, has a most melancholy effect on the mind', and then the Warm Bokkeveld, less barren but still surrounded by 'horrid mountains'. From here the only way down to Rood Zand was by the dangerous pass of Mosten's Hoek, where, says Masson, they 'looked down with horror on the river, which formed several cataracts inconceivably wild and romantic', and which they had to cross four times by fords made dangerous by large stones – 'but we thought our labour and difficulties largely repaid by the number of rare plants we found here'. They got safely to Rood Zand and the waggons, and spent a day climbing the nearby mountain of Winterhoek, hoping for hardy plants from its cold altitudes, but found nothing on the summit save a few grasses. Thunberg, however, discovered *Disa caerulea* in a hollow in the mountains, into which he took a short cut by jumping, he tells us, some 20–24 yards down, his fall being safely broken by bushes. It was near Rood Zand that they found the beautiful, now rare, *Ixia viridiflora*.

They followed the Breede River down to Swellendam, where they stopped for five days to rest the oxen. They resumed their journey on 11 November, and shortly afterwards met with a mishap in the flooded Duyvenhoek's River, when, as Masson says, 'The Dr imprudently took the ford without the least enquiry, when on a sudden he and his horse plunged over head and ears into a pit' – from which he was saved by the 'strong exertions' of the horse. Thunberg's version of the incident was rather different. 'I, who was the most courageous of the company,' he explains, and consequently always in the lead, had the misfortune to plunge into a deep hippopotamus-wallow, which might have proved

[1] Thunberg in his four-volume *Travels*, and Masson in the *Philosophical Transactions* LXVI 1776

fatal 'if I, who have always had the good fortune to possess myself in the greatest dangers, had not with the greatest calm and composure, guided the animal . . . and kept myself fast in the saddle.' (Thunberg strongly resembled the engaging Mr Toad.) Quite undeterred by this misadventure, and by another ducking when a Hottentot drove the waggon with Thunberg inside, through the deep instead of the shallow part of a muddy river, we find the incorrigible botanist on 1 December wading in the shallow estuary of the Zeeko or Sea-Cow River 'as well for the sake of bathing, as of collecting insects and shrubs that grew on the banks, with nothing but a handkerchief about my waist, not suspecting that the sunbeams would have any bad effect upon me'. He got so badly sunburned that he was confined to bed for several days, and the journey was not resumed till 9 December. Fortunately they were then enjoying the hospitality of a settler, Jacob Kock, who supplied them with provisions and fresh oxen. On 10 December they crossed the Gamtoos River, which formed the boundary of Cape Colony – settlers not being allowed in the Kaffir region beyond. Across the river was a fine country of woods and groves, which, Masson said, 'could not but charm us, who for upwards of three months had been climbing rugged mountains and crossing sultry desarts', and the local Hottentots, though wild in appearance, 'behaved very obligingly, and slept at our fire all night'. Six days later, they reached Sunday River, and their guides and servants refused to go any farther for fear of the Kaffirs; their waggons were in need of repair and some of their oxen sick with the hoof-distemper, so they returned again to the Zeeko River and the hospitality of Jacob Kock. On the way home, Thunberg and Masson planned to make a detour on horseback over the barren plains of the Karroo, but they got lost and had to spend a night in the open, by a brook where there was still some water, without food except for some biscuits and sugar-candy that Thunberg had put into his shooting-bag. When, in the morning, they found that their horses had strayed, the position began to look serious; but the animals were recovered and the travellers eventually found their way back to the waggons. By 29 January 1774 they were back at Cape Town, ready to pack and despatch their extensive collections.

During the autumn and winter the botanists kept in touch, for both accompanied Lady Anne Monson (who was on a visit to the Cape) to outlying farms, and contributed to her natural history collections. They probably made short excursions, together or apart, for Thunberg says that he climbed Table Mountain fifteen times in the three years of his stay, and it was on Table Mountain that Masson found *Nerine sarniensis*, previously thought to be a native of Japan. In the spring they set out on another extensive journey, although Masson, as Thunberg reported, 'was not much inclined to make any long excursions this year'. Could it be that Masson was finding the company of the volatile Swede a little wearisome?

This time they ventured into new country, where neither of them had been before. Masson left Cape Town on 26 September 1774, and joined Thunberg, who had preceded him, at Paarle Kerk, whence they climbed Paarle Berg and Paard Berg. On 8 October they reached Riebeck Castel, and left the waggons and oxen below while they made a wide circuit to get to the summit of the mountain, immediately above the waggons but separated by seemingly impassable precipices. While searching for plants Thunberg 'stumbled upon a very near but at the same time dangerous way. . . . This was a chink of a few fathoms length, and so narrow as to be capable of admitting a middle-sized man only. Through this I ventured to crawl on my hands and feet, and was fortunate enough to get safe over to the other side, from whence it was only a musquet shot to our waggons. My fellow-traveller together with his dog, stood astonished at my adventurous exploit, the one howling, the other almost crying, and at the same time vexed to think he should be obliged to go alone the long way about, without once daring to take the direct path. My courage was rewarded with a small plant which I got in the chink, and which I afterwards sought in vain in other places'. Needless to say Masson does not mention this incident; he says only that they here collected 'many remarkable new plants, in particular a hyacinth with flowers of a pale gold colour'.

They forded the Berg River with difficulty and went on to Piquet Berg, then followed the coast from the mouth of the Verloore River to the mouth of Oliphant's River, finding the going heavy on account of loose sand. After a few days' rest they struck inland, crossed the river, and embarked on the passage of part of the Karroo. Here they found 'great treasure' of new succulents, but dared not stop to botanize; to save the lives of their oxen they were obliged to press on to the next water, nearly three days' journey away. Even so, they picked up 100 new species by the wayside, and on 1 November found *Aloe dichotoma*. Then they went northwards along the plateau, and on the tenth reached the last Dutch habitation, about 350 miles north of the Cape. After another rest they made south-east through uninhabited country to Rhinoceros River and the Roggeveld mountains. It was cold among the mountains, with frost at night, and they missed the ericas and proteas to which they had become accustomed, but such plants as they found were new. The waggons were much shaken by the rough tracks and the horses and oxen sore-footed, so on 3 December they turned homewards, and made the steep and difficult two and a half hour descent to the Karroo. After four hot and thirsty days they reached the Bokkeveld Mountains and a spring of pure cold water, where they 'spent the remainder of the night and part of next day in great luxury' before returning by way of Hexen River, Breede River and Rood Zand to Cape Town on 28 December.

Two months later they parted, Thunberg sailing on 2 March 1775 for Java and Japan, and Masson shortly afterwards returning to England. Each of them afterwards ranged far afield, but for each the Cape provided his first experience of collecting and established his reputation. Masson's methods of packing and sending home his seeds and plants were evidently highly successful; he wrote to Linnaeus that he had been the means of adding 'upwards of 500 new species to His Majesty's collection of living plants'. 'I am confident that the famous journey to the Levant made by Monsr. Tournefort . . . at an enormous expence,' Banks wrote later in a memorandum to the King, 'did not produce so great an addition of Plants to the Paris Gardens as Mr. Masson's Voyage to the Cape only, has done to that of Kew. . . . His Majesty's appointment of Mr. Masson is to be accounted among the few Royal bounties which have not been in any degree misapplied.' In *Hortus Kewensis* Masson's name recurs monotonously on almost every page concerned with Cape genera – 50 pelargoniums, nearly 70 mesembryanthemums, 88 heaths, 48 oxalis, 40 stapelias, 16 diosmas, 11 lobelias, 12 arctotis, and many others (some the products of his second visit) and if some of his introductions proved short-lived, this was hardly felt, 'so abundant and repeated were his supplies'.

After barely a year in England, Masson sailed for the Azores, Canaries and West Indies, and it was ten years before he returned to the Cape; in the interval, much had changed. In 1772, 'a French officer, though dressed to the best advantage, and frequently wearing a star on his breast . . . had but little respect paid him, whereas an English mate of a ship, with his hair about his ears, was much esteemed on account of his being flush with money, and his nation's being in alliance with Holland'.[1] But the War of American Independence intervened, and in 1780 Holland joined France and Spain against Britain; and although peace was restored in 1783, the Dutch East India Company ordained that no stranger should be allowed beyond a forty-mile radius from Cape Town. Masson, in particular, was warned not to 'excite the jealousy of the inhabitants' which had been aroused by the activities of a mysterious person called Patterson,[2] supposed to be collecting for Lord Strathmore. Thunberg met this man on his return from the East in May 1788, and says 'he professed to travel at the expence of certain individuals, and possessed some small knowledge of Botany, but was in fact a mere Gardener', and Forster[3] reported to Banks in January 1786 that 'Mr. Patterson made an ill use of the liberty that was given him, and an ungenerous return of the great kindness that was shown him'. The Dutch suspected him of spying, and even Masson came under the same aspersion.

Nevertheless, Masson was kindly received by the Governor when he arrived on 10 January 1786; he obtained permission from the Council to go about his business – within limits – and despatched sixty packets of seeds to Banks within ten days of his arrival. In March he was allowed to make a five-day visit to the Hottentot Holland mountains, and sent Banks a further consignment of seeds from there – which was little appreciated. 'These expensive journeys up the country', Banks wrote in June 1787, 'seldom produce an adequate return in really ripe seeds . . . one rare described Plant is worth two nondescripts.' It would be far better, Banks thought, if Masson would keep to one or two rich regions – False Bay and Hart Bay are mentioned – and work them thoroughly; till these are exhausted 'I trust you will remain quiet; afterwards you may propose excursions'.

In 1788 therefore, Masson dutifully made two trips to False Bay; but

[1] Thunberg *Travels*
[2] Not to be confused with Colonel Paterson, at the Cape from 1777–81, and afterwards Governor of New South Wales
[3] Mr. George Forster of the Civil Service, not the Forster who sailed with Cook

Banks could no more control his collectors than a hen a brood of ducklings,[1] and despite his reproofs and the restrictions nominally in force, Masson's accounts at Kew reveal that he made a 200-mile trip to Oliphant's River the same year, and another, twice as long, north-west to Kamiesberg in the year following, while in 1790 he visited Kleine Roggeveld and the Karroo. Unfortunately Masson's second sojourn at the Cape is very scantily documented and we have no details of these journeys. The flora had lost much of its novelty and there was little inducement for Masson to write about his travels; and there was no Thunberg to report on their common adventures.

At Banks' suggestion Masson established a garden in which he could keep his plants until an opportunity arose for sending them home; it was about two miles from the town in 'a small enclosed recess under the Table Mountain', and Masson paid a man to look after it during his absences. We get occasional glimpses of the garden and its owner through the eyes of botanists who visited the Cape from time to time. Hove met him when on his way home from India in May 1789, and the two tried to climb Table Mountain together, but were defeated by bad weather. Masson gave Hove some seeds to take home, and wanted him to take plants also, but Hove had already lost many of his own plants through shortage of fresh water on board ship, and dared not take charge of any more, as 'the Captain was very particular concerning it'. In March 1793 James Main called on his way out to China, and had a 'most interesting interview' with Masson; there was all the news from England to be told, and Masson's latest acquisitions to be admired, including the new *Nymphaea caerulea* and many stapelias and heaths.

That year a new war broke out in Europe – the French Revolutionary War. French troops overran Holland; the Prince of Orange took refuge in England, and authorized a British expedition to forestall any French attempt to capture the Cape. But the Cape was not willing to be taken over by the British, and there was some local resistance; Masson feared to lose his collections, and asked to be recalled. He left early in 1795, and the colony capitulated to Britain in September of the same year. A scandalous report subsequently arose, that Messrs Lee and Kennedy were regularly supplied by Col. Gordon with seeds collected by Masson when at the Cape in the service of Kew. This story was sent to Banks by Viscount Valencia in December 1802 – seven years after Masson had left the Cape and when he was far away in Canada, and five years after Colonel Gordon's death by suicide in February 1797. The older James Lee died in 1795, so there was nobody to refute the allegation. Admittedly Masson was on intimate terms with the Lee family, with whom he lodged when living in London between journeys, but he seems always to have been punctilious over money-matters; he never exceeded his expense allowance, and when Main had tried to buy some of his seeds 'for a small packet of which we offered a bag of dollars . . . he was honourably proof against the temptation'.[2] Banks, to do him justice, seems to have ignored the report, which was probably due to misunderstanding or malice.

[1] Hove and Kerr gave him similar trouble [2] Main *Reminiscences*

In England, Masson was given a year's well-earned leave, to work on his book *Stapeliae Novae*, which was published in two parts, in 1796 and 1797; it was illustrated by forty-one plates drawn 'in their native climate', and many, though not necessarily all, by Masson himself. He was now fifty-six, and had been a collector for twenty-five years; he knew no other way of life. There were no pension-schemes in those days, and in the foreword to his book (which was dedicated to the King) Masson shows a pathetic anxiety about his future. He expresses himself as 'anxious to recommend my employment as a collector, and still enjoying, though in the afternoon of life, a reasonable share of health and vigour, I am now ready to proceed to any part of the globe to which your Majesty's commands shall direct me . . . all of them are equal to my choice'. The unfortunate result was a decision to send him to Canada, where the climate must have proved a severe trial after a quarter of a century in warm regions.

Thunberg, who was a prolific writer of inaccurate and now discredited botanical works, has been called the Father of South African Botany. Masson was a man of action, not a writer, and except for the stapelias left the description and publication of his plants to others; but the number of his discoveries and introductions must far have exceeded Thunberg's. His work had a tremendous influence on British horticulture, especially in stimulating the fashions for Cape pelargoniums and heaths.

Other collectors besides Masson were working in South Africa at this time. In 1786 the Emperor Joseph II sent two gardeners, **Franz Boos** and **Georg Scholl**, to obtain plants for the Imperial Botanic Garden of Schönbrunn near Vienna, which Banks called 'The only rival to Kew I am acquainted with'. Boos was the senior and superior, and had returned not long before from a collecting-trip to North America and the Bahamas. They arrived at Cape Town six months after Masson, on 1 June 1786, and Boos left on 17 February 1787 for Mauritius and Bourbon, the main object of his whole journey being to bring home a collection of plants assembled for the Emperor by M. Céré, the director of the Mauritius botanic garden. He was back at the Cape on 20 January 1788 with 280 cases of tropical plants, and possibly some live animals. He was unable to find a ship that could take so many cases, and when he sailed to Europe on 5 February he left some of the plants behind in charge of Scholl, to wait for another opportunity. Scholl's repatriation was delayed for more than ten years. During this time he travelled extensively, sometimes with Gordon or Masson, sometimes alone; and sent to Vienna consignments of bulbs and seeds, but not of living plants.

In 1791 an attempt was made to send out two more gardeners from Schönbrunn, Bredemeyer and the younger van der Schot; they were to collect another consignment of plants from M. Céré and to pick up Scholl on the way back – but they never got so far. The captain of the ship in which they had embarked 'made sail for Malaga, where they discovered in good time that his intentions regarding them were not of the fairest. This circumstance obliged them to return to Vienna without having been able to fulfil their commission'.[1] (To be sold as a slave to the Moors was a

[1] Koenig and Sims *Annals of Botany* 1806

possibility that might daunt even the most intrepid botanist.) Scholl got safely home in 1799 with a large cargo of living plants, some of which, however, were severely damaged on the voyage. It is believed to have been Boos or Scholl who introduced to Schönbrunn another of Africa's extraordinary plants, *Fockea capensis*,[1] whose large water-storing roots may attain six feet in diameter and weigh over 50 lbs. The specimen grown in the Imperial Garden before 1800 was still alive and not noticeably increased in size in 1935.

Unfortunately, little seems to have been recorded of the activities of another collector, **James Niven** (1774?–1827), who arrived at the Cape in 1798, three years after Masson's departure. Niven was a Scot who came south about 1796 and was employed by the Duke of Northumberland at Sion House, till engaged by the wealthy amateur George Hibbert of Clapham to collect in South Africa. His first visit lasted five years, during which he mastered the Dutch and Kaffir languages, and sent home many new plants and a 'valuable herbarium of native specimens'. He returned home in 1803, but within three months was re-engaged by a syndicate whose members included the firm of Lee and Kennedy and the Empress Josephine – regardless of the fact that the Napoleonic Wars, temporarily concluded by the Peace of Amiens the previous year, had broken out again. This time Niven stayed at the Cape for nine years, but 'had difficulties to encounter and embarrassments to contend with which quite deranged his intended excursions into the interior'. The colony, restored to Holland in 1802, was again taken by the British in 1805; but there followed wars with the Kaffir nations on the borders, and the experienced Niven was pressed into service as interpreter and guide, having to submit to all the inconveniences and privations of a soldier's life and receiving in return many encomiums, but few emoluments. On his return to England in 1812 he gave up botany and gardening and went into business with his brother in his native Pennicuik, where he died in 1827, aged about fifty-two. Loudon, to whose obituary notice in the *Gardener's Magazine* we owe most of the foregoing particulars, says that his wife 'died at the instant her husband's corpse left the door of the house, leaving five orphans!'; but the pathos of this incident is diminished by the reflection that one of the orphans (Ninian, b. 1799) was then twenty-eight years old.

Niven seems to have been an industrious and efficient collector – as well as being 'a most affable and friendly-hearted man' – and served his employers well. Hibbert was particularly interested in proteas, of which he possessed the most extensive collection of living plants that had ever been formed; Niven must have been responsible for most of these, besides adding five new species to the genus. The Empress Josephine was keen on heaths; her collection at Malmaison rose from fifty species in 1805 to 132 by 1810. Niven found thirty ericas not previously known to science (including *elegans*, *pellucida*, *blandfordiana*, *hibbertii* and *nivenii*) which, considering the thoroughness with which Masson had already combed the area, was no inconsiderable feat. His miscellaneous introductions included

[1] It was not found again in the wild till 1906

William Burchell trading from his waggon. From Burchell *Travels in the Interior of Southern Africa* 1822 *By courtesy of the British Museum*

Gazania pavonia, *Lobelia gracilis*, *Moraea longiflora*, and *Nivenia corymbosa*, named by Robert Brown in 1809 in honour of the 'intelligent observer and indefatigable collector'.

It seems that Niven did not explore the more arid parts of the country, for his introductions (as recorded in *Hortus Kewensis*) did not include a single new mesembryanthemum or other succulent. This omission was remedied by **William John Burchell** (1781–1863) one of South Africa's most remarkable travellers. He was the son of a nurseryman of Fulham, and lived for five years on St Helena, first as merchant, then as schoolmaster and finally as official botanist; but he experienced various frustrations, became restless, and, giving up his post, sailed for the Cape on 16 October 1810. As an explorer, Burchell was a veritable one-man-band, combining geographer, ethnographer, zoologist, artist and botanist in one protean individual. He travelled at his own cost and from sheer curiosity, thinly disguised as zeal for the advancement of science; playing the flute in the evening to the appreciative Hottentots, and trading from a waggon flying the British flag.

Burchell landed at Cape Town on 26 November, and set himself at once to learn the language. Although it was now the dry season, he was enraptured by the flora. 'I could not for some time divest myself of feelings of regret that at every step my foot crushed some beautiful plant; for it is not easy, at one's first walks in this country, to lay aside a kind of respect with which one is accustomed, in Europe, to treat the *Proteas*, the *Ericas*, the *Pelargoniums*, the *Chironias*, the *Royenas*, etc.' Like Masson, he made a short preliminary tour the following April to the Hottentot Holland Mountains, Zwart Berg, Brand Vallei, Tulbagh, Paarl and Stellenbosch; and on 19 January 1811, in company with a party of missionaries, set off on a journey that was to last nearly four years. His waggon was so heavily loaded (his equipment included a library of fifty books) that he had to add a second lighter vehicle, but he refused the offer of a European companion, fearing 'the disagreeable, and often fatal, consequences arising from the want of harmony between the members of such an expedition' – a decision he afterwards regretted.

For three and a half months the party travelled NNE; once past the Middel Roggeveld the country was uninhabited and almost trackless, but they were following a regular route to a well-established mission-station, and at various stages of the journey were joined by other waggons, to the number eventually of eighteen. At Zand Vlei in the Prieska district Burchell picked up what he thought was a pebble, but which turned out to be a stone-mimicking mesembryanthemum – *M. turbiniforme*, on which the Lithops section of the genus was afterwards founded. On 15 September they reached and crossed the Gariep or Orange River (where he discovered South Africa's only poppy, *P. aculeatum*) and on the 30th reached the caravan's destination, the mission-station of Klaarwater. Here Burchell decided to make a long stay, while his oxen were rested and his waggon repaired; meantime he joined a hippo-hunting party and was the first European to explore the junctions of the Riet and Vaal and of the Vaal and Orange rivers.

At Klaarwater he was unable to find any Hottentots willing to travel farther into the interior, and the missionaries refused to help him, as they regarded any such journey as frankly suicidal. Against their wishes Burchell decided to make an equally dangerous trip south-east to Graaf Reinet in search of men and supplies. He left with a party of eight on 24 February 1812, and steered by compass over entirely unknown Bushman country, among tribes that were reputed dangerous, but which he found only oppressed and pathetically poor. After some vicissitudes and hardships he reached the town safely on 25 March, but labour was at a premium and he was only able to recruit the sweepings of the jails. He started the return journey on 28 April, escorted for the first part of the way by a party of friends, in a carriage with six fast horses; but he disliked the unaccustomed speed, preferring his plodding oxen 'whose steady pace seemed to have been measured exactly to suit an observer and admirer of nature'. He was back at Klaarwater on 24 May, but received no welcome from the missionaries, chagrined because their prophecies of doom had not been fulfilled. He was glad to get there – and equally glad to leave.

Burchell resumed his journey northwards on 6 June 1812, through country unknown to Europeans; and on 13 July reached the considerable native town of Litákun. Up to this point we have his own detailed narrative of his travels in two handsome and very readable volumes which tell everything needful, including the best way to cook ostrich eggs; but here his story stops, and no subsequent volumes were ever published. It is known that he penetrated into what was later British Bechuanaland, reaching Chue Lake (Honing Vlei) in the country between Kuruman and Vreiburg on 4 October. He then returned to Klaarwater, and left it again in February 1813 by a different route for Graaf Reinet; from there he continued to Grahamstown and the Great Fish River, and thence returned to Cape Town along the coast. He sailed for England in a small trading-vessel on 25 August 1815 – with a cargo almost sufficient to freight her; forty-eight packages containing 40,371 beautifully-prepared plant-specimens, all numbered to correspond with notes in his carefully-compiled *Catalogus Geographicus*; nearly 500 drawings (not all botanical), more than 2000 species of seeds and 276 bulbs, which he cultivated on his return in his father's garden at Fulham. Several were subsequently illustrated in the *Botanical Register* and in Herbert's *Amaryllidaceae*, but few seem to have remained in cultivation. It is recorded that he kept the more ornamental species and gave the others away – 'at heart', says a disapproving botanist, 'he was a horticulturist'. He was generous in his gifts to those with special interests, and provided A. H. Haworth with specimens and living plants from which he was able to describe eleven new species of mesembryanthemum.

Haworth was indebted for many of the succulent plants in which he specialized to another South African collector of the period, **James Bowie** (1789?–1869). The son of a London seedsman in modest circumstances, Bowie went as gardener to Kew in 1810, and four years later was engaged by Sir Joseph Banks to collect plants at Rio de Janeiro in company with Allan Cunningham. In 1816 they were sent further orders; Cunningham was to proceed to New South Wales and Bowie to the Cape, where the Governor had been instructed to supply him with a waggon and oxen.

He arrived on 1 November 1816, and for the first eighteen months of his seven-year stay collected only in the vicinity of Cape Town; afterwards he spent approximately nine months of every year in travel. (He did not return to Cape Town between his two last excursions, but made his summer base at Uitenhage.) With one exception, all his journeys were along the coastlands to the east, the farthest points reached being Bathurst, Grahamstown[1] and the Great Fish River. His only inland journey took place in the last four months of 1821, when he went north from Algoa Bay to Graaf Reinet, and beyond to Erste Poorte in the Coleberg district.

This predilection for the coastlands may have been in accordance with his instructions, but it may also have been influenced by the friendship which sprang up between Bowie and that remarkable character George Rex, a natural son of George III, who had been persuaded to leave his country for his country's good, and had removed from Cape Town to Knysna, eastward along the coast, when the colony was restored to Holland in 1802. Bowie probably met him during his first visit to Knysna, about July 1818, and afterwards stayed with him several times, including a six-month winter sojourn in 1820 and a four-month stay in 1822. Rex was noted for riotous living and the number of his tinted descendants; and his companionship, though doubtless congenial, was perhaps not very salutary for a man whose later career was marred by his intemperate habits. It was on his estate that Bowie found *Streptocarpus rexii*; but perhaps his most important introduction from this period was a striking plant of the Quagga Flats by the Great Fish River, which he called *Imatophyllum aitonii*, though by a fortunate botanical debacle it came by the simpler official name of *Clivia nobilis*.

After the death of Sir Joseph Banks in 1820, Kew entered on a period of decline; the money allocated for plant collecting was reduced by half, and Bowie was summoned home. He left Cape Town on 20 May and arrived

[1] A new settlement, which at that time was attracting many English immigrants

in London on 5 August 1823. Up to this time his conduct seems to have been good; in Rio he was reported to have been 'very assiduous', and Hooker considered that 'Every friend to Science must regret that this indefatigable Naturalist, after sending the greatest treasures both of living and dried plants to the Royal Gardens . . . has, by a needless stretch of parsimony, been recalled'.[1] But Bowie found it hard to settle down to life in London, and is said to have spent much of his time 'among the free and easy companions of the bar-parlours, recounting apocryphal stories of his Brazilian and Cape travels, largely illustrated with big snake and wildebeeste adventures'.[2] At last he decided to return to the Cape as an independent collector, and advertised in *The Gardener's Magazine* (1826) to the effect that he would receive and execute orders for seeds, bulbs, plants and dried specimens, and that he was also willing to examine, free of charge, consignments of growing plants on ships touching at the Cape, and advise on their care. He returned to Cape Town in 1827, but his affairs did not prosper. He had no tact, perseverance, or business ability, and the competition was great; there was even 'an officer of the army who has sometimes 40 soldiers at a time told off to collect for him'. A letter that he wrote to *The Gardener's Magazine* in 1829 is querulous in tone, and complains of the backward state and 'wilful and scandalous neglect of this colony by the British Government'.

In 1835 another botanist came to the Cape – a lovable young Irishman, **William Henry Harvey** (1811–1866). He was not at liberty to travel and collect, for he succeeded his brother the following year in the post of Colonial Secretary, but he worked at his duties by day and at botany by night until 1841, when his health broke down and he had to return to Britain. (Two years later he succeeded Thomas Coulter as curator of the herbarium at Trinity College, Dublin, and in 1854–6 made a journey to Australia, in the course of which he visited the aged James Drummond.) While at Cape Town Harvey compiled his first work on the South African flora, *The Genera of South African Plants* (1838), in which he acknowledges help received from Bowie, who in the meantime (1837) had obtained the post of superintendent of the fine garden established near Cape Town by the German Baron C. F. H. von Ludwig. About 1841 Bowie gave up this position and resumed collecting; he wrote to Harvey that he was about to make a journey of 180 miles into the interior, but that it was necessary first to establish a garden-depot for the plants he sent back, with a man to look after it; to defray the expense of this, the hire of horses and the cost of tools and equipment, he had been 'giving instruction and inspection in horticulture, a science which . . . is making a decided advance in this country'. He complains of the difficulty of collecting succulents, many of the species being sparingly distributed over a great tract of country, and almost unrecognisable when shrivelled in the dry season. 'Even when closed in tight tin boxes', he wrote, 'I have known them, especially the more fleshy specimens, melt away entirely from the heat'.

In later years Bowie seems to have led an aimless irregular life, often in

[1] *Botanical Magazine* 1827 [2] *Journal of Botany* XXVII 1889

great poverty, always complaining of lack of recognition of his great services to science; 'his habits' as one biographer neatly puts it 'were such as to interfere with his prospects'.[1] At last he was charitably employed as a gardener by R. H. Arderne of Claremont in the environs of Cape Town, where he died in 1869.

Bowie was the last of the great Cape collectors. There were others, but their visits were relatively short; and the work of discovery was subsequently taken over by the colony's own resident botanists, who have been numerous and brilliant. It is a pity, however, that so little is recorded of the work of **Thomas Cooper** (1815–1913), who visited the Cape during Bowie's lifetime. The ground that he covered, and the plants he introduced, entitle him to be ranked among the most important collectors, but only the most meagre facts are available. He was engaged in 1859 by W. Wilson Saunders, who had a renowned garden at Reigate; and worked in South Africa till 1862. He made his headquarters inland at Worcester, not at Cape Town, and ranged much farther afield than any previous collector (Burchell excepted) – to Natal, the Drakensberg Mountains and the Orange Free State. His plants include the popular greenhouse Asparagus Fern (*Asparagus plumosus*) and the handsome and hardy *Galtonia candicans*; *Aloe cooperii* and *Cyrtanthus mackenii* var. *cooperi* are among the species that bear his name. He lived to be nearly ninety-eight, and by the time his obituary notices came to be written, the exploits of his collecting days seem to have been forgotten.

Information is also tantalizingly scanty about the expedition in 1927 of four gardener-botanists so distinguished that one expects all the flowers of South Africa to have bowed down to them as they passed. Three of the four appear elsewhere in this book – Collingwood Ingram, George Taylor and Laurence Johnstone of Hidcote. The fourth, Reginald Cory, had a fine garden at Dyffrin near Cardiff, and is gratefully remembered for the bequest of his considerable fortune to Cambridge University for the benefit of the Botanic Garden, and of his magnificent botanical and horticultural library to the Royal Horticultural Society.

No time was wasted, though much was spent; the two cars in which the trip was made were waiting at the quayside on the September day when the ship drew in, but it took three months to accomplish the 6000-mile journey north-east to Pretoria, with delays and diversions wherever plant-hunting seemed likely to be profitable. (One such diversion was a day spent botanizing with General Smuts at his farm in the Transvaal.) On this journey Ingram concentrated chiefly on gladioli, and collected corms of about a hundred species and subspecies, several of which he was able to distinguish as new. Among those previously known was another of Africa's oddities, *Gladiolus lilaceus* (=*versicolor*), the Brown Afrikander, which at nightfall changes from rusty brown to pale glaucous blue, resuming its brown in the morning. (This, though a rarity, is still in cultivation.) Ingram also introduced *Cyrtanthus flanaganii*, but his great gift to gardeners was the hardy and elegant little red-hot-poker, *Kniphofia galpinii*.

[1] *Dictionary of National Biography*

North America

THE EASTERN STATES

1637	Tradescant II	1748	Kalm
c. 1680	Banister	*c.* 1764	Young
1712	Catesby	1785	Fraser
1722	More	1785	Michaux
1730	Bartram	1802	Lyon

THE WESTERN STATES

1804	Lewis and Clark	1842	Fremont
1809	Nuttall	1846	Hartweg
1809	Bradbury	1849	Lobb
1825	Douglas	1850	Jeffrey
1831	Drummond		

CANADA

1797	Masson	1825	Drummond

'. . . much of the soil that is now traversed by the locomotive, and gladdened by the joyful sound of the gospel and civilisation, was then the hunting-field of the Indian, and the scene of many a bloody conflict.' R. HOGG in *The Cottage Gardener* VIII 1852

'I like the free range of the woods and glades; I hate the sight of fences like the Indians.' C. RAFINESQUE *New Flora of North America* 1836

North America

The Eastern States

The United States of America are cracked almost from top to bottom by the mighty Mississippi, which divides so conveniently the Eastern States from the West. The river also marks as it were a visible date-line, for botanical exploration hardly crossed its barrier before 1804, more than two centuries after it had started in the East. The traffic in plants began with the first attempts at American colonization, for there was an eager interest at home in all the products of the unexplored New World; but very little is recorded of the earliest collectors. Among the colonists brought back by Drake in 1586 from Raleigh's unsuccessful attempt to establish a settlement in Virginia, was the mathematician and astronomer, Thomas Hariot, who brought home a number of plants, some of which were later described in Gerard's *Herball*; and Clusius is said to have made his third visit to England in 1581 in order to see the collections of Drake and Raleigh.

The elder John Tradescant was a subscriber to the Virginia Company, and though he did not himself cross the Atlantic he obtained a number of American plants; about forty are named in his garden-list of 1634, and while some were earlier introductions, Tradescant is credited with being the first to grow the Virginia Creeper (*Parthenocissus quinquefolia*), *Aquilegia canadensis, Aster tradescantii*, *Rudbeckia laciniata, Tradescantia virginica* and, possibly, *Robinia pseudo-acacia*. His son visited the country three times, in 1637, 1642 and 1654, but nothing is known of what he did there. A further thirty or forty American species appear in the catalogue of garden-plants incorporated in the *Musaeum Tradescantianum* of 1656, and it seems fair to assume that some at least of these were the younger Tradescant's importations. They included the Red Maple, the Tulip

Tree, the Swamp Cypress and the Occidental Plane; the vines *Vitis labrusca* and *V. vulpina*; *Adiantum pedatum, Anaphallis margaritacea, Lonicera sempervirens, Smilacina racemosa* and *Yucca filamentosa*.

John Watts, the curator of the Chelsea Physic Garden, sent a gardener called James Harlow to collect in Virginia in 1686–7, but received from him only three or four parcels 'most of all which are either trees or arborescent plants or bulbs, these being the most obvious; I do not believe he has got above one hundred in all'.[1] The most successful importer of American plants at this time was Henry Compton, whose position as Bishop of London and head of the Church for the American Colonies gave him special opportunities. One of the missionaries he sent out was **John Banister** (1654–1692) who had already shown great interest in the study of plants when an undergraduate at Oxford.

After a stay of uncertain duration in the West Indies, Banister arrived in 1678 at Charles Court County in Virginia, where he 'industriously sought for plants, described them, and himself drew the figures of the rarer species'.[2] Within two years he sent John Ray a 'Catalogue of Plants observed by me in Virginia', which was published in Ray's *Historia Plantarum* Vol. II (1688), and is regarded with great respect as the first printed account of the American flora. Banister was an all-round naturalist; he seems to have made a particular study of freshwater snails, and a colleague, writing about the backbone of a whale, referred to him as a 'Gentleman pretty curious in those things'. He planned to write a natural history of Virginia, which should include 'a more particular Acct. of the Plants of this country: viz. such as are Cultivated and manured, or Wild and spontaneous'[3] – and some of the fine drawings he sent home may have been intended to illustrate it; but, owing to his early death, it never materialized. He went for several excursions, on one of which he accompanied a party that ascended forty or fifty miles above the falls of the James River; and it was on some such excursion, in 1692, that he died. Pulteney says that 'he fell from the rocks, and perished', but according to Professor Ewan he was accidentally shot by one of his companions.

Banister sent seeds to several correspondents – to Ray, to Compton and to Bobart at the Oxford Botanic Garden. His introductions included *Magnolia virginiana, Rhododendron viscosum, Echinacea purpurea, Mertensia virginica*, and possibly *Dodecatheon meadia*, though this has also been attributed to Tradescant and to Catesby.

With **Mark Catesby** (1682–1749) we come to the first professional full-time collector whose movements in America are reasonably well known. His Suffolk home was not far from that of the aged John Ray, and it was Ray who inspired him with that 'passion for natural history' by which he was 'very early allured from the interesting pursuits of life' – meaning, one imagines, occupations that might have been more conducive to his *financial* interest. His father died when he was twenty-three, and left him a modest but sufficient patrimony; and, as he wrote later, 'Virginia was

[1] Berwick, E. *The Rawdon Papers* 1819
[2] Pulteney *Sketches*
[3] Quoted by Britten and Dandy

the place (I having Relations there) suited most with my Convenience to go'. The relations were his elder sister Elizabeth and her husband Dr William Cocke, whom she had married against her parents' wishes. The couple had emigrated about the turn of the century, and Dr Cocke was already becoming a person of note in the colony.

Catesby reached Williamsburg on 23 April 1712; but rather as an observant visitor than the professional natural-history collector he afterwards became. Soon after his arrival he became acquainted with Col. William Byrd, also a keen amateur naturalist, whom he visited several times and accompanied to a gathering of Indians at Pamunkey. Seeds of Catesby's collecting were sent to Bishop Compton by a correspondent in the spring of 1713, and Catesby himself sent seeds to the nurseryman Thomas Fairchild in London, and specimens to Ray's friend, the botanist Samuel Dale. The following year he went with some others up the James River to the foothills of the Appalachian mountains, where he discovered the Rose Acacia (*Robinia hispida*) but was unable to introduce it, for when he returned to get seed he found the country had been burnt by the 'ravaging Indians' and the trees destroyed. The same year (1714) he made a voyage on a cargo vessel to Jamaica, and again sent botanical specimens from there to Samuel Dale. Of the rest of his stay in Virginia little is known; possibly he was a good deal occupied with personal matters, looking after his sister and her family while his brother-in-law was absent on a visit to England. He himself returned home in the autumn of 1719.

Catesby's own estimate of the results of these years was a modest one. He claims only to have sent home a few collections of dried plants, 'and some of the most specious of them in Tubs of Earth, at the Request of some curious Friends', and to have made a few observations of the country. But he had also made many friends, who afterwards became valuable correspondents, had learned much about the conditions of American life, and had made notes and drawings which were used later in his great book. Above all, his specimens had been shown by Dale to William Sherard and others, and when Catesby returned to England his reputation as a promising naturalist was already established.

It was Sherard who organized the despatch of Catesby to America as a natural-history collector, on behalf of himself, Sir Hans Sloane, Dr Richard Mead, the Duke of Chandos and other subscribers. It was arranged that he should accompany the Governor, Col. Francis Nicholson, who was about to be sent out to the new colony of Carolina, and who agreed to pay 'Mr. Keatesby' an allowance of £20 a year. Catesby's departure was delayed for a year, however, owing to a counter-scheme put forward by Chandos to send him to Africa instead, which eventually came to nothing; so that he did not embark for America until February 1722. He arrived at Charleston on 3 May, and was kindly received by Governor Nicholson and introduced to the best of the local society. By 22 June, Catesby had already made a journey of forty miles 'up the Country' and was beginning to appreciate the time-wasting difficulties of packing and despatching his collections in a place where all sorts of containers were expensive and difficult to procure. Boxes cost '20d. that is 4d sterling' and even at that price the 'makannics' of Charleston made a favour of supplying them. He had previously made a practice of sending seeds, well sealed, in the shell of a gourd, and had found this very successful. However, few seed-containers were required that first season, for about the middle of September Catesby developed a swelling in his cheek (perhaps an abscess from a tooth?) which, though twice lanced and treated with 'injections' every day, kept him indoors until December. Meantime, a hurricane had hit the district, which 'disrobed' the trees of their leaves, cones and seeds, and was followed by floods; so that even if he had been well, there would have been nothing for him to collect. He was very concerned that he had been able to do so little, but Sherard was quite satisfied with his explanation, and the excellent dried specimens that were sent.

As soon as he had recovered, Catesby visited several plantations in the neighbourhood of Ashley River, and in March and April ascended the Savannah to the frontier post of Fort Moore, near the present city of Augusta. This he found a delightful spot – 'the inhabited part of the Country is a Sink in comparison of it; it is one of the Sweetest Countrys I ever saw'; and he returned to it twice more, to observe its vegetation at different seasons. The Indians in the neighbourhood were hospitable and friendly, and he employed one of them to carry his painting and collecting materials.

The Savannah forms the northern boundary of Georgia, and Catesby is supposed to have visited both Georgia and Florida – perhaps only on the strength of their place on the titlepage of his book, for no such journey is mentioned in his letters to Sherard, which are the only surviving documents concerning his travels. Indeed, his time seems fully accounted for, by the local excursions that he reports. The winter of 1723–4 was spent in the neighbourhood of Charleston; in the spring he embarked on an expedition of 400 miles into the 'Apalatheans', thinking that as the Cherokees were busy fighting a neighbouring tribe, they would have no leisure for the slaughter of occasional white travellers; but in this he may have been over-optimistic, for he was back within a month. He paid another visit to Fort Moore, and in August suggested to his patrons that he should make a long trip to the 'remoter parts', and ultimately to Mexico. He asked them to try to obtain the necessary Spanish permits, and while waiting for the reply (which never came) filled in the time with other local excursions. He was getting restless, for by this time he had more or less exhausted the country accessible from Charleston, and could find little that was new; for any change in the flora, he thought it would be necessary to go at least 300 miles farther south. Instead, he went in January 1725 to the Bahamas, staying first at Nassau on New Providence, and then visiting the islands of Eleuthera, Andros and Abaco, while engaged on the study of fishes. He returned to England in 1726.

Catesby always showed a practical interest in gardening. He had hardly been in Virginia a month before he was advising his new friend Colonel Byrd on the improvement of his garden, and soon after his arrival at Charleston he wrote asking Sherard to send bulbs for the gardens of his colonial friends. He also introduced to them some of the plants he found in the interior, such as the catalpa and perhaps the spice-bush (*Calycanthus floridus*), which in his time was found 'in the remote and hilly parts of *Carolina*, but nowhere among the Inhabitants', but which thirty years later was common in Charleston gardens. The plants he introduced to Britain included, besides these, *Callicarpa americana, Coreopsis lanceolata*, and the American wisteria, then called *Glycine frutescens*. After his return to England he lodged for a time at Fairchild's nursery in Hoxton; later he had a garden of his own at Fulham, to which he often refers, and was closely associated with the nurseryman Christopher Grey, who specialized in American plants. His correspondents across the Atlantic sent him species that he had been unable to procure, or that had subsequently died; in this way he was the means of introducing the first stewartia.

It is unfair, however, to represent Catesby only as a botanist and gardener. His reputation as a pioneer ornithologist stands high, and he made valuable contributions to other branches of natural history. The plates in his *Natural History of Carolina, Georgia, Florida and the Bahama Islands* (1730–47) are predominantly of birds, with appropriate plants in the background, and established the precedent later followed by Audubon. In the production of this amazing work Catesby spent almost the whole of the remaining twenty-three years of his life. Finding that it would be very expensive to have his drawings professionally engraved, he took lessons in etching, and prepared his plates himself. Each of the two serially-published volumes contained a hundred illustrations, coloured by hand under his supervision, and the work was concluded with an appendix of a further twenty plates. It was a remarkable achievement for a man with little or no scientific training or financial backing, and shows great tenacity of purpose. He celebrated its completion by getting married – at the age of sixty-five; at the time of his death, two years later, he had mysteriously acquired two children, the elder of whom was a boy of eight. Catesby was described as being 'tall, meagre, hard-featured and [with] a sullen look . . . extremely grave and sedate, and of a silent disposition; but when he contracted a friendship was communicative and affable'. Only on one occasion, after a convivial evening with Colonel Byrd in Virginia, was he known to sing.

Some of the enthusiasts who promoted Catesby's journey to Carolina also sent a collector to New England. This was **Thomas More** or Moore, an eccentric character with a high opinion of himself, who patronized his patrons in a way that must have been extremely irritating, in view of his deficiencies as a collector. Although the date of his birth is unknown, he was probably between fifty and sixty at the time of his visit to America. In 1704 he was present at a meeting of the Royal Society, and demonstrated a set of tables, designed, as one observer put it, 'to give us an account of the whole creation, from an Angell to an Attome'. He was then designated a Philosophical Pilgrim, and the name of the Pilgrim Botanist stuck to him ever after. When Sherard met him in 1718, he was making fresh copies on parchment of his tattered 'Tables', which looked like 'so many tailor's measures joined at the top and rolled up'. He seems to have been employed as a natural history collector around the coasts of Britain, and possibly also abroad, as he refers to having visited Egypt and other hot countries.

Sherard busied himself in collecting subscriptions, and More left England in June or July 1722, and after a passage prolonged by adverse winds, reached Boston about mid-September. He lost no time in impressing the inhabitants with the extent of his previous travels ('over the greatest part of the world') and the benefits he expected the colony would reap from his researches. He brought with him a quantity of English seeds to try in the new conditions 'if he could gett a suitable Spott of Ground of some hundred acres', and was aggrieved that this was not immediately granted to him, free of charge. He embarked at once on enthusiastic if rather undiscriminating collecting, anxious to supply some returns for his patrons in spite of the lateness of the season, but it had been an unusually hot, dry summer and such specimens of herbaceous plants as he was able to obtain were 'but faint and languid'. He sent his first consignment home on 27 October, with the assurance that 'what I now send is but a fleabite to whats a comeing', but unfortunately the carelessness of his packing rendered the contents almost useless. 'Mr. More has been very diligent for the short time he was in the country before the ship came away', wrote Sherard the following February, 'but most of what he gathered were common, and spoiled in coming over by his fault, for he put in the same box with the dried Plants, Fruits and Seeds, Limes, Gourds and such-like trash, and to fill it up, put seaweeds atop. I took him for a greater philosopher, and shall give him orders how to pack for the future'.

But there was little occasion for such instructions, for the Pilgrim Botanist was already distracted by other interests. He became embroiled in local politics, and in the spring of 1723 sent a long letter to the Secretary of State, making complaints about the state of the colony, suggesting the steps that ought to be taken, and applying for the vacant post of Forest Ranger. In the summer he embarked on a visit to an Indian Chief whom he had previously met in England,[1] and who had invited him 'very loveingly up to his Country', which he thought would 'make my range exceedingly the larger'. Sherard did not altogether approve. 'I have heard nothing yet from the Pilgrim Botanist, which I admire at', he wrote in August; 'Col. Dudley wrote word he was gone up into the country, to visit his old acquaintance, the Indian Kings that were in England. I had rather he would first send what grows near Boston, but they have all a notion that the farther they go, the more rare things they find.'

The results of this season's work were despatched to England on 12 December, and consisted largely of tree-seeds – it seems, at Sherard's particular request. This time More included nothing soft – he was endeavouring to follow instructions 'as nicely as I can' – which was as well, for the barrel lay a month in the Customs before Sherard was advised of its arrival. This was at the end of February 1724, and More himself came home shortly afterwards, largely to press his claims for the post he coveted. He expected to return to America early in the summer, and promised Sherard

[1] Probably in 1710, when five chiefs visited the court of Queen Anne

to follow his instructions better in the future; but from that time onward, the Pilgrim Botanist is no more heard of. In the beginning, Sherard had expected more from him than from Catesby; but he is recorded to have introduced only the American ash (*Fraxinus americana*) and the Poison Tree (*Rhus vernix*), of which Catesby also sent seeds.

One of Catesby's staunchest friends and supporters was Peter Collinson, a Quaker linendraper of London and a garden enthusiast. He pestered his business correspondents to send him seeds and plants, until one of his American contacts contrived to get rid of his importunities 'By recommending a Person whose Business it should be to gather seeds & send over plants. Accordingly John Bartram was recommended as a very proper Person for that purpose, being a native of Pensilvania with a numerous Family – the profits arising from Gathering Seeds would Enable Him to support it'.[1] The Quaker **John Bartram** (1699–1777) – farmer, amateur physician and self-taught botanist – married for the second time in 1729, and his family ultimately numbered eleven; at this time (about 1732 or 1733) he had recently moved to a property on the Schuylkill River near Philadelphia, where he built a house with his own hands and laid out a botanic garden.

Within a year or two the correspondence between Bartram and Collinson had become a 'settled trade and business'. Bartram sent over boxes containing 100 species of seeds (chiefly of trees) which Collinson distributed to various patrons at five guineas apiece. At first he and Lord Petre were the only customers; then they were joined by Philip Miller and the Dukes of Richmond, Norfolk and Bedford, and afterwards by others, until Bartram had orders for about twenty boxes a year. All this involved Collinson in considerable trouble and expense, but he 'willingly undertook it without the least grain of profit to myself in hope to improve or at least adorn my country'. But this was only a small part of the traffic that flowed back and forth across the Atlantic, in spite of many hazards – rats, damp and decay, wrecks, war and piracy – and the resulting losses. Besides plants and seeds, Bartram sent maps, drawings, journals, shells, minerals, turtles' eggs – some arrived on the point of hatching and caused great excitement – even live animals, but few zoological specimens, as he was averse to taking life. Through Collinson he received paper from Dillenius, seeds from Philip Miller, a silver cup from Sloane, and from Collinson himself plants, books and tools, lengths of material and articles of clothing, letters of introduction and a great variety of instructions; he was told how to study fossils and mosses, how to mount butterflies for safe transport and how to make a rhubarb tart. The correspondence thrived for thirty-five years, ceasing only with Collinson's death.

Bartram is regarded as the first American botanist, and as such his every footstep has been traced by assiduous American scholars. The wanderings of more than thirty years are too long to record in detail here. They began modestly enough with short trips up the Schuylkill River[2] when the harvest on the farm was over; Collinson warned his friend not to neglect his

[1] Collinson, in *Journal of Botany* 1925
[2] In 1735 Bartram traced and mapped the Schuylkill to its source

farming for his collecting, for 'the main chance must be minded'. As his family responsibilities diminished and the number of his subscribers increased, Bartram's journeys took him farther and farther afield; but of the details of many of them we know little. Only two of his trips were recorded in print – that of 1743 and his final triumphant progress in 1765.

The first of Bartram's longer journeys took place in 1738, when he went south through the coastal areas of Delaware, Maryland and Virginia to Williamsburg on the James River, then up the river to the Appalachians and '... over and between the mountains, in many very crooked turnings and windings, in which ... I travelled 1,100 miles in five weeks time; having rested but one day in all that time, and that was at Williamsburg'. It proved a profitable journey, for it was an exceptionally good season for seeds. In July 1743 he took the opportunity of accompanying an interpreter, Conrad Weiser, who was travelling to attend a conference between the colonists and the Indians at the Onondago Long House (near the present Syracuse, NY). The journey north, by Reading, Shamokin (Sunbury) and the Susquehanna River to Owego and Cortland, took four weeks to accomplish. Bartram kept a journal, three copies of which were sent to England; the first two were taken by French privateers, but the third arrived safely, and Collinson published it[1] 'without the author's knowledge, at the instance of several gentlemen, who were more in number than could conveniently peruse the manuscript'. It was on this journey that Bartram found and introduced *Magnolia acuminata*.

A short journey northwards to Newburgh in the autumn of 1753 is noteworthy because he was then accompanied for the first time by 'my little botanist' – his fifteen-year-old son William, afterwards to become a considerable naturalist in his own right. Another memorable tour was that made early in 1760 to Charleston, when he went to stay with Dr Alexander Garden (whose acquaintance he had made when Garden was passing through Philadelphia five and a half years before), and also met for the first time the celebrated woman horticulturist, Martha Logan, who afterwards sent him many seeds and plants. (The two elderly Quakers were very arch in their correspondence about this 'fascinating widow' of fifty-eight). The following year, the English captured Fort Duquesne and changed its name to Pittsburgh; Bartram was 'all in a flame' to go there, and down the Ohio 'as far as I can get a safe escort', in spite of the fact that he had been rather badly hurt that summer by a fall from a tree. So far as is known, the trip down the river did not materialize, but he got to Pittsburgh in the autumn, and found there his first pecans. Another long tour in North and South Carolina was made in 1762.

These are only a few of the travels of the indefatigable naturalist. His principal hunting-ground was the 600-mile valley of the Shenandoah, between the Alleghanies and the Blue Ridge mountains, which he called 'my Kashmir'; but he also made several visits to the Catskill mountains, and many local excursions about New Jersey. Age did not diminish his enthusiasm. 'Oh! if I could but spend six months on the Ohio, Mississippi and Florida in health', he wrote to Collinson in 1763, 'I believe I could find

[1] *Observations ... Made on his Travels, from Pensilvania to ... the Lake Ontario* 1751

more curiosities than the English, French and Spaniards have done in six score of years.' His opportunity was soon to arrive; at the age of sixty-six, his greatest journey still lay before him.

Florida became a British colony in 1763, and in April 1765 Collinson at last managed to procure for his friend the official recognition he had long been seeking for him – his appointment as King's Botanist to George III. But Collinson's utmost influence could not obtain a salary of more than £50 a year, nor even letters of introduction to the governors of the new colony; the King had no interest in botany, and the Queen's patronage was fully engaged by John Hill and his protégé William Young. 'Thou knows the length of a chain of 50 links', Collinson wrote to Bartram; 'go as far as that goes, and when that's at an end, cease to go any farther.' He advised the naturalist to save up for a year or two before embarking on so ambitious a project; but Bartram dared not wait, in case some accident should deprive him of his long-cherished ambition.

Meantime, 'Billy', Bartram's fifth son William, had been causing his father and friends considerable concern. He was an excellent artist, and wished to establish himself as a natural-history draughtsman; but his father thought there would not be sufficient employment for him in the Colonies, and wanted him to take up some steadier and more lucrative career. The result was that William drifted from one job to another, always starting out with great enthusiasm but never evincing any staying-power. At the time of the Florida journey he was restless again, and delighted to join his father, as his assistant.

The elder Bartram left Philadelphia by sea on 1 July 1765; the passage to Charleston took a week, and he was seasick all the time, so 'came to my worthy dear friend Dr. Gardens very faint'. Garden was inclined to patronize Bartram on account of his lack of education, and to be incredulous and even a little resentful of his official appointment as King's Botanist, but he had a great admiration for the naturalist's cheerful, intelligent and industrious personality, and always gave him a warm welcome. From Charleston Bartram made a preliminary journey northwards to Cape Fear River and up the river to Asheville, where William was employed on a trading-venture, under the eye of a land-owning uncle. Father and son returned together to Charleston, and set off from there on their southward journey on 29 August.

They went by Jacksonberg to Savannah, and up the river to Augusta, where they stayed from 10–21 September; then back to Ebenezer and south to Fort Barrington on the Altamaha River. The four days that they spent there were botanically important, for it was there that they found *Gordonia altamaha* and *Nyssa ogeche*. Then they proceeded to St Augustine on the Florida coast (11 October) where Bartram came down with perhaps the most severe attack of malaria he ever had; on 23 October he records 'I am so very weke can hardly stand without reeling'. It was a month before he was well enough to resume his journey, though in the meantime he made short trips to the lighthouse on Anastasia Island and to attend a congress of Creek Indians at Fort Piccollata on the St John. At last, on 19 December, they set out for the principal object of their journey

– the exploration of the river St John. (The journals give no details of the party, but refer to a hunter and a pilot; there would also be some boatmen.) For nearly a month they made their way upstream, till on 13 January 1766 they reached, not the source, but the point at which the boat could go no farther, the water being only three feet deep and the channel choked with water-plants. They returned on the west or 'Indian' side of the river, and were back in St Augustine by 13 February, having traced the stream for nearly 400 miles. The Governor of St Augustine 'ordered A room & conveniency to draw ye courses of ye river for him', and this map-making and some local excursions occupied Bartram until 17 March, when he took ship to Charleston, and thence about three weeks later to Philadelphia – without William. Nothing would do for Billy but to become a planter in Florida; 'this frolic of his, and our maintenance', wrote his father, 'hath drove me to great straits'. Billy's subsequent attempt to raise indigo was, of course, unsuccessful.

This was the last of Bartram's major journeys; Collinson died in 1768, and although Bartram continued to send seeds to some of his other correspondents, they were probably from plants in his own garden, or collected locally. A threat to this garden is said to have hastened his death; the War of American Independence had broken out, and troops were ravaging the neighbouring countryside after the battle of Brandywine. General Howe's progress up-river to Philadelphia troubled Bartram greatly, and as Howe entered (as one of his biographers puts it) the old botanist departed.

The value of Bartram's work can hardly be overestimated. One botanist has assessed the number of American plants he introduced to cultivation at 200 species – most of which were credited not to him, but to their recipients. They included, among shrubs, *Chionanthus virginica*, *Epigaea repens*, *Leucothoe racemosa*, *Rhododendron maximum* and *R. nudiflorum*; and among herbaceous plants *Cimicifuga racemosa*, *Iris cristata*, *Phlox divaricata*, *P. maculata* and *P. subulata*; *Lilium superbum* and *L. philadelphicum*, *Monarda didyma* and *Veratrum viride*. He also introduced the climbing Dutchman's Pipe, *Aristolochia macrophylla*, and the Ostrich Fern, *Onoclea struthiopteris*, besides many important trees. Quite apart from his new introductions, the sheer quantity and bulk of his collections were of great importance to the widespread distribution and secure establishment of American species. He had a wonderful eye for a new plant, and boasted that in thirty years of travel, revisiting perhaps twenty times the same locality, he never found a species 'in all after times, that I did not observe on my first journey through the same province'. Often, indeed, plants that he found on early trips disappeared later; the most famous example being the Franklinia (*Gordonia altamaha*) of which he found only one small stand, which was afterwards exterminated partly by the clearing of the land for agriculture, but partly, it must be confessed, by the rapacity of later collectors. It now survives only in cultivation. Unfortunately, Bartram seems to have been reluctant to put pen to paper except in correspondence; Kalm, who knew him well, lamented that his book harmed rather than benefited his reputation, as he did not put into it 'a thousandth part of the great knowledge which he has acquired'.

Pedr Kalm (1715–1779), the son of a poor Finnish pastor, had renounced a career in the Church to become a pupil of Linnaeus. He entered the service of Baron Bjelke, who was interested in the improvement of agriculture, and collected economic plants on his employer's behalf in Sweden, Finland and Russia. In 1747 he was appointed Professor of Economy at the University of his home-town of Åbo, and in the same year was chosen for a mission to study the botany of North America. It was chiefly an economic venture, with the emphasis on the benefits to be expected from plants growing in a similar latitude to that of Sweden. Linnaeus believed that 'all Lapland could be rendered fertile by the introduction of appropriate American plants'; botanical discovery was only to be incidental.

Sufficient funds were with difficulty scraped together, with contributions from the Swedish Academy and three universities, and Kalm set off, accompanied by a gardener called Lars Yungstroem. On the way he made a six-month visit to England, where he met all the leading agriculturists, botanists and gardeners, including Philip Miller, Peter Collinson, the aged Sir Hans Sloane and Mark Catesby.

It was natural that Kalm and Yungstroem should make first for Philadelphia, for that part of America had originally been colonized by Sweden, and there were still large Swedish communities both in the town and in the surrounding villages. They arrived on 15 September 1748, and Kalm found himself in a world where everything was new. 'Whenever I looked to the ground, I everywhere found such plants as I had never seen before . . . I was seized with terror at the thought of ranging so many new and unknown parts of natural history.' He made industrious notes on the plants, birds, quadrupeds, insects, fish and molluscs; on minerals and climate, and on the manners, customs, diseases and architecture of both natives and immigrants. Three days after his arrival he paid his first visit to Bartram, who had, he found, 'the great quality of communicating everything he knew'; of this he took full advantage, for he quotes Bartram on almost every page of his journal during his stay in Philadelphia, and often afterwards.

Towards the end of October Kalm sent home his first consignment of plants and seeds; then made a short trip to New York, and after his return, removed in early December to the Swedish village of Raccoon, not far away on the New Jersey side of the Delaware, which was his base till the following May. There may have been a special attraction there, for during his stay in America Kalm married a pastor's widow, and eventually took a wife back with him to Sweden. He was to have returned home in 1750, but wrote to ask if his time could be extended. Linnaeus was delighted, and urged him, not for the first time, to go farther north – if possible, to Hudson's Bay; he could not be convinced that nothing was to be found in those regions but sub-arctic tundra. Kalm dutifully set out at the end of May 1749 on a long-planned visit to Canada.

Accompanied by Yungstroem, he went first to New York, and then up the Hudson River by yacht to Albany, where they hired a canoe and two guides, intending to continue up the river to the site of the deserted Fort

Nicholson; beyond Saratoga all houses had been burnt in the recent wars, and people were only just beginning to return to their abandoned farms. Owing to the lowness of the water, however, they had to leave the boat earlier than they intended, and tramp some fifty miles, first following the Hudson, then branching off ENE to the Woodcreek River – camping in the woods, oppressed by heat and mosquitoes, apprehensive of snakes and Indians, and increasingly short of provisions. At the Woodcreek River they made a bark boat to proceed downstream (much hampered by fallen trees and beaver-dams), but were delayed by losing their way and rowing for twelve miles up a tributary stream; so that by the time they reached Crown Point on 2 July they had been twenty-four hours without food. They were received 'very politely' by the French governor, and had over a fortnight to recover while waiting for a boat to Fort St Johns.

The Governor of Quebec, the Marquis de la Galissonière, was himself a keen naturalist, and had received orders from France to assist Kalm in every way he could. As Kalm explained to Bartram, when France had sent a team of scientists to Sweden fifteen years before to measure a degree of latitude from the Pole, the Swedish crown had afforded them free hospitality; Kalm was now to receive hospitality in return – 'It is not permitted to me to pay anything, but the French King he pays that all'. Accordingly the Marquis sent horses to St Johns to meet him, and also two small casks of wine. There was feasting that night (22 July) at the fort, and the healths of the Kings of France and Sweden were drunk, to the accompaniment of salutes of cannon. Next day they were in Montreal, where the people crowded to see the first Swedes who had ever come there, and by 5 August were in Quebec.

From Quebec Yungstroem returned almost immediately to Pennsylvania, to collect seeds under Bartram's guidance, while Kalm remained in Canada for another two months, making excursions in company with the Governor's physician, Dr Gaulthier. Kalm's published journal ceases abruptly on 5 October, when he was on his way back to Philadelphia, and his subsequent travels are not recorded in any detail. He sent home a new consignment of seeds, plants and curiosities in the late autumn, and in 1750 made another northern trip, following the Mohawk River to the country of the Iriquois and visiting Lake Ontario and the Niagara Falls. (Collinson was relieved to hear that he had returned safely in October to New York, as he 'was afraid of some Wild Indians doing him a mischief'.) Yungstroem in the meantime was collecting in New Jersey.

Early in 1751 they sailed for Europe – as Garden disgustedly put it, 'loaded with British treasures' for which Sweden would take all the credit. They had another six-week sojourn in England, during which Kalm saw American plants thriving in the gardens of his English friends, from seeds he had sent them, and reached Stockholm in June. When he got to Uppsala, Linnaeus was unwell, but hearing that Kalm was back from America with his harvest of new plants, 'he rose from his bed and forgot his troubles'. Yet the herbarium which Kalm brought home was not very large; it contained about 325 species, many of which Linnaeus subsequently described in the *Species Plantarum*. Kalm himself was aware that much more could have been done, but he had conscientiously spent most of his time on his agricultural researches. In 1751, Kalm resumed his professorship at Åbo, where he established a small experimental garden in which he grew 'many hundreds' of American plants. He was much impoverished by his journey, having spent on it all his own savings as well as his salary; and he published the story of his travels (*En Resa til Norra America*, 1753–61), 'by intervals', on account of the expense. It was never completed, and the manuscript was subsequently destroyed by fire; but its author is effectively commemorated in the beautiful genus of shrubs which Linnaeus 'conformable to the peculiar friendship and goodness he has always honoured me with' named Kalmia in his honour.

The fame of Bartram's achievements inevitably produced rivals, especially towards the end of his career; for a short time one James Alexander seemed a potential danger, but he soon transferred his obvious business abilities to other fields. Bartram's most serious competitor was **William Young** (d. 1785) who is first mentioned in a letter of 25 July 1761, from Dr Garden to John Ellis in England. 'I have at last met with a man who is to commence nursery man and gardener, and to collect seeds and plants, etc. for the London market. He is a sensible careful man, and has a turn for that business. I must beg your interest in his favour. . . . His name is Young, and any letters to him enclosed to me, will be taken care of.' It seems odd that Garden should say he had 'at last' found someone to collect seeds and plants for England, when Bartram, whom he had known for six years, had for nearly thirty years been doing just that; it is also odd that Garden, in Charleston, should be at such pains to patronize someone who lived in Philadelphia, and who had possibly been introduced to him by Bartram himself.

Still more astonishing was the news, announced in the summer of 1764, that this upstart, on the strength of three years' experience as a nurseryman and collector and a consignment sent to London of a few familiar and common plants, had been appointed Queen's Botanist – and this, before he had so much as paid a visit to Britain. The letter in which Bartram comments, with some consternation, on the appointment was written on 23 September, and when he wrote again, three weeks later, Young had left for England, and Bartram was wondering how he would succeed at Court. He generously surmised that 'if he is put under Dr Hill's care, he will make a botanist, as he is very industrious and hath a good share of ingenuity'. In the same letter he hopes that Collinson will find some means of forwarding to the King a box he has sent, containing choice specimens of *really* rare plants; 'not that I depend on having any such preferment as Young had', he writes with heartbreaking humility, 'but chiefly as a curiosity . . .'. His own ill-paid appointment did not materialize till the following April.

Young's rapid promotion may have been due to the fact that whereas Bartram and Collinson had only merit to recommend them, Young had a friend at court. His father, a small farmer in the neighbourhood of Philadelphia, augmented his income by the sale of a nostrum called Hill's Balsam, for which he employed agents up and down the country. It seems at least possible that this was one of the productions of the versatile 'Sir' John Hill, who was a quack doctor and compiler of herbal medicines, as well as a journalist, playwright, actor, botanist and gardener. If William Young senior had been acting as Hill's colonial agent and sending him a substantial commission – and his accounts show the sale of the balsam to have been considerable – that might account for Hill's patronage of the younger William; and Hill was a protégé of Lord Bute, who was a favourite at Court and well-placed to obtain any preferment. A remark of Bartram's seems to imply that the Youngs were of German stock, and likely therefore to be acceptable to the German Queen.

Arrived in London, Young did, apparently, take up the study of botany under Hill, who later referred to him as his pupil; but he did not display the qualities of common-sense and carefulness with which Garden had credited him. The following May he called on Collinson 'so new-modelled and grown so fine and fashionable, with his hair curled and tied in a black bag, that my people, who have seen him often, did not know him. I happened not to be at home, so could not enquire what scheme he is upon'. By December 1766, earlier than was expected, he was back in Philadelphia, cutting a great figure with his sword and gold lace, and boasting of his salary of £300 a year[1] and his intention to winter in the Carolinas. Nevertheless, there were rumours that he had left England under a cloud, having been imprisoned for debt, and escorted on board ship by two prison officers. Kindly Bartram would not believe this story without confirmation, but Collinson assured him it was probably true, as he knew Young had been living above his income, and had warned him many times of the consequences of his extravagance and folly.

However, when he recovered from his head-turning visit to court, Young seems to have settled down to work soberly enough. In 1767 he sent Dr Fothergill a collection of 302 rather crude figures of plants, with accompanying specimens, and he returned to England in the summer of 1768, bringing with him a cargo of plants which he established in a garden taken for the purpose at Isleworth. The most sensational of his introductions was the Venus' Fly-trap, or Tippitiwitchet (*Dionaea muscipula*, originally Bartram's discovery), which had never before been seen alive in Europe. This curiosity aroused great interest in the botanical world and much enhanced Young's unmerited reputation. By 1771 he was back in America and again sending over plants and seeds. His plants were well-packed and gave satisfaction; not so his seeds. 'W. Young has been very diligent', wrote Fothergill to Bartram, 'but has glutted the market with many common things.... But, contrary to my opinion, he put them into the hands of a person who, to make the most of them, bought up, I am told, all the old American seeds that were in the hands of the seedsmen here, and mixed them with a few of W. Young's, to increase the quantity. Being old and effete, they did not come up; and have thereby injured his reputation. I am sorry for him; have endeavoured to help him, but he is not discreet.'

Young was more fortunate in the relations that he established with the

[1] Compare to Bartram's £50

firm of Vilmorin in Paris. A '*Catalogue d'Arbres, Arbustes et Plantes Herbacées d'Amerique*, par William Yong, jr. Botaniste de Pensylvanie' was printed in 1783, listing 173 species of trees and shrubs, and 145 herbaceous plants, which could be ordered, either as plants or seeds, through the agency of Vilmorin. It also included a list of forty-one species that could not be supplied, owing to the length and expense of the journeys that would be necessary to procure them; most were natives of Carolina or Florida, but a few were from Hudson's Bay, showing the wide range of American plants that was then in demand.

Young continued to send plants to Kew up to the time of his death in the winter of 1784–5. He is credited with some twenty-five first-introductions, none of them of great horticultural importance. One would think that this unstable personality had been born under a watery planet; his best introductions were a moss (*Lycopodium obscurum*), a bog-plant (the dionaea, a native only of the Great Dismal Swamp on the border of Virginia and North Carolina) and a water-lily (*Nuphar advena*); and he met his death by drowning.

The year of Young's death saw the advent to America of two collectors of much greater importance. Just as in South Africa Thunberg and Masson arrived in the same year and made some of their earliest journeys together, so in America Michaux and Fraser joined forces at the start of their travels. Michaux, like Thunberg, was the better qualified as a botanist; he was three or four years older than Fraser and had already had considerable collecting experience; but Fraser's introductions, like Masson's, were of greater value to gardeners. Unfortunately there are gaps and discrepancies in the memoirs of both collectors that make it hard to trace their movements with any certainty, Fraser's biographers in particular differing from each other and from the few particulars given by himself.

John Fraser (1750–1811) was a Scot who came to London about 1770, married, and established himself as a linendraper and hosier in Chelsea, where he soon became acquainted with William Forsyth (also a Scot) then Director of the Chelsea Physic Garden. Having been delicate in youth, and being threatened with tuberculosis, Fraser was persuaded in 1780 to take a voyage to Newfoundland for the sake of his health, with his friend Admiral Campbell, in command of the Newfoundland station. Here he stayed for some four years, developing at the same time his taste for botany, and a habit of restlessness which unfitted him for the pursuit of the hosiery trade. On his return he showed his specimens to Forsyth, who told him they were of value, and encouraged him to take up professional collecting. According to the account given by the younger Forsyth in Loudon's *Arboretum et Fruticetum Britannicum*, Fraser made his first journey to Charleston in 1784,[1] sending his plants to Frank Thorburn, a nurseryman of Old Brompton. He returned early in 1785 and 'expected to receive the rewards of his labours, but was told that all his valuable plants had died, and that those remaining were common, and not very saleable'. This led eventually to a long and costly lawsuit.

[1] No one else mentions this journey, not even Fraser himself, but it sounds convincing

In the following year Fraser returned to America, this time with the support of a number of subscribers. He arrived at Charleston on 20 September 1786, and met Michaux, it would seem, almost at once. Although Michaux had a 'very liberal establishment' of 12,000 livres a year and Fraser was travelling 'solely at my own expense and risk', he at once 'took up a determination, which may be thought rather presumptuous, of endeavouring to excel Mr. Michaux, or at least to share with him the honour of extending the knowledge of my countrymen over the vegetable kingdom in this part of the world'.[1] Their rivalry did not prevent them from 'pursuing (their) botanical researches for a considerable time together'.

André Michaux (1746–1803) with his fifteen-year-old son and a gardener called Paul Saulnier, had arrived at New York on 1 October the previous year (1785). He had instructions to send all seeds and plants to the park of Rambouillet (where they were received and distributed by the Abbé Nolin) except for two parcels a year for his first patron, M. le Monnier, and two for the Jardin des Plantes. He was also to arrange a depot at New York for the establishment of plants before despatch. He duly organized the purchase of land and the laying out of a nursery-cum-botanic garden; but by December he had already realized that the climate of New York was too cold for the successful cultivation of plants such as magnolias, and by January he was planning a journey to Carolina, where botany and gardening could be carried on all the year round. Before 12 May 1786, he had bought a piece of ground at Ten Mile Station, north of Charleston, without waiting for authorization from France; and leaving the New York garden in charge of Saulnier, henceforward he made Charleston his headquarters.

It was not yet the season for a major plant-hunting expedition, but in the November after his arrival Fraser seems to have gone on an excursion with two companions, a Dr Porter and 'young Mr. Brooks'. They went first to Savannah (Porter by sea, the others by horse and chaise) and leaving his companions to look after his collections in his absence, Fraser went on to Sunburry, where he joined a hunting-party of seventeen for a ten-day trip to Saplo and other islands off the Georgia coast. On his return to Savannah he was obliged to chase after Brooks, who had boarded a ship for England, taking with him Fraser's seeds, and without paying his share of the expedition; he gave as his excuse that he 'could not think of staying with such a disagreeable young man as Porter'.

Fraser had other consignments to despatch from Charleston, and Michaux, hearing that among the collections that had already been put aboard the *John* were some plants he had not seen, set out with Fraser in a boat for the ship, which was anchored five miles from the harbour. A storm blew up while they were on board, with much lightning, which 'terrified the poor Frenchman so much, that it was with the greatest difficulty he could be persuaded to get into the boat in order to get ashore again, notwithstanding he was almost like to go to sea for England next

[1] Fraser *Agrostis Cornucopiæ* 1789

morning, as the ship was then a-going'.[1] At this time (January 1787) the two had already arranged a combined operation. 'My friend Mr. Michaux and I have altered our plans, and are now going to the Indian country, for had we gone to the Bahamas, the Spring would have been spent before we could have returned. The Frenchman seems very courageous, and means to go as far as the Mississippi, and I have every reason to believe that I shall not be the one that will give out first.'[2]

The journey up the Savannah River to the 'Indian country' began in April 1787, but the two botanists did not remain long in company. Between Savannah and Augusta Michaux's horses were stolen,[3] and after a delay of several days during which he tried in vain to recover them or to purchase others, Fraser got tired of waiting and went on by himself. He ascended the river to its beginnings, crossed the Blue Mountains and entered the Cherokee country, three or four days' journey beyond. He found the Indians thinly dispersed and peaceable, and was enchanted by the beauty of the country, which he prophesied would become 'the Italy of America'. The date of his return is unknown; but it seems to have been on this trip that he found *Galax aphylla* 'at the foot of the mountains at the back part of the state of Georgia'.

During his stay at Charleston Fraser formed an intimate friendship with Thomas Walter, who was preparing a flora of Carolina. Fraser added a further 420 species to the 640 already collected by Walter, including several new genera, and a grass[4] which they hoped to develop as a valuable contribution to agriculture. Unfortunately Walter died in January 1788, and when Fraser sailed for home soon afterwards, he brought with him, besides 30,000 plant specimens and a great variety of living plants, the manuscript of his friend's unpublished work. He reached England in March, after an absence of nineteen months, during which he had made journeys amounting to some 4000 miles.

One of Fraser's subscribers complained that he and his friends had received small returns for their expenditure, and that most of the plants and seeds he brought back were treated as his own property; but the traveller was hard pressed at this time by the expenses of his unsuccessful lawsuit, and had, he says, 'nothing to save me and my family from ruin, but by making an immediate sale of the Collections I had brought along with me in the ship'. He was much occupied that summer seeing Walter's *Flora Caroliniana* through the press, and with the distribution and advertising of the new grass, on which he published a monograph (with some biographical notes in the foreword) in 1789. To obtain further subscriptions for his grass-seed he paid a visit to Paris, where he met, among others, the American Minister Plenipotentiary, Thomas Jefferson.

Fraser made at least two more voyages to America between 1789 and 1795, but of these no record seems to have survived; he intended to publish an account of his journeys, but it never materialized. About 1795 he was back in England and had started a nursery of some twelve acres for

[1] and [2] Fraser, letter to Forsyth (*The Cottage Gardener* 1852)
[3] Michaux was consistently unlucky with horses
[4] *Agrostis perennaus*

the reception, propagation and distribution of his American plants, in the neighbourhood of Sloane Square. A catalogue was published in 1796 in which he refers to having returned from his fifth voyage to America,[1] and to the purchase, with his brother James, of a plantation near Charleston. Forty-seven kinds of seeds are listed and forty-two named plants 'which he is convinced will enrich any Botanical Collection in Europe', besides a tantalizing general entry: 'A great variety of Herbaceous Plants, new.'

In 1796 Fraser made his first visit to Russia, taking with him a collection which was purchased at his own price by Catherine the Great. She died the following year, but her successor the Empress Maria ordered a further consignment, and in 1798 Fraser was appointed botanical collector to their Majesties Paul and Maria, with instructions to 'furnish such other rare or novel plants as he should recommend for the completion of the Imperial collections'. Accordingly he embarked again for America in 1799, this time accompanied by his elder son, John.

Fraser seems to have gone farther afield on this journey, if it is true that he visited Tennessee, Kentucky and Ohio. Late in 1800 he made arrangements for a visit to Cuba. Spain being then at war with Britain, the Frasers procured passports as American citizens, not realizing how difficult it was to enter any Spanish colony without express Royal permission. In the event, the boat in which they travelled was wrecked on a reef, about eighty miles from Havana and forty from the American mainland.[2] After six days of severe suffering the survivors – eighteen in all – were rescued by a fishing-boat and taken to the Cuban port of Matanzas. The American consul got permission for the Frasers to proceed overland to Havana, where they arrived, quite destitute, about the beginning of February 1801. They were befriended by Alexander von Humboldt, recently arrived from Venezuela, who supplied them with money and lodged them in his own house; but he would not take the responsibility of concealing from the Governor the fact that the newcomers were British. The Marquis de Somourclos, however, after the gentlemanly fashion of the period, waged no war on scientists, and gave them passports to move freely about the island.

Humboldt seems to have taken a liking to the younger Fraser, whom he invited to accompany him to Mexico; but the lad was timid about going among Spaniards (with whom his nation was at war and whose language he did not speak) and moreover he was in a hurry to return to England with the new Kentucky plants. It was arranged therefore that young Fraser should take charge of two large chests of Humboldt's collections, which he brought safely via Charleston to London in June 1801. Meantime his father stayed in America for another season and left for England in 1802. Again he was unfortunate in his vessel, which developed so bad a leak that it took the combined efforts of crew and passengers at the pumps to get her as far as Port Nassau in the Bahamas. Eventually he reached home, only to find that the demented Czar Paul had been mur-

[1] It is not clear whether he counted his preliminary voyage to Newfoundland as one of the five

[2] According to Forsyth, in the Cayos – a group of small islands at the entrance to the Old Channel

dered, and that his son (and some hint, assassin), Czar Alexander, had repudiated the agreements made by his predecessor. Fraser made three trips to Russia between 1802 and 1807 to try to recover his dues, but received little recompense, though the gentle Empress Maria did what she could, and gave him a valuable diamond ring. On Cuba he had discovered a silver-leaved palm (*Coccothrinax miraguama*) from the leaves of which the natives wove hats 'all in one piece without sewing, in a new and peculiar manner'; Fraser tried to introduce this manufacture to Britain, under the superintendance of his sister, Christina; but although patronized by Queen Charlotte, this venture, too, was a failure.

Back, then, to America – perhaps with less enthusiasm now, but there was nothing else to be done. Little is known of this last trip, which lasted from 1807 to 1810; it seems to have followed the pattern of the one before, and to have included another visit to Cuba. It was during this journey, however, that Fraser crowned his life-work with one of his most triumphant finds – *Rhododendron catawbiense*, which he found on the summit of Great Roa or Bald Mountain of the Apalachicola Ridge. His son John was with him, and returned with a cargo of plants, leaving his father, as before, to work for another season. But on his return to Charleston from the mountains the elder Fraser had a fall from his horse and broke several ribs, 'the distance from surgical aid aggravating the consequences'. He never quite recovered from his injuries; after he got back to England he was in bed for several months, and died on 26 April 1811, aged sixty, lamenting that 'Providence had cut him off in the midst of his labours'.

Fraser was a kindly and amiable character – obviously no businessman, but a born collector, undergoing hardships that 'nothing but the ardent zeal of the man combined with the dauntless hardihood of the Highlander, could have enabled him to support'.[1] The list of his introductions contains an unusually high proportion of good garden-plants. *Hortus Kewensis* (published before the conclusion of his last journey) credits him with forty-four new species; Hooker lists 215, counting those subsequently brought by his son, but probably not all of these were new. Besides the species already mentioned we owe him *Aesculus parviflora, Hydrangea quercifolia, Rhododendron calendulaceum, R. speciosum* and *Zenobia pulverulenta*; among herbaceous plants, *Oenothera tetragona fraseri*, and *Lepachys* (*Rudbeckia*) *pinnata*; the discerning gardener will also remember him kindly for *Allium cernuum, Euphorbia marginata* and *Uvularia grandiflora*.

We left Michaux stranded near the Savannah River. Eventually he resumed his journey, following much the same route as Fraser, and penetrating, with an Indian guide, as far as the River Tennessee – the first of a long series of botanical explorations, which extended from Florida to Hudson's Bay. The Florida trip took place the following year (February 1788) when Michaux and his son, after exploring the river Tomaco, followed the route taken by the two Bartrams up the St John River to Lake St George, and thence along one of its smaller tributaries. Florida by then had reverted to Spain, so Spanish passports were necessary, and Michaux

[1] Hooker *Companion to the Botanical Magazine* II (1836)

sent a collection of seeds to Madrid at the conclusion of the journey. Similarly, when he visited the Bahamas in the spring of the following year, he presented the governor with some seeds for Sir Joseph Banks; they probably included *Illicium parviflorum*, the only plant credited to him in *Hortus Kewensis*.

On his return to Charleston from the Bahamas in May 1789, Michaux heard for the first time the news of the French Revolution. He expected soon to be recalled to France; and anxious to make the most of what he thought would be his last opportunity, he set off at the end of the month to visit the 'highest mountains' in North Carolina.[1] He went to Morganton, and from there plunged into the forests with his son and a guide; but the latter was badly injured by a bear, causing Michaux to reflect on the necessity for having two guides, as it would be almost impossible for a European – even himself, who was respected by the Indians for his woodcraft – to make his way through the forests alone. Finding the Indians temporarily hostile, they revisited New York and Philadelphia, and returned to Charleston after an absence of five and a half months.

The recall to France never came – but neither did any further remittances; and for the remaining seven years of his stay in America Michaux was engaged in a perpetual struggle to make ends meet. For a year or two he concentrated on the nursery, adding trees from Europe and Asia, the latter from seed brought by ships trading to China; during this period he introduced to America *Ginkgo biloba*, *Lagerstroemia indica*, *Albizzia julibrissin* and many other valuable plants. But he was eager to make further studies in the botanical geography of America, and in 1792 he made arrangements for a long journey to the north, pledging his own patrimony in France in order to raise the necessary funds.

He left Charleston on 18 April (without his son, now studying medicine in France) and followed the usual route by Philadelphia, New York, Albany, Saratoga and Lake Champlain to Montreal and Quebec, which he reached on 16 July. From there he went by Lorette and Montmorency to the fur-trading station of Tadoussac, a 'miserable village', where he hired two canoes and four Indians and proceeded up the Sagueney River to the lake and trading-post of St John. This was the farthest station inhabited by Europeans; from there on, his way led through the wilderness. He navigated the River Mistassen as far as the falls, where he lodged in a cabin, lived on beaver and cranberries, and botanized on a rock between the two arms of the cascade; then leaving the river, he struck across the mountains to a small stream that descended to Lake Mistassen, which he reached on 4 September. By this time the weather was turning bitterly cold, but he pressed on northwards down the Marten to its junction with Rupert River. At this point his guides refused to go farther, in view of the dangers of the rapidly-advancing winter, which would soon make travel impossible; and he was obliged to turn back within a relatively short distance of Hudson's Bay, and return by the way he had come. He was back in Tadoussac by 1 October, and dismissed his Indians, who had served him

[1] Probably the Mt Mitchell area

with honesty and zeal; and he reached Philadelphia on 8 December, having traversed some eighteen degrees of latitude from south to north in an almost straight line, and back again. He had ascertained the northern limits of many species of trees, besides obtaining much geographical information, and had found *Primula mistassinica* and other new plants.

Shortly after his return Michaux proposed to the American Philosophical Society that he should make an expedition up the Missouri to the undiscovered West. This plan was 'exceedingly well received' by the Secretary of State, Thomas Jefferson, who had had something of the sort in mind for at least ten years, and who had contemplated sending only a single individual with a servant 'to avoid alarming the Indians'. A subscription was set on foot, and although the total raised was only $\$128\frac{1}{2}$ – a modest sum even for those days – it is said that Michaux had already set out, when he was recalled by the arrival in Philadelphia of Citizen Genêt, the new Republican Minister from France, who claimed his services to go on a diplomatic mission to Kentucky. As a loyal Frenchman and staunch Republican Michaux could not refuse, but he must have been disappointed. His errand was to carry letters and messages to certain French generals, concerning a projected campaign against Spanish Louisiana; but the generals were reluctant, Genêt was recalled to France and the invasion of Louisiana was abandoned. Michaux, after a five-month tour (July–December 1793), was free to resume his botanizing; he had, of course, made some collections in the meantime, and stayed in Philadelphia till February 1794 putting these in order, before returning to Charleston. But the Kentucky countryside had aroused his interest, and in April 1795, having mortgaged the last of his inheritance, he set off on his final American journey.

He started up the Catawba River – Cambden, Lincolnton and Morganton, finding five new plants in the first week – and botanized for ten days among the Lineville mountains. Then he descended by way of the Doe and Holsten rivers to Knoxville, where he had to wait till a sufficient party could be assembled to cross the Wilderness, a stretch of uninhabited country subject to marauding Indians. Fifteen well-armed men and more than thirty women set out on 4 June, and slept the first night at the frontier post of West Point on the Clinch River; they crossed the Cumberland mountains without incident and arrived safely at Nashville. After a week at Nashville, Michaux turned north, crossed the Barrens – another dangerous area – and reached Danville in Kentucky, which had been his principal centre when he was on his diplomatic mission two years before. Here he stayed for several weeks, revisiting also Bardstown and Louisville. On 9 August he left Louisville for Vincennes on the River Wabash; this turned out to be one of his most difficult journeys, owing to the density of the brushwood in the forests and the number of fallen trees. He arrived on the 13th, 'almost ill' from the effects of a fall; his horse had come down when trying to jump a tree-trunk, and Michaux was thrown, bruising his chest on the trigger of his gun. After a few days' rest he resumed his botanizing, and on the 23rd set off again with 'a Savage and his wife' as guides. Progress was delayed by the inordinate appetites of these Indians, who, when game was obtainable, stopped to eat five times a day; moreover, it rained daily. On 30 August they reached Fort Kaskaskia, near the mouth of the Kaskaskia River; and Michaux at last stood beside the Mississippi.

During the next three and a half months he made the fort his base for excursions which ranged from the Illinois River in the north to the Cumberland River in the south. On 14 December he started the return journey by boat – down the Mississippi, then up the Tennessee and the Cumberland rivers to Clarksville, on 11 January 1796. Here he bought a horse – which promptly ran away – and continued to Nashville, which he left on 23 January, to go north to Louisville and back. (One wonders what strong inducement drove him to make this detour, during which he suffered from frostbite and many other hardships – eighteen days' winter travel for the sake of five days at Louisville.) Back at Nashville, he rested for ten days before finally starting for home (25 February) by much the same route as he came. From 24 March his journal makes frequent references to pulling 'shoots of shrubs . . . to transfer them to the Garden of the Republic in Carolina';[1] several hundred, including rhododendrons and azaleas, were collected in the Lineville mountains. He arrived at the nursery on 11 April.

No class of men suffered more from the French Revolution than the harmless and hard-working botanists. Adanson, Dombey and L'Héritier were among those ruined by it; and Michaux was no exception. His private means were now exhausted; he did not feel entitled to sell the nursery, which he regarded as the property of the French Government; and he was disinclined to take other employment. He had no resource but to return to France, and leaving the nursery in charge of his son, he embarked in August 1796, with a large consignment of plants and specimens, on a ship bound for Amsterdam. But a storm blew her out of her course, and she was wrecked (early in October) on the coast of Holland, near the little village of Egmond. Crew and passengers were rescued; Michaux, lashed to a yard, was unconscious when brought ashore, but was ill only for two days. He lost all his personal property, but nearly all his collections were saved; and after spending six weeks drying and re-papering his specimens (the first review, he says, took sixteen days, working from 4.0 a.m. to 8.0 p.m.) and washing the salt water from his plants, he set off in a ricketty cart for Paris.

He was received with honour and distinction – but not with cash. The Government took no responsibility for the engagements of its predecessors, and the exchequer being reduced by war, granted him only a small proportion of his seven years' arrears of salary. He was given a piece of ground at Rambouillet for his new collections, but found few remaining of the 60,000 living trees (not to mention ninety consignments of seeds) he had previously sent home. Many, inadequately packed, had perished on the journey; the Abbé Nolin had been 'worn out' by demands for the survivors from various members of the aristocracy, some of whom had

[1] as he had patriotically named his nursery

cared for their trees, and some had not; many had been sent by Marie Antoinette to her father's gardens at Schönbrunn, and the Rambouillet plantations had been laid waste during the Revolution. Disheartened, Michaux settled down to live as cheaply as possible and to work on his two books, the *Histoire des Chênes d'Amérique* (1801) and the posthumously-published *Flora Boreali Americana* (1803) but authorship was temporarily abandoned while he did all he could to help his former patron, M. le Monnier, now ill and impoverished. Michaux could not afford to return to America, and the Government would not send him there; so he accepted an offer to join Captain Baudin's expedition to Australia, left it (as his contract permitted) to explore the Mascarene Islands, and died from fever on Madagascar.

A third collector was working in much the same area and at the same time as Michaux and Fraser. Indeed, **John Lyon** (d. 1814) called on Fraser in Charleston in 1809, and received from that friendly and unworldly character information about his collecting localities. Little is known of Lyon's early life, except that he was a Scot, born at Gillogie in Forfar, and that by 1796 he was employed by William Hamilton of Philadelphia, whose garden, 'The Woodlands', when under Lyon's care, housed one of the finest collections of native and foreign plants in America. Hamilton was a difficult and domineering employer, and Lyon was restless for some years before he finally broke away and started his first major collecting trip in December 1802, having already been on two small journeys on Hamilton's behalf.

Professor Ewan has rescued and reprinted Lyon's journals, so his travels are well-documented. His first journey was much the longest, as though he was spying out the territories, to the best of which he was afterwards to return again and again. He went by sea to Charleston, then by stage to Savannah, and spent the first six months on or near the Georgia coast, with a short excursion up the St John River. In June he struck inland, via Augusta, Fort Wilkinson, and the Oconee and Flint rivers, for a venture into Creek Indian territory, when he accompanied Colonel Hawkins to a large meeting of Indian Chiefs on the Chattahoochee River. 'While out on a Botanical ramble . . . at some distance from the camp' he was bitten by a mad dog; he cauterized the wounds, 'three in number', and bathed them for two or three days with 'volatile Caustic Alkali' which he also took internally. Further treatment consisted of a night's camp in heavy rain without fire or shelter, followed by a long, fruitless search for a lost horse; and he experienced 'no ill effects' from his accident. He was back at Hawkins' Springs near Fort Wilkinson by 18 August, and stayed there till early October, except for a five-day trip to Bath (near Augusta), to collect an expected remittance from Hamilton – which was not forthcoming.

In October Lyon made his way north to visit a friend near Lincolnton in North Carolina, where he stayed for about three weeks, making excursions in the neighbourhood, and returning to Charleston by a different route at the end of November. In January 1804, he revisited Savannah, Cumberland Island and other places, collecting and packing plants previously deposited, and on 13 March embarked for Philadelphia, which he reached on the 24th after an absence of more than fifteen months. During this tour he had established four reserve gardens on the properties of different friends, where his plants could be accommodated until he was ready to pack and despatch them to Philadelphia; and at Philadelphia he had an arrangement with the famous nursery firm of David Landreth, with whom he lodged between journeys, and who looked after his plants in his absences until he had accumulated sufficient stock to be worth a voyage to England.

On this occasion, however, he did not stay long in town; in May he made a short trip to Harper's Ferry to collect plants of *Jeffersonia diphylla*, and at the end of the month he was once more on his way south – this time by way of the Shenandoah valley, a more direct route to his chosen territory. Again he visited his friend Colonel Hawkins, on the Flint River in Georgia, and on the way back stopped near Athens to gather seeds of the poisonous *Rhus pumila*. Four or five days later (7 September) 'a braking out all over me commenced with considerable fever, much alarmed' and by the time he reached Asheville, with difficulty, on the 13th, he was 'almost in one continued blister all over my body'. It was not till 8 October that he was able to resume his journey; he returned by the way he came, and reached Philadelphia on the last day of the month.

Lyon then packed his collections for a voyage to London, but for some reason was 'disapointed' of his expected passage and had to unpack them again and replace all the plants in the nursery. It was another year before he actually sailed, on 8 December 1805; the ship had a stormy passage and reached Liverpool in January 1806 'in distress'. After a month's visit to Dublin Lyon arrived in London early in March, and established his plants in a nursery he rented at Parson's Green. He then set about organizing a sale, the catalogue of which ran to thirty-four closely printed pages; the lots included quantities of one-year-old seedlings in pots, and collections of fifty varieties of seeds. Few of the species were actually new, but it was thought to be the largest collection of American trees and shrubs ever brought over by one person at one time. After deducting his expenses and the costs of his journeys, Lyon estimated the 'Neat Profits' at £923 9s 0d.

He returned to Philadelphia, and three more collecting-trips followed, all covering approximately the same area of the southern Alleghanies. On the first, which lasted from April till October 1807, after revisiting Asheville and Knoxville, he went west as far as Nashville, and procured a passport for a return through Cherokee Indian territory, by the Hiwassee, Conasanga and Cosawattee rivers; then by Franklin and French Broad River to Asheville and Lincolnton again. On the way back he stayed two nights at Winston and climbed Mt Pilot, finding on the top of it *Pieris floribunda*. On the second tour, from August 1808 till April 1809, he was laid up for a month at a friend's house on the Seneca River by an inflammation in his left leg. He then went on to Colonel Hawkins', by way of Milledgeville, 'now the capital of Georgia containing about 900 Houses, an elegant State House, etc. which 4 years ago was woods in a state of nature and which I considered good botanising grounds'; but again was held up by inflammation, this time in his other leg. He returned by way of Charleston, where he met Fraser and visited Michaux's derelict garden. The third trip lasted from July to November of the same year (1809), when he once more visited Knoxville, Nashville and Lincolnton, and returned to Philadelphia by the Shenandoah valley.

In May and again in August 1810 Lyon made short trips to New York, and he spent the following year cultivating and preparing his collections. He sailed again for England in December 1811, and reached London via Liverpool in mid-February. But the sale that followed in due course was less successful than the previous one – the expenses were higher and the takings less, and he took back to Philadelphia only a 'Neat Amount' of £613 6s 9d for six years' work.

From 1812 to 1814 Britain and America were at war, and Lyon stayed prudently at home; but in July 1814, with peace concluded, he set off again for his familiar collecting-grounds in the south. About a month after his arrival at Knoxville, he fell a victim to a local typhoid epidemic, and died at Jonesburgh near Asheville on 13 September, after less than a fortnight's illness, 'amidst those savage and romantic mountains that had so often been the scene of his labours'. A Dr Macbride who had vainly travelled a long distance to treat him, thought he had been bled to death through well-meaning solicitude.[1]

John Lyon was no pioneer; ten years after Lewis and Clark had crossed the Continent to the Pacific, he penetrated no farther west than Nashville, and the greater part of his territory had previously been explored by Bartram, Michaux and Fraser. Although Pursh called him 'a gentleman through whose industry and skill more new and rare American plants have lately been introduced into Europe than through any other channel whatever', this applies to quantity rather than variety; we read of him collecting 3600 plants of *Magnolia macrophylla* at one time. (It is to be feared that he was of the type that contributes to the extermination of rare species.) His attitude was commercial; in all his journals he never expresses pleasure in a plant, but he almost invariably notes the mileage covered and the cost of the journey. On the other hand, he never complains of the hardships of his life, and he seems to have had many friends. The younger Landreth, to whom he acted as tutor between journeys, described him as 'an amiable, well-bred, intelligent man, of most sterling worth, and a loyal Briton'. In person he was of medium height and stocky build, not handsome but with strong features and bright blue eyes.

With three collectors working in one area at one time, it is natural that finds should be duplicated, and many 'first introductions' are credited by different authors, in turn to Lyon, Fraser or Michaux. *Magnolia macrophylla* and *Rhododendron calendulaceum* appear on all three lists, *Illicium parviflorum* is credited both to Michaux and to Fraser, while Fraser and Lyon overlap with *Pieris floribunda*, *Jeffersonia diphylla*, *Oenothera tetragona fraseri* and several other plants. A number of species, however, were undoubtedly Lyon's alone, including *Chelone lyoni*, *Dicentra eximia* and *Iris fulva*.

[1] Lyon referred several times to treatment by bleeding, both for his Sumac poisoning and for his inflamed legs

A passion for plants does not seem to be a heritable trait; Lyon was unmarried, but neither Bartram's son, nor Michaux's, nor Fraser's, became a collector of the stature of his father. **William Bartram** (1739–1823) was a good ornithologist, with considerable literary and artistic talent. In 1772 Peter Collinson's friend Dr Fothergill commissioned him to go on a collecting journey, for a fee of £50 a year and expenses, drawings to be paid for separately; and the next year William 'sat off', as he puts it, on a series of travels which might rather be called wanderings, which occupied five years. At first he sent back reports, specimens and drawings, though little in the way of living plants or seeds; but afterwards he seldom wrote, even when communications were possible, and had long been given up for lost when he returned in January 1778, to find his father dead and America and Britain at war.

For the rest of his life, with gentle determination, William did nothing whatever – nothing, at least, that he did not want to do. He never married, but lived with his brother and sister-in-law, who had inherited his father's house and botanic garden. Four years after his return he was offered the professorship of botany at the University of Philadelphia, but refused on the grounds of ill-health – though he lived to a hale eighty-four, content, as one author puts it, 'to dig, barefoot and coarsely clad, in the ancient garden'. (How is it possible to dig barefoot?) He did, however, compile a valuable check-list of American birds, and wrote a book of travels, though this was not published until 1791. William Bartram was a romantic, through and through; and not even Reginald Farrer could write a more glowing description of a plant in flower. His book, a second edition of which was printed in England the following year, had a profound influence on Wordsworth, Coleridge and Southey, all of whom recognizably drew inspiration from its pages. Puc Puggy (or Puc Puggy Faya, the flower-hunter) as the Indians called him, died on 22 July 1823 a month too soon to receive a visit from David Douglas, who called at the garden on 22 August of the same year, but found no-one at home. It is odd that two eras of American collecting history – and John Bartram's first journey and Douglas' last were a hundred years apart – should so nearly have made contact.

François André Michaux (1770–1855) seems to have followed his father back to France about two years later, in 1798, leaving the Charleston nursery to look after itself, with some casual supervision from a neighbouring planter. The elder Michaux having sailed with Baudin late in 1800, François completed the editing of his book on American oaks and saw it through the press, and then returned to America, commissioned by the French Government to sell what was left of the nursery. From June to October 1802 he went on a collecting journey, starting from Philadelphia and making a wide circuit through Pittsburg, Marietta, Lexington, Knoxville, and Charleston. Most of the route had already been traversed by his father, but there had been many changes, and he was now able to make the first part of the journey by stage-coach. (It was still advisable, however, to have an escort across the Wilderness.) He returned to France in March 1803 with a great variety of seeds and plants, but many of his trees, like those of his father, were suffered to die through accident or neglect. The next three years saw the publication of three more works – his father's *Flora Boreali Americana*, his own *Relation d'une Voyage*, and a memoir on the acclimatization of American trees. This last procured him a commission to study the forests of America; he left Bordeaux on 5 February 1806, but the American ship in which he sailed was arrested by the British frigate *Leander*, and he was detained for a time in the Bermudas, before being allowed to continue his journey to Philadelphia, which he reached on 5 July. He went north to visit Saulnier, still in charge of the New York garden, and is reported to have travelled extensively in Maine, Ohio and Georgia, returning to France in 1808. This was his last visit to America; his *Histoire des Arbres Forestiers d'Amérique* was published from 1810 to 1813, but its botany has been subjected to considerable criticism.

Fraser's son, who might have won fame as Humboldt's companion, ended life as a 'respectable nurseryman at Ramsgate'. He made at least two more trips to America after his father's death in 1811, for his catalogue of 1813 refers to 'seeds collected by him, and brought home under his own care' and Hooker dates his final return at 1817. (This 1813 Catalogue was the one in which many of Nuttall's plants were offered for sale, for which he afterwards complained that he was never compensated.) The nursery, which John Fraser ran in conjunction with his younger brother James Thomas, was sold up in 1817 and Fraser moved to another called The Hermitage near Ramsgate, which he managed till his retirement in 1835. The date of his death, like that of his birth, is uncertain.

The Western States

Botany turned over a new leaf when Lewis and Clark made their celebrated trans-continental journey in 1804–6. Michaux's projected venture up the Missouri with a single attendant had now developed into a Government-sponsored expedition of more than forty well-equipped men, charged with finding a practicable route to the Pacific. Neither of its two leaders – Captain **Merriwether Lewis** (1774–1809) and his chosen companion, Captain **William Clark** (1770–1838) – was a botanist, though both were skilled in hunting and woodcraft and had served in frontier wars against the Indians; but Lewis was sent to Professor Barton in Philadelphia for some preliminary coaching in natural history and Indian lore. The expedition was organized in 1803, but the actual start was delayed till the conclusion of the Louisiana Purchase, when Jefferson bought from Napoleon (who had had it from Spain) the whole of Lower and Upper Louisiana, which then stretched from the Gulf of Mexico to Canada, and for an indefinite extent into the unknown west – surely the most profitable package-deal ever made.[1]

The area they were about to traverse being now American territory, the party waited only till the ice was out of the rivers, and left the village

[1] In 1819 Bradbury estimated the cost at 1½d an acre

of St Louis in May 1804. For six months they worked their way up the Missouri, and camped for the winter near the settlements of the Mandan Indians, about fifty miles above the present Bismark in North Dakota. On 7 April 1805 they sent home despatches and collections before resuming their journey; and that was the last that was heard of them till their unheralded return eighteen months later. By the end of June they were making a difficult eleven-day portage past the Great Falls, where they cached some skins and plant-specimens which were afterwards spoiled by an unexpected rise in the river. At Three Forks (July) they chose the branch of the river now called the Jefferson and followed it to the limit of navigation (August) when they changed from canoes to horses hired from the Indians. For some time they searched the mountains for a practicable crossing-point, and it was here that Lewis, scouting ahead for provisions as well as for passes, found some roots being dried in an Indian encampment which were 'bitter and nauceous to my pallate', and named the range the Bitter Root Mountains. At last they crossed by the Lolo Pass, considerably farther to the north, and gained the upper reaches of the Clearwater. Here they built canoes, and navigated the Clearwater and Snake rivers to the Columbia, which they descended to the coast, sighting the Pacific on 7 October. For the first time, the continent had been crossed from east to west.

Fort Clatsop was built for winter-quarters, about three miles up a small tributary to the south of the Columbia estuary; and in the spring of 1806 the homeward journey was begun. The same route was followed as far as the Lolo Pass, where they had to wait for the snow to melt before they could proceed; then the party divided and Lewis explored north to the rise of Maria's River, which he followed down to the Missouri, while Clark turned south and south-east along Big Hole River to the Jefferson. He followed the latter to its junction with the Madison, then struck across country to the Yellowstone River and descended it to its junction with the Missouri, where the party reunited on 12 August. They then had a quick passage down to St Louis, arriving in triumph on 23 September.

The surveying instruments used by the expedition were primitive, and their estimates of heights and distances often inaccurate; their geological knowledge, too, was sketchy, and they crossed the Montana gold-fields without noticing the sparkles in the sand or the nuggets in the river bars. But they had opened the door to the West, and many were stimulated to follow them – including botanists. The collection of seeds and specimens made by Lewis and Clark on their homeward journey was a relatively small one, but almost everything in it was new to science.

This collection afterwards underwent many vicissitudes. Jefferson entrusted the seeds, including a large part of his own personal share, to William Hamilton (Lyon's employer) and the Irish-born nurseryman MacMahon, both of Philadelphia; the latter was also charged with Lewis's herbarium, with instructions that it should be given to the German botanist Frederick Pursh, then absent on a collecting trip for Barton.[1] On

[1] Pursh may have been intended to pass on the specimens to Barton

his return, Pursh stayed with MacMahon for eighteen months, preparing drawings and descriptions to be included in the account of the expedition that Lewis was expected to publish; nor would MacMahon distribute any of the plants he had raised, for fear of depriving the explorers of their hard-earned priority. But nothing was heard of Lewis, now Governor of Louisiana, till October 1809 – and then it was the news of his death, apparently by suicide. The duty of publishing a report of the expedition now fell on Clark, but for various reasons its appearance was delayed till 1814. Meantime Pursh, who on leaving MacMahon's had gone to take charge of Dr Hosack's botanic garden at New York, had sailed for England, taking Lewis's herbarium with him; later to be incorporated, with much other material (to some of which his title was equally doubtful) in his *Flora Americae Septentrionalis* (1814). It was in this work that Lewis's Bitter Root was named Lewisia in his honour, and Clark commemorated in the Clarkia. Other plants discovered on the expedition were *Gaillardia aristata* and *Mimulus luteus*; *Calochortus elegans* and *Erythronium grandiflorum*; *Holodiscus discolor* and *Philadelphus lewisii*; while from Lewis's seeds MacMahon grew the now over-familiar snowberry (then thought a great acquisition) and the Oregon Grape or Mahonia.

The first botanists to follow the call of the West were Bradbury and Nuttall. **Thomas Nuttall** (1786–1859), an impecunious journeyman-printer from Liverpool, came to Philadelphia on 23 April 1808; he had already acquired a love of books and plants, but had had little opportunity for the study of botanical science. The story goes that on his first walk abroad Nuttall found a flower that neither he, nor anyone in his boarding-house, could identify; he was advised to show it to Professor Barton, who lived nearby, and the meeting 'made Nuttall a botanist and Barton his friend and patron'. Barton found him a brilliant and impassioned pupil, and being himself prevented by poor health from going far afield, sent the young newcomer on two collecting trips in 1809. The first was to the coasts of New Jersey and Southern Delaware, where Nuttall got so badly bitten by mosquitoes that he was 'all but driven away' from a house he approached, on suspicion of smallpox; and the second took him to the shores of Lake Erie and the Niagara Falls.

Nuttall had obtained a printing job, and financed these two expeditions out of his own savings; but the ambitious trip planned for 1810 was to be made at Barton's cost. The Professor gave him a salary of eight dollars a month and expenses, and a certain amount of equipment, including a veritable arsenal of weapons – a double-barrelled gun, a pistol and a dirk; Nuttall afterwards found the gun-barrels useful for digging up plants. The itinerary planned by Barton was quite impossible to realize; it extended to Winnipeg, but poor Nuttall, racked by recurrent bouts of fever, had considerable difficulty in getting as far as Detroit. Himself no traveller, Barton had probably little idea of the extent of his demands on a youth who could neither swim nor shoot, and who had no experience of woodcraft or of frontier conditions.

On 12 April 1810 Nuttall left Philadelphia by stage-coach for Pittsburgh; the rest of his journey was accomplished on foot or by boat. He had already had several attacks of malaria by the time he reached Franklin at the junction of the Allegheny and French Creek rivers, and during the week of his stay there (7–14 May). He went on for a further two days, but learned at Le Boeuf that his trunk, which was to be brought by water, had not arrived owing to the lowness of the river, and might be delayed for another month; so he had to return to Franklin to retrieve his possessions. Again he was laid up with fever; and started a week later 'very weak and burthened' with the contents of his trunk on his back. He struggled through to Erie, on the shores of the lake, and having waited two days in vain for a boat, set off to walk to Detroit, leaving some of his luggage to be sent on later. He followed the south shore of the lake, past the newly-founded settlement of Cleveland, to the mouth of the Huron River, where he stayed for a week, making forays into the prairie country towards Sandusky Bay, until he at last got a boat for Detroit, and arrived on 26 June.

After a month's needed rest, Nuttall resumed his journey by birch-bark canoe up the coast of Lake Huron, with a surveyor bound for Michili-Mackinac. He found it was impossible to proceed farther to the north or north-west: the Canadian shores of Lake Superior were prohibited by the 'sinister regulations' of the North-West Company, and the British military posts were in general unfavourable to the movements of Americans.[1] Michili-Mackinac was a fur-trading centre and the headquarters of John Jacob Astor's fur-company; the Astorian Expedition was nearly ready to leave, and Nuttall decided to join it. This party of thirty-four was to follow Lewis and Clark's route to the Pacific, establishing trading-posts by the way, and meeting at the coast another party which was being sent round by sea. On leaving Mackinac their route led through Green Bay on Lake Michigan, then by the Fox and Wisconsin rivers to the Missouri, and down-stream to St Louis. And there Nuttall met a compatriot.

John Bradbury (1768–1823), like Nuttall, was a north-countryman, born at Stalybridge near Manchester, and had started his working life in a cotton-mill. His researches into local botany brought him into contact with William Roscoe and Sir Joseph Banks, and he was sent to America (with a salary of £100 a year for three years) by the Liverpool Philosophical Society and the Liverpool Botanic Garden, for the dual purpose of collecting plants and of finding new cotton-growing areas to satisfy the increasing demands of the Lancashire weaving industry. He was considerably older than Nuttall and left a wife and eight children behind when he quitted England in the summer of 1809. Roscoe had given him a letter of introduction to President Jefferson and he stayed for a couple of weeks at Monticello, spending 'every day in the woods, from morning till night'; Jefferson found him 'a man of entire worth and correct conduct' and a first-class botanist. In person he was 'swarthy, broad-shouldered, and of medium height, amiable yet stubborn in disposition, temperate in his habits and a most excellent marksman. He was fond of music, active on his feet, and determined in his methods and opinions'. At Jefferson's sugges-

[1] Although British, Nuttall was travelling with an American passport

tion he abandoned his intention of going to New Orleans, and went instead to St Louis; cotton had been shelved in favour of botanical exploration.

He reached St Louis on 31 December 1809, and spent the next nine months on relatively short local excursions. When Nuttall arrived at the end of the following September, Bradbury had gone down the river to St Genevieve, to oversee the despatch of his collections, which included a large number of living plants. Soon after his return he was invited to join the Astorias expedition, which was now encamped and waiting for the spring; in the meantime, he and Nuttall went botanizing together, and he was able to show his young colleague plants he had found earlier in the season, which were now out of flower.

Nuttall and Bradbury left St Louis on 13 March 1811, with the leader of the Astorians, Wilson P. Hunt, to join the main body of the expedition, encamped about 450 miles up the Missouri under its auxiliary leaders, Ramsay Crooks and Donald McKenzie. No journal of Nuttall's for this period seems to have survived, but we have Bradbury's account, and also a narrative by Henry Brackenridge which covers part of the trip. At first the weather was bad; when the incessant rain permitted Bradbury and Nuttall sometimes left the boats and walked along the bank, and it was on one such occasion, faced with a creek swollen by the rains, that Bradbury discovered to his surprise that Nuttall could not swim. He offered to swim over with Nuttall on his back, but Nuttall prudently declined, and they went upstream till they found a raft of driftwood held by a fallen tree, and were able to get across.

On 8 April they joined Crooks and his party at Fort Osage, and on the 17th, the rest of the expedition with McKenzie, near the mouth of the Nodoway River. From there the group, now numbering nearly sixty, proceeded in four boats, equipped with both oars and sails; Bradbury was invited to go with McKenzie, and Nuttall was probably in one of the other boats, since Bradbury makes no further mention of him for more than two months. Bradbury took every opportunity to botanize on shore, and on 2 May started a 200-mile overland journey with Crooks and two Canadians to the Maha Indian village on the Platte River, where they rejoined the boats on the 11th without mishap. The Indians here were friendly, especially to Bradbury, whom they regarded as a medicine-man; but farther on the Sioux were hostile, and had to be approached with great caution. A large war-party was encountered and gave rise to some alarm, but a palaver was held at which peace was eventually concluded and presents exchanged.

About this time, on 2 June, their numbers were increased to ninety by the arrival of a Spanish fur-trader of doubtful probity, Manuel Lisa, whose party included Henry Brackenridge. This young man had joined chiefly for the adventure, and was much looking forward to making excursions with Bradbury,[1] whom he had met at St Louis and to whom he was very much attached. They had left the town a month later than the Astorians,

[1] On such occasions Brackenridge hunted or kept guard, while Bradbury botanized

and Lisa had driven his men almost to mutiny in his endeavour to overtake the larger party before entering the dangerous Sioux country. But although united for defence, Lisa represented the rival Missouri Fur Company, and neither party trusted the other; three days after their meeting, he and Hunt had a violent quarrel over an interpreter and only the intervention of Bradbury and Brackenridge prevented bloodshed. Thereafter the two parties kept their distance and Bradbury saw less of his friend than he could wish. It is from Brackenridge that we learn that the Canadian boatmen called Nuttall 'le fou', because of his preoccupation with botany to the exclusion of all else, including his personal safety and the convenience of his companions. Bradbury was more of an all-rounder, always ready to join in a hunting or fishing party, and was consequently more popular.

On 12 June they reached the settlement of the Aricara Indians, about opposite the present town of Campbell. Here the Astorians left the river and set about bargaining for horses to continue their journey by land, and for the first time Bradbury was able to collect living plants and prepare a piece of ground for their reception. A week later he accompanied an overland party sent to fetch some horses from Fort Mandan, now a post of the Missouri Fur Company; they got there on 22 June, and Bradbury found the botanizing so good that he decided to stay for a time and return with Lisa, who arrived by boat (with Brackenridge) three days later. Nuttall, too, was at the fort, though it is not said how he got there, and they all attended a party given by Lisa on 4 July – including two Indian Chiefs, who behaved 'with much propriety'. Two days later, leaving Nuttall behind, Bradbury and Brackenridge returned with Lisa to the Aricara settlement and rejoined the main party on 7 July.

Nearly eighty horses had now been assembled, and Hunt was ready to start on the second part of his journey. He invited Bradbury to go with him; but Bradbury realized that it would be impossible to take his collection of 'several thousands' of living plants across the Rockies, and even when the Pacific was reached, he could not be sure of getting a passage by sea to the USA 'or even to China'. He therefore arranged to embark himself and his collections – they filled seventeen chests – on two boats loaded with furs that Lisa was sending down the river under Brackenridge's charge. He had bargained with the trader that he should be allowed to stop and botanize on the way, but the Sioux were again troublesome, and Lisa privately instructed his boatmen to press on day and night whenever the weather was favourable; so opportunities of going ashore were few. They sailed on 17 July, and on the 20th were overtaken by a severe storm in a part of the river where the rocky banks afforded no shelter. They tied up to an inadequate shrub (Bradbury notes that it was *Amorpha fruticosa*) and endeavoured to raise the height of the gunwales by extra boards and blankets, which they had to hold in place as there was no time to lash them securely. Even so, they were nearly swamped by the waves and the drenching rain that followed. Bradbury was much concerned for Brackenridge. 'Poor young man, his youth and the delicacy of his frame ill suited him for such hardships, which nevertheless he supported cheerfully.' Brackenridge was equally concerned for Bradbury. 'Poor old man, the exposure was much greater than one of his years could well support. His amiable ardour in the pursuit of knowledge did not permit him for a moment to think of his advanced age . . .' (Bradbury was then forty-three). They reached St Louis safely on 29 July.

Lisa himself returned with the remainder of the season's furs on 21 October, bringing Nuttall with him. All this time Nuttall had been living at Fort Mandan, and going on botanical excursions; there is a tradition that he was found in a starving condition many miles away and brought back into camp by a friendly Indian. He is known to have penetrated about a hundred miles farther up the Missouri, where he found *Yucca glauca*; but he was not, as he fondly imagined, practically among the Rockies.

Although Nuttall returned to St Louis so much later than Bradbury, he was the first to leave, by about a month; and thereby escaped the misfortunes that befell his colleague. Nothing, for Bradbury, seemed to go right. The day after his arrival he received a letter from his son, informing him that his patrons in England had 'determined to withhold any further supply' – perhaps because of his failure to fulfil their requirements concerning cotton. With much labour, he planted his collections in a piece of ground provided by a friend, and then collapsed with a fever so severe that for a time he had little hope of recovery. At the end of November he was cheered by a letter informing him that his remittance was restored, and that Shepherd at the Liverpool Botanic Garden had received his previous consignment, 'out of which he had secured in pots more than one thousand plants, and that the seeds were already vegetating in vast numbers'. On 5 December, he embarked, still shaky, for the journey home; on the way down the river he experienced the violent earthquake which on 14–15 December destroyed the town of New Madrid and did much damage elsewhere. He reached New Orleans on 13 January 1812, and soon afterwards embarked for New York, but was prevented from sailing for England by the outbreak of the Anglo-American War. He entered into business, but sustained losses through a dishonest partner; and on the conclusion of peace made another journey into Illinois, of which no record seems to have survived. Meantime, the herbarium specimens he had sent home to Liverpool were, very unwisely, lent by William Roscoe to Frederick Pursh; and Pursh published descriptions of all Bradbury's new plants – some forty-one of them – in an appendix to his *Flora Americae Septentrionalis* (1814). Bradbury was justifiably embittered by this piracy, 'depriving me', as he said, 'both of the credit and profit of what was justly due to me'; and he never made another collecting expedition. He returned to England to publish his book of travels (at his own expense) in 1817; then brought his family with him to America and settled at St Louis, where he became director of the Botanic Garden. He died in Kentucky in 1823.

Nuttall, meanwhile, had sailed direct to England from New Orleans – either because he guessed war was imminent and wanted to get home while it was still possible, or simply because a homeward-bound ship was about to leave port. Before sailing, he despatched bulky collections to Barton, but he took with him a large quantity of seeds and living plants, which were temporarily established in his uncle Jonas Nuttall's garden at Nutgrove, near St Helen's, Lancashire. He made generous donations of these to the Liverpool Botanic Gardens and to other institutions and friends, and arranged for the remainder to be marketed by the Frasers, who specialized in American plants. Their 1813 catalogue (as previously mentioned) was drawn up by Nuttall, though his name does not appear. Many of these plants were illustrated in the *Botanical Magazine*, and Nuttall therefore got credit for a number of species which, almost certainly, were also introduced by the neglected Bradbury – for example *Oenothera missouriensis*, *Ribes aureum* and *Shepherdia argentea*. Nuttall's plants included *Camassia fraseri*, *Lepachys* (*Rudbeckia*) *columnaris*, *Mentzelia decapetala*, *Oenothera caespitosa*, *O. nuttallii* and *Penstemon glaber*.

The Missouri expedition of 1811 was only the beginning of Nuttall's long series of botanical travels. He returned to Philadelphia in May 1815, and in 1818 published his *Genera of North American Plants*, setting a large part of the type with his own hands. This was the third American flora – Michaux had written the first, and Pursh the second – and by far the most complete; a remarkable feat for a man with little previous knowledge of botany, who had only set foot on the continent ten years before. Shortly afterwards, with the aid of some subscriptions from friends, Nuttall set out on another adventurous trip to the west.

He left Philadelphia on 2 October 1818 by stage-coach, which as yet only went as far as Lancaster. From there he travelled on foot to Pittsburgh, bought a skiff for six dollars and hired a boatman to take him down the Ohio. He hoped to get a steamboat at Louisville, but the river was too low for the boats to ascend so far; he therefore purchased a flatboat and a cargo 'nearly at my own cost' whose loss 'would probably have plunged me into penury and distress' – but he seems to have had no difficulty in selling his whisky, salt and flour at a good profit farther down the river. On 17 December he reached the mouth of the Ohio and proceeded down the Mississippi – a dismal voyage as the scattered settlements had been abandoned owing to earthquakes and inundations, and all was 'irksome silence and gloomy solitude, such as to inspire the mind with horror'. On 16 January he began the ascent of the Arkansas River – the real object of his journey – and by tedious towing arrived at Arkansas Post on the 26th. He had now been nearly four months on the way, and the spring flowers were beginning to appear.

At Arkansas Post he had difficulty in making arrangements for further transport as his appearance 'in the meanest garb of a working boatman, and unattended by a single slave', did not command respect. Eventually he got a passage on a skiff bound for Bairdstown, and there transferred to the 'large and commodious travelling boat' of a trader called Drope, to whom he had a letter of introduction. When Drope went ashore to trade at the settlements Nuttall was able to botanize, but the settlements were few, and consisted at most of three or four families; seventy miles of river would go by without a single house. At the Dardanelle Hills he parted from Drope and set off in a pirogue with two French boatmen for 'the garrison' at Fort Smith, where he arrived on 24 April. This was to be his base for the next six months.

The fort was under the command of Major Bradford, who was friendly and helpful, and on 16 May, Nuttall was allowed to accompany a contingent of troops that was being sent overland by the Poteau and Kiamichi valleys to Red River, to clear white settlers from an area that had been allotted as an Indian reserve. The prairies were now 'enamelled with innumerable flowers . . . and charming as the blissful regions of fancy'; these flowers included *Coreopsis tinctoria*, *Oenothera speciosa*, *Penstemon cobaea* and *Rudbeckia maxima*. On the return journey Nuttall (with the undisciplined enthusiasm previously noted by Brackenridge)[1] stayed behind for two hours to botanize. He intended to overtake the troop at its next encampment, but lost his way, and eventually had to stay for nearly three weeks with a hospitable farmer before he could join a party of hunters returning to the Arkansas; without money or equipment he dared not venture on the journey alone. Even so, his party took a wrong route and had a hard struggle to reach Fort Smith on 21 June.

An agent of Drope's, M. Bogy, came up the river early in July, and Nuttall accompanied him upstream to his trading-station at the mouth of the Verdigris River. After an excursion up the Neosho River to the Osage salt-works, and a week's prostration with fever, he set out again on 11 August with a hunter and trapper called Lee, on a cross-country journey to the Cimarron River, which almost proved fatal. The weather was oppressively hot, provisions went bad, and they were plagued with flies; Lee's horse proved quite unfit for travel, which imposed a double burden on Nuttall's; they were apprehensive of attack by Indians and Nuttall's fever returned with such severity that he fainted on horseback when he tried to proceed. With Nuttall occasionally delirious and unable to eat, they advanced at a rate of three to six miles a day, and reached the Cimarron on 3 September. Here they made a canoe and began the descent of the river, Lee (his horse abandoned) in the boat, and Nuttall, now somewhat better, riding along the bank – and still botanizing. On 15 September he got back to Bogy's trading-post, more dead than alive; when sufficiently recovered he returned to Fort Smith.

The homeward journey began on 16 October and was accomplished without special incident. Nuttall refers to collecting living plants in the prairies near Arkansas Post, though the season was by then (January) so advanced that it was hard to find what he wanted. He got a passage down the Mississippi to New Orleans, which he reached on 18 February 1820, and returned to Philadelphia by sea.

In 1822 Nuttall was appointed Curator of Harvard University's Botanic Garden at Cambridge (Mass.), a post that he held till 1834. During this period he made several less ambitious journeys, none of which took him west of the Mississippi. He revisited England in 1823–4, 1827–8 and 1832–3, and on the first occasion took with him seeds of a number of his

[1] In this he resembled Richard Cunningham, and was lucky to escape a similar fate

Arkansas plants, which were raised by various friends and nurserymen. He also made an intensive study of ornithology and in 1832 published a text-book on American birds. J. J. Audubon, who visited him the same year, and found him 'a Gem . . . after our own heart', thought his descriptions of bird-song particularly apt.

In 1833 Nuttall's friend Captain Nathaniel Jervis Wyeth brought him a collection of plants gathered in Oregon, from which he had just returned. On each of his two previous Western expeditions Nuttall had vainly hoped to reach the Rockies; but he neither had the skills needed for survival in the wild, nor the financial means to equip himself with adequate helpers and guides, so the mountains had remained an unattainable dream. He now resigned his post at Harvard in order to be able to accompany Wyeth on his next trans-continental expedition, the object of which was to establish trading-posts beyond the Rockies for the newly-formed Columbia River Fishing and Trading Company. Again we have no account directly from Nuttall, but are able to see him through the eyes of others – the ornithologist J. K. Townsend, and later, the temporary seaman, Henry Dana. Nuttall was now an experienced traveller, an established botanist, and forty-eight years of age, but Townsend's account shows him to have lost none of his earlier ardour.

Nuttall and Townsend were able to make the journey from Pittsburgh to St Louis by steamboat; and after a short cut overland, another steamboat took them from Boonville to Independence on the Missouri, where the party was to assemble. Seventy men – missionaries, hunters and traders – and 250 horses left Independence on 28 April 1834; but instead of following up the Missouri they turned west along the Nebraska River, and crossing the South Fork of the Platte River on 24 May, proceeded up the North Fork. Nuttall was finding dozens of new species daily, and both naturalists had already jettisoned all their superfluous clothing, to make room for the transport of their collections. In the narrow gorges through the bluffs of La Platte, Nuttall 'rode on ahead of the company, and cleared the passages with a trembling and eager hand, looking anxiously back at the approaching party, as though he feared it would come ere he had finished, and tread his lovely prizes underfoot'.

The junction with the Laramie River was reached on 1 June and the company followed up the Sweetwater River to its source and crossed the watershed to the Green River or Siskadee, where they stayed for a fortnight to rest. Then they proceeded by way of Bear River to the upper waters of the Snake River (in the neighbourhood of Pocatello), where the main body stopped to build a trading-post (Fort Hall) while Townsend and a small party were sent to find provisions – a foray made dangerous by the hostility of the Blackfoot Indians. They returned after ten days, well-loaded with buffalo-meat, but Townsend found that in the meantime Nuttall 'had become so exceedingly thin that I should not have known him' – the result of living on grizzly-bear meat, and very little of that. On 6 August the journey was resumed, over arid plains and difficult mountain passes, in one of which they found 'an abundance of large yellow currants', most welcome after months of exclusively animal food; 'some of our people became so attached to the bushes that we had considerable difficulty to induce them to travel again'. By 18 August they were almost out of provisions, and their horses were dying. Shortly afterwards they reached the Boisée River, full of salmon which they had no means of catching; but when at last they got some fish from the Indians the men were sick from the change of diet. All this time they were following the Snake River valley, and as its junction with the Columbia drew near, the men bathed, shaved and changed in preparation for the arrival at civilization; but Townsend dined on rosebuds, and Nuttall and another on an owl that Townsend had shot as a specimen. Nevertheless, they were all in high spirits when at last they reached the Columbia, the fort of Walla-Walla[1] – and food.

The hardest part of the journey was over, but there was still a long trail ahead. The march down the Columbia began on 5 September, and at the Dalles they embarked on canoes, glad of a change of transport, but were nearly swamped by an adverse gale and had to struggle back to land and camp while waiting for better weather. Nuttall's collection of plants got wet, and he spent much time in drying them, with 'a degree of patience and perseverence which is truly astonishing'; he believed he had added nearly a thousand new species to the American flora. They arrived at Fort Vancouver on 16 September, and received a great welcome; almost immediately afterwards Nuttall found *Cornus nuttallii*, and collected seeds which were later raised in England. On 29 September the two naturalists accompanied Captain Wyeth on an excursion up the Willamette River to look for a site for a farm; and on their return they went to live on Wyeth's brig, the *May Dacre*, which was anchored at the mouth of the Willamette River, and made their excursions from there. In early December they sailed to the Sandwich Islands for the winter, returning to the Columbia in the same vessel on 16 April 1835, in time for the salmon-fishing, the migratory birds and the spring flowers. Nuttall spent the next six months at various settlements on the Columbia and Willamette rivers – ground that had previously been covered by David Douglas, who had died the year before; and returned to the Sandwich Islands in October.

Early in 1836 Nuttall started his homeward journey, by way of California and Cape Horn. He reached Monterey in March, and then coasted down to Santa Barbara, San Pedro, and San Diego, where he waited for the arrival of the brig *Alert*, on which he was to make the remainder of the voyage. This was the ship on which the former Harvard student Henry Dana was serving his celebrated two years before the mast;[2] and Dana recognized Nuttall with astonishment. 'I had left him quietly seated in the chair of Botany and Ornithology in Harvard University, and the next I saw of him, he was strolling about San Diego beach in a sailor's pea-jacket, with a wide straw hat and bare-footed, with his trousers rolled up to his knees, picking up stones and shells. . . .' Like the Canadian boatmen years before, the sailors did not know what to make of the white-haired naturalist; they called him 'Old Curious', but thought his madness harm-

[1] Then situated on the Columbia, about nine miles below the confluence of the Snake River
[2] See his book of that title

less. The passage round the Horn was long and stormy, and everyone was relieved when Staten Island was sighted on 22 July; 'even Mr. N –, the passenger, who had kept in his shell for nearly a month . . . and who we had almost forgotten was on board, came out like a butterfly, and was hopping about as bright as a bird'. He begged to be allowed on shore to botanize, if only for a few hours, but one gathers that the Captain's reply was unprintably nautical. By September he was back in Boston.

In 1841 Nuttall fell heir to the estate of his Uncle Jonas (the printer to whom he had been apprenticed as a boy) which was bequeathed to him on condition that he resided there for at least nine months of every year. He was most reluctant to leave America, which had become his home; but since he had resigned his post to go on Wyeth's expedition, his finances had been precarious, and he could not afford to renounce his inheritance. He sailed at the end of the year and lived in England for the remainder of his life, except for a six-month visit to Philadelphia – the last three months of 1847 and the first three months of 1848, thus complying with the terms of the bequest. In the garden at Nutgrove he cultivated nostalgically his favourite American plants, and also the Indian rhododendrons and orchids brought to him by his nephew, Thomas Booth.[1] He suffered much from bronchitis, and died in 1859.

Although he introduced a number of important and valuable plants to cultivation, Nuttall was primarily a botanist; in the field of horticulture he was far excelled by his contemporary, **David Douglas** (1799–1834). The son of a village stonemason at Scone, Perthshire, and apprenticed at the age of ten as gardener's boy at the local 'big house', young David yet contrived by his own efforts to pick up a good education, especially on natural history subjects, and in 1820 obtained a post at the Glasgow Botanic Garden, where he attended the lectures on botany of Dr W. J. Hooker. When Joseph Sabine, the secretary of the Horticultural Society was looking for a collector in 1823, Douglas was warmly recommended; and after three months in the Society's gardens he set out on his first mission to America on 6 June – an expedition which served as a curtain-raiser to the more famous and prolonged travels that followed.

He arrived in New York on 3 August, and though most of his time was spent visiting gardens and nurseries in the neighbourhood of New York and Philadelphia, he made one longer excursion, during which he sampled for the first time the joys and hazards of collecting in the American wild. A month after his arrival he left New York for Albany, and from there travelled to Buffalo and along Lake Erie almost to Detroit, visiting Amherstberg and Sandwich on the Canadian side of the water. (He refers to his day in the woods at Amherstberg, rather oddly, as 'My first day in America'.) At Sandwich, the day being hot, he took off his coat to climb a tree, and his rascally guide made off with it – and the money and other possessions in the pockets. Douglas had to hire a man to drive him back, for 'the horse only understanding the French language . . . placed me in an awkward situation'. After returning to Buffalo he went to see the Niagara

[1] See p. 166

Falls, then back by the same route to New York. On a second visit to Philadelphia he met Nuttall, whom he found 'very communicative' (the highest encomium one naturalist can bestow on another) and also saw Landreth's nursery and Bartram's garden. He sailed for England on 12 December and arrived on 10 January 1824.

The Horticultural Society was satisfied with the results Douglas obtained, and quickly planned to send him on a more ambitious journey. China was suggested, but was considered too dangerous. Instead, the help of the Hudson's Bay Company was requested, and willingly granted on condition that the mission was for scientific purposes only; and after little more than six months at home Douglas set sail on the *Mary and Anne* for the mouth of the Columbia River. He was happy to have the company on board of a fellow-student from Glasgow, John Scouler, who had joined the ship as surgeon for the sake of the opportunities it would afford for the study of natural history,[1] and with whom Douglas was on 'the strictest terms of friendship'.

The voyage was an expedition in itself; it lasted more than eight months, with stops at Madeira, Rio de Janeiro, Juan Fernandez and the Galapagos Islands. By 12 February 1825 they were off the Oregon coast, but owing to adverse winds and storms, were unable to cross the bar of the Columbia River till 8 April, Scouler and Douglas helping to take soundings. (This crossing was always a tricky operation, and on her next voyage, in 1829, the *Mary and Anne* was wrecked on the bar with the loss of all on board.) Next day Douglas was able to go ashore and gather his first Western American plant, *Gaultheria shallon*, specimens of which he had previously studied in Menzies' herbarium in London, and which ever afterwards was one of his favourites.

At this time the Hudson's Bay Company's post of Fort Vancouver-which had originally been established near the coast, was in process of being transferred to a site ninety miles farther up the river, near the farthest point reached by Vancouver's expedition. Here Douglas was received with every kindness by the factor, John McLoughlin, and this became his base for the next two years, where he lived between journeys first in a tent, then in a deerskin lodge and finally in a hut of thuya bark. He spent the summer exploring the lower Columbia and its tributary the Multnomak, with two arduous trips to the Cascade Mountains on each side of the Grand Rapids.

When Douglas returned to the Fort on 13 September he found Scouler there; the *Mary and Anne* had completed her cruise to the north and would soon be returning to England. Douglas set to work to pack his collections, to be sent home under Scouler's care – seeds and specimens of nearly 500 plants, journals, letters and skins of birds and quadrupeds; he could not himself see them stowed on board, for on 3 October he injured his knee on a rusty nail, causing an abscess which incapacitated him for some weeks and was not completely cured till Christmas. Nevertheless, hearing on 22 October that the ship had been delayed by contrary winds

[1] Scouler's interests ran to zoology and bryology, rather than to phanerogamic botany

and was still in the bay, he set off by canoe with a few last seeds and letters hoping to overtake her, but was delayed in his turn by bad weather, and reached the river-mouth an hour after she had sailed.

In spite of his knee injury, Douglas then made an excursion north up the coast to the Chehalis River, which he had not been able to visit in the summer because the local tribes were fighting among themselves, and the sight of their dances and death-songs had caused in him 'a most uncomfortable sensation'. Now he found the villages deserted; and in consequence, could get no food. It poured with rain; he lived on berries and roots, and when at last he shot some ducks, he found his appetite had fled. At last he came to the house of a friendly Indian guide, who gave him help and food. He reached the Chehalis on 7 November and went up it for about sixty miles, still in continuous rain; then struck across country for two days to the Cowlitz River, which he descended by canoe, using his blanket and cloak as sails. By 15 November he was back at Fort Vancouver, after twenty-five days 'during which I experienced more fatigue and gleaned less, than in any trip I ever made in this country'. Shortly afterwards the annual 'express' (two boats and forty men) arrived by the overland route from Hudson's Bay, bringing news and letters – but none for Douglas; the party had started on 21 July, and at that date the ship that left England in May had not yet arrived, so any mail addressed to him would have to wait on the other side of the continent for another year. And for another year, Douglas with much heart-searching decided to stay; he was not, he thought, costing the Horticultural Society much to maintain, and there was still a great deal to be done. On his first season in the country he had travelled a mere 2000 miles; during his second, he nearly doubled that amount.

His first journey in 1826 was with the returning 'express' party (leaving on 20 March) whom he accompanied up the Columbia as far as the Spokane River, where he joined another party bound for Kettle Falls, a trading-post situated at the junction of the Columbia and Colville rivers. This was his base from 22 April to 5 June, one of his excursions being to a point on the Spokane River, ninety miles away, where a man lived who could repair his gun. In June he returned down the Columbia as far as Wallawalla, where he stayed some weeks; the river being too rough for boating or salmon-fishing, he was reduced to living on ground-rats and a little boiled horse-flesh. From Wallawalla he made two trips into the mountains with a rather unsatisfactory Indian guide, but on the second was 'haunted continually by the thought that our people, who were daily expected from the coast, would have arrived and brought my letters', so he returned to Wallawalla on 3 July, and thence downstream to Dalles, where he met the party coming up and at last received some mail. 'I am not ashamed to say', he wrote, 'I rose from my mat four different times during the night to read my letters; in fact, before morning I might say I had them by heart – my eyes never closed.'

It was during this season that Douglas began to experience trouble with his eyes, complaining to Hooker in April of 'dimness of sight'. In desert country near Wallawalla in June his eyes became sore and inflamed from blowing sand and the reflection of the sun from bare ground; and only five days later he was suffering from snow-blindness in the mountains. This was later to have serious results. Meantime, he joined another party of thirty men bound for the Forks of Lewis and Clark's River, the weather being so hot that it was necessary to camp during the middle of day, and Douglas was obliged to resort to the 'certainly enfeebling' practice of bathing, night and morning, in the river. From the Forks he went with one companion on a trip to the mountains, visiting Munroe's Fountain and breakfasting on the spot where Lewis and Clark's party had built their canoes, twenty-one years before. He reported to Sabine that the month's gleanings were 'three bundles of dry plants (ninety-seven distinct species) forty-five papers of seeds, three *Arctomys* and one curious rat, which I hope you will receive safe'.

From Lewis and Clark's River Douglas and three others went across country and by the Spokane River to Kettle Falls; and hearing that a ship, probably the last of the season, was to leave for England on 1 September, Douglas made for Fort Vancouver with all possible haste, to see his precious collections safely despatched. The river was now so low that water-transport in the higher reaches was impossible, and he had to go the 250 miles to Okanagan by land, travelling mostly by night to avoid the heat, but even then 'parched like a cinder'. At Okanagan he was able to transfer to a canoe, and reached Wallawalla on 26 August; though exhausted, he dared not linger, but with a larger canoe and fresh guides arrived at Fort Vancouver on the 29th, having traversed 800 miles of the Columbia valley in twelve days. The captain of the *Dryad* was still at the Fort, and Douglas saw his chests safely stowed on the boat under the captain's personal charge.

When Douglas reached the Fort he looked so weary, ragged and desolate that the inhabitants thought some great disaster had occurred, of which he was the sole survivor; but within three weeks he was off again, this time taking advantage of a party headed by A. R. M'cLeod, which was going south to the Umptqua River. Douglas was particularly anxious to explore the region, in order to find a magnificent pine of which he had heard from the Indians; but this autumn journey turned out to be as miserably uncomfortable and almost as unproductive as his northern one the autumn before.

They set off on 20 September, following the valley of the River Willamette; much of the country had been burned by the Indians and the game was scarce and shy. From 13 October it rained heavily, and they reached the Umptqua three days later, worn out by hard going and insufficient food. The exhausted horses, including the animal that carried Douglas's bedding and scanty comforts, did not reach camp till the following day – but his plant-collections he carried on his own back. The coveted pines still lay some way to the south-east, and after a day's rest Douglas started, with an Indian guide, to seek them; but he blistered his hands so badly trying to make a raft that he had to send the Indian back for help, and then, left to himself, fell down a ravine when hunting a buck and lay unconscious for five hours. Some Indians rescued him, and he struggled back to camp. A new attempt was made a few days later, with the son of a local chieftain for a guide; this time they met with nothing worse than an alarmingly severe nocturnal thunderstorm, and at last reached the region where grew 'this most beautiful and immensely large tree' *Pinus lambertiana* – second only to the redwood for size and splendour. The branchless boles soared gloriously upward (one fallen monster measured 215 ft) and of course the unreachable cones grew right at the top. Douglas shot some down as if they were pigeons – and the noise brought down upon him eight hostile Indians. He had a bad half-hour before he managed to get rid of them, by a mixture of boldness and strategy, and return to his camp with the three cones which were all he had managed to secure.

On 29 October he rejoined M'cLeod's party, now encamped well down the Umptqua River towards the sea; and on 7 November started with two companions on the return to Fort Vancouver, anxious to be back at his base in case anything should prevent his journey home the following spring. Four days later he got lost, and had to spend a night drenched with rain and sleet, numbed with wind and without food or fire; 'on such occasions', he mildly remarks, 'I am very liable to become fretful'. Swimming the Sandiam River a few days afterwards, his collections were swept away and lost, but fortunately the three pine-cones were saved. He reached the Fort after 'twelve days of extreme misery' on 23 November.

One is forced to conclude that Douglas had a certain relish for tales of hardship; as he said himself, 'when my people in England are acquainted with my travels, they may perhaps think I have told them nothing but my miseries'. But the hardships were real and unavoidable. Except on boat expeditions he was obliged to travel light, for much of the country was unsuitable for horses, nor were the Indians amenable as porters; he was rarely able to use a tent. Food supplies were uncertain; the local tribes grew no crops, but lived on fish, game, wild roots and berries; they had seasons of abundance, but also long periods of dearth. Douglas was continually beset by the dangers of starvation, exposure and Indian hostility. Yet when the time came to leave Fort Vancouver he did so with regret, having spent there 'not many comfortable days, but some such happy ones'.

He left on 20 March 1827 by the overland route, joining the well-organized Hudson's Bay Express, which travelled that way every spring – though only Douglas tried to take live plants and live eagles on the arduous journey. The baggage went by canoe, but for the first twenty-five days Douglas walked, taking to the canoe only when his feet were sore. At Fort Colville (12 April) he dug up roots of *Fritillaria pudica*, *Erythronium grandiflorum* and *Claytonia lanceolata*, which he successfully preserved alive until he reached York Factory in August; he afterwards regretted that he did not bring his favourite *Gaultheria shallon* as well. On 27 April they reached Boat Encampment, beyond the two Arrow Lakes, where the canoes were abandoned and the crossing of the Rockies begun on foot. The first day the winding, icy river had to be forded fourteen times in ten and a half miles, Douglas afterwards 'skipping with my load to recover my heat among the hoar frost'. The next day there were seven more fordings and Douglas had to learn to manage unaccustomed snow-shoes; and on the third, they nearly lost the track, as six feet of snow had hidden the blazes on the trees. On 1 May they reached the Athabasca Pass, the highest point of the journey, and Douglas made a side-trip to climb the nearest peak, which he named Mount Brown, but could not spare time to climb the one he named Mount Hooker.[1] Crossing the watershed next day, the temperature rose in the afternoon to 57°F, 'a heat which we found dreadfully oppressive'. On 3 May they were met by a Canadian guide with horses for the next stage of the journey, and at Rocky Mountain House were able to transfer to canoes again.

The miles and the months went by. By the middle of May Douglas had left the Athabasca and crossed by way of the Pembina River to Fort Edmonton on the North Saskatchewan River. Here he collected one of his eagles, a young Calumet; he does not mention where he got the other, a white-headed eagle, but unfortunately both died before he got them to the coast. From Fort Edmonton the river-route led almost due east, across half Alberta and the whole breadth of Saskatchewan to Lake Winnipeg. Douglas, like all botanists, was bored by river-travel, but landings could not often be made as the local tribes were unfriendly. The monotony of the journey was disagreeably broken when one of the party, F. McDonald, was attacked by a wounded buffalo during a hunt and seriously injured before he could be rescued. The medical treatment administered by Douglas consisted of some (surely unnecessary) bleeding, twenty-five drops of laudanum and binding of the wounds, and the victim eventually recovered.

The American continent is vast – but set two naturalists on it, and they will draw together like globules of quicksilver. As Michaux met Fraser and Nuttall met Bradbury, Douglas was to meet Drummond, and, later, Coulter. Douglas had known for some time that Thomas Drummond, 'my old botanical acquaintance',[2] who had accompanied Sir John Franklin's second expedition in the capacity of botanist, was spending the winter on the other side of the Rockies, in the neighbourhood of the Peace River. In October 1826 Finan McDonald, whom Douglas had charged with the transport of some plants and seeds, met Drummond on the Athabasca River, to Douglas's dismay – 'Hope my box is safe', he wrote, when he heard of the meeting. '(Do not relish botanist coming in contact with another's gleanings)' – but actually Drummond saved some of Douglas's specimens from ruin by changing the papers for him. The two met on 3 June 1827, at Carlton House, where Drummond was staying for a month or two on his homeward journey; and Douglas could not withhold his grudging approbation, both for the quality of Drummond's specimens and the liberality with which he showed them and shared his information. They were not long together, for Douglas resumed his journey a day or two later, to Cumberland House and finally to Norway House at the

[1] In Australia, Allan Cunningham named a Mt Brown in 1817, and Mt Hooker in 1828

[2] They probably met when Douglas was helping to collect material for Hooker's *Flora Scotica*, to which Drummond also contributed

northernmost point of Lake Winnipeg. Finding that he had some weeks to spare before the departure of the ship to England, he started an excursion southwards down the lake, travelling as far as Fort Alexander in company with Sir John Franklin and Dr Richardson, and then hiring a canoe and guide to explore the Red River.

Throughout his stay on the Columbia Douglas seems to have been as short of clothes as he often was of provisions; when packing to come home, he had only four shirts (two linen and two flannel), three handkerchiefs and two pairs of stockings, and the list of his personal possessions was pitifully small. But in all his vicissitudes he managed somehow to preserve a suit of Stuart royal tartan – not one of the quieter patterns – and this he now wore. He seems surprised that it attracted considerable notice; 'scarcely a house I passed without an invitation to enter, more particularly from the Scottish settlers',[1] who thought he was fresh from their homeland. He must have been a gay figure – 'a sturdy little Scot, handsome rather', McDonald described him, 'with a head and face of a fine Grecian mould; of winning address, and withal the most pious of men'.

Douglas reached Fort Garry (now Winnipeg) and explored the neighbourhood, before catching a boat to return to Fort Norway; from there he made a rather laborious journey down the River Nelson to York Factory, where he arrived on 28 August, finding the ship for England already in the bay. Here he again met Drummond, and on 1 September the two, with three other passengers and eight crew, went out in a boat to visit the ship, anchored five or six miles offshore. On the way back a violent storm arose; their dismasted boat blew out to sea and only continual bailing kept it afloat during the night and the tempestuous day that followed. It was not until the following night that they were able to use their oars, by which time they had been carried sixty or seventy miles out to sea; but with the aid of a favourable current they managed to get back to the ship. Douglas became 'dreadfully ill' and had hardly regained the use of his limbs before the vessel, after an exceptionally quick passage, reached England on 11 October.

His return was something of a triumph; never before had a collector introduced so many hardy and ornamental plants. It is said that the botanical world was 'literally startled by the number and importance of his discoveries', and that the Horticultural Society had difficulty in distributing his seeds, so large was the quantity. One of his introductions, the flowering currant (*Ribes sanguineum*) was considered to be worth in itself the whole cost of the expedition.[2] Many of our most familiar plants, which had previously been known only to botanists, were now brought into general cultivation; *Cornus alba* and *Mahonia aquifolium*, *Mimulus moschatus* and *Lupinus polyphyllus*, *Clarkia elegans* and *Eschscholtzia californica*, besides the Douglas fir(*Pseudotsuga taxifolia*) and so many other conifers that, as he wrote to Hooker, 'you will begin to think that I manufacture Pines at my pleasure'.

[1] Fifteen years before, Lord Selkirk had established in this area a settlement of evicted Highland crofters

[2] Less than £400

In London Douglas enjoyed being lionized for a time, but he soon got bored and irritable, until his friends began to wish him back in America as heartily as he wished it himself. He suggested to the Horticultural Society that he should be sent to California, but this at first they were disinclined to do, thinking that the day of the plant-hunter was over and that seeds from every accessible country could now be obtained by correspondence. Douglas occupied himself with the preparation of his journals for publication (an uncongenial task which was never finished) and latterly in the study at Greenwich of geographical observation and surveying, under Sabine's brother, Captain Edward Sabine. He was also called in to advise the Oregon Boundary Commission, on account of his knowledge of the terrain. On 31 October 1829 he sailed for the third time for America, with a commission from the Horticultural Society and a load of surveying instruments supplied by the Colonial Office. On this last journey he seems to have buzzed about like a maddened fly – to and fro between Columbia and California, Hawaii and Honolulu; south to Santa Barbara, north to Fort James; but there was a reason for each bewildering change of direction. He was engaged on a survey of the Columbia valley (his map of the river is still considered among the best) he was fascinated by the flora of the Sandwich Islands, and the winter climate of California was kind to his much-tried constitution.

The *Eagle*, which brought him out from England, anchored for a month at the Sandwich Islands, and reached the mouth of the Columbia on 3 June 1830; but for several weeks Douglas was unable to leave the coast, owing to unrest and tribal warfare among the Indians. The lower part of the country had been almost depopulated by an epidemic of fever which had wiped out whole villages, but which Douglas fortunately escaped. Eventually he managed to penetrate to one of his old hunting-grounds at the head of Lewis and Clark's River, in a trip involving '60 days of severe fatigue', and to send three chests of seeds by the *Eagle* when she sailed for home in mid-October. Late in the year he got a passage down the coast to California, and arrived at Monterey on 22 December.

California then belonged to Mexico, and Douglas had to apply for a permit to travel and botanize. This took so long to obtain, that although the Californian spring had already begun at the time of his arrival, it was the end of April (1831) and the flowers were almost over, before he was able to make any but the most restricted local excursions. He then went south down the Salinas valley to Santa Barbara, stopping at the Roman Catholic missions – Nuestra Senora de Soledad, S. Antonio de Padua, S. Miguel Arcangel, S. Luis Obispo de Tolosa, La Purissima Concepcion, Santa Inez Virgen y Martyr – on the way. Like the Hudson's Bay Company's trading-posts farther north, these were the only spots of comparative civilization in the wilderness; and although in general 'no friend to Catholicism', Douglas had nothing but praise for the missionaries, whom he found uniformly helpful, well-educated[1] and disinterested.

He returned to Monterey in June, and set out again on a journey north

[1] Douglas, having no Spanish, conversed with them in Latin

up the Sacramento valley, the object being to locate more Sugar-pines, and if possible to attain from the south the spot near the Umptqua River he had previously reached from the north. He turned back at Fort Ross, having failed of his aim, according to his own estimate, by about sixty miles; actually it must have been nearly four times as far. He may have been in a hurry to return to Monterey, because a Hudson's Bay Company's ship was expected there in November, on which he hoped to get a passage back to the Columbia. But no ship arrived; and Douglas stayed at Monterey all the following spring and summer (1832) never daring to venture far from the coast, for fear of missing the chance of a vessel. The tedium was mitigated by the company of Thomas Coulter, whom he met unexpectedly towards the end of November and who remained in California till the following March. Zealous, amiable, an excellent shot and fisherman, Douglas found him a most congenial associate, and told Hooker 'it is a *terrible pleasure* to me thus to meet a really good man, and one with whom I can talk of plants'.

Tired of waiting, on 18 August Douglas took a passage for the Sandwich Islands on a very small American vessel which took nineteen days to make the crossing. From there he sent home his Californian collections – about 670 species, many of them new, including *Garrya elliptica*, *Dendromecon rigidum*, *Delphinium cardinale*, and whole handfuls of annuals – godetia, limnanthes, mentzelia, nemophila, phacelia, platystemon and a host of others. He was unable to do any botanizing in the islands on this visit, owing to a severe attack of rheumatic fever. About this time, he received news that his patron Joseph Sabine had resigned his post as Secretary to the Horticultural Society, owing to the pecuniary embezzlements of a subordinate, for which he was held responsible. Douglas, though said to be quarrelsome, was intensely loyal to friends who had helped him,[1] and with sympathetic indignation he resigned likewise. Nevertheless he continued to send collections to be shared between the Society and Hooker.

By 14 October Douglas was back on the Columbia River, where he spent the winter mainly on astronomic observations; it was for surveying purposes, too, that he made a laborious but successful early spring excursion to 'New Georgia' (probably the Olympic Mts and Puget Sound) fixing the heights of peaks and the position of capes and headlands. He was preparing, meantime, for a more ambitious journey. During his stay in London he had met the Russian Dr Mertens and Captain Lutke, fresh from their voyage of circumnavigation; and they had suggested that he should return from America by way of Alaska and Siberia. They gave him letters of introduction to Baron Wrangel, the Governor of Alaska; and these, when forwarded, brought a cordial response, and the offer of a passage to Siberia on a Russian vessel. Although warned that Siberia was like a rat-trap 'which there is no difficulty in entering, but from which it is not so easy to find egress', Douglas thought this was 'a glorious prospect', and that the Russians were 'a sort of people whose whole aim is to make you

[1] Michaux also had this quality

happy'. Full of excitement at the possibility of making new geographical and botanical discoveries, he left Fort Vancouver for the north on 20 March 1834.

In his next letter to Hooker, Douglas reported the total loss of sight in one eye. The damage begun by ophthalmia and snow-blindness in 1826 had been completed by the blazing sun of California, and his understandable refusal to wear 'purple goggles' when botanizing. Those who would pity him for his 'half-blind' condition are, however, mistaken. He told Hooker that the sight of his remaining eye seemed to have become all the more delicate and clear, and a companion on the Columbia remarked on the quickness of his sight, especially 'in the discovery of any small object or plant in the ground over which we passed', or in noticing plants on the banks when travelling by boat.

From the Columbia Douglas travelled up the Okanagan River and Lake, where he joined a cattle-party for Fort Kamloops on Thompson's River, and thence up the east bank of Fraser's River to Fort Alexandria. Here he quitted the party and went on in a canoe with one attendant to Fort George, then by the Stuart River to Fort St James, at the south end of Stuart Lake, 1150 miles from his start. His aim was to cross from there to the Skeena River, from the mouth of which he could go to Sitka by sea. Precisely why the journey was abandoned hardly beyond the halfway stage is not known; reference was made to 'a series of disasters', but the worst of them occurred when he was well on his way home by the same route. On 13 June his canoe was dashed to pieces on some rocks in the Fraser River;[1] his astronomical and barometrical instruments and notes were saved, but his botanical notes and collections and all his personal possessions, including his journals, were lost. Douglas himself is said to have spun in a whirlpool for an hour and forty minutes; it is a miracle that he emerged alive.

Damaged in health and much disheartened, Douglas set about retrieving his losses as best he could. He did not return immediately to Fort Vancouver, but stayed for a time at Wallawalla, and from there made another excursion to the Blue Mountains; he also attempted to climb Mt Hood in the Cascades. He was back at the Fort by September, and on 18 October went on board the brig *Dryad* for the Sandwich Islands. It had a stormy passage, and on 4 November put in at San Francisco; but though Douglas went ashore and pitched his tent on the hill of Yerba Buena,[2] he could not venture far, as the captain was only waiting for a favourable wind to continue the voyage. It was resumed on the 30th and the *Dryad* anchored at Woahoo on 23 December; from there Douglas crossed to Hawaii a few days later.

In January 1834 he climbed two of Hawaii's celebrated volcanoes, Mauna Kea and Mauna Loa; then revisited Honolulu from the end of March till early July, when he returned to Hawaii. Shortly afterwards, on 14 July, he met his terrible death – trampled and gored in a pit-trap, by a wild descendant of the cattle landed by Captain Vancouver, on the island

[1] Between Quesnel and Fort George; since blasted away to afford passage for steamboats
[2] Telegraph Hill, now part of the city

that had been fatal to Captain Cook. It seems that he approached too close to the edge of the pit, to observe the animal already trapped there, and that the bank caved in beneath him, throwing him inescapably beneath the bullock's feet; a suggestion that he had been pushed seems to have no justification. His Scots terrier, the companion of all his travels, was found guarding his bundle a short distance away.

Douglas has been blamed for impatience, and for missing important plants that he might have found, in his eagerness to explore new ground. Nevertheless, he must rank as one of the greatest collectors, as well as a notable explorer, geographer and alpinist; and considering his short life and humble beginnings, these were remarkable achievements. Much, however, would have more; and Douglas's abundance, far from satiating the market, only stimulated the despatch of other collectors to the treasure-house of the west.

Thomas Drummond has already been mentioned.[1] After his return home with Douglas in October 1827, he became Curator of the Belfast Botanic Garden (1828–31), and during this period published a work on Scottish mosses which caused the American botanist John Torrey to refer to him later as 'the great Scottish muscologist'. He left to become an independent plant-collector, supported by the Glasgow and Edinburgh botanic gardens and by private subscribers, who were to receive plant-specimens (distributed by Hooker) at the rate of £2 a hundred. He is said to have been 'allured' to Texas by the sight of some plants collected there by the Swiss botanist Jean Louis Berlandier, who accompanied the Mexican Boundary Commission in 1827–30. Drummond, however, did not go immediately to Texas, but worked his way gradually in that direction after two seasons elsewhere.

He arrived in New York on 25 April 1831, and there met Torrey, who reported that Drummond had sent *two tons* of paper round by New Orleans – 'In that, you may judge how extensively he means to collect'. By 7 May he was in Philadelphia, and sent Hooker 'an excellent set of spring plants of Pennsylvania'. Then he went south, probably by stage-coach, by Baltimore and Washington to Frederickstown, and set out to cross the mountains on foot, with his essential luggage in a waggon; having been accustomed to the Rockies he found the Alleghanies 'mere ridges'. The waggon made about twenty-five miles a day, and Drummond found his botanical activities restricted by the necessity for keeping in touch with it. He reached Wheeling in seventeen days, and intended to purchase a small boat to proceed down the Ohio, but found that 'since steam-navigation has become so common', small boats were not to be had. This, too, hindered botanizing, for the steamboat on which he travelled only made short stops, to take in wood. At Louisville he was held up for ten days by an attack of fever, and when he reached St Louis five days later, he had a relapse with complications, which reduced him to skin and bone. Owing to this illness he missed an opportunity of going with a fur-trading party to Santa Fé in the mountains of New Mexico. Nevertheless, he managed to do some collecting, and sent Hooker shells and living

[1] See pp. 309, 325.

plants – reserving some two to three hundred kinds of seeds for the next consignment – before descending the river to his base for the next year, New Orleans.

The excursions he made from there during the following year (1832) included one to Natchitoches on the Red River and three to the northern side of Lake Ponchartrain and to Covington in Louisiana. He sent home three cases of plant-specimens in January and two in August, as well as living plants and seeds, including those of *Penstemon cobaea*, found, but not introduced, by Nuttall. It was not till 1833 that he ultimately got to Texas – and he could hardly have chosen a worse time; it turned out to be a year of floods (afterwards known as the Great Overflow), of political unrest and of a cholera epidemic.

He took ship from New Orleans to Velasco at the mouth of the River Brazos, and went up the river as far as Brazosia, then returned to Velasco to explore the coast. Having made arrangements to go by sea to Galveston Bay, he was seized by cholera; he dosed himself with opium and pulled through, but the ship's captain, his sister, and seven others out of a 'town' of four houses died in two or three days, and Drummond, slowly recovering, almost starved because the survivors were too weak to supply food. Nevertheless he contrived to pack and send home about a hundred species each of plants and birds, with some snakes, land-shells and seeds, including those of two handsome coreopses and *Oenothera drummondii*. He then returned by boat to Brazosia, over prairies flooded to a depth of nine to fifteen feet; the floor of his boarding-house was a foot under water. He continued by boat up the Brazos to Bello and then on foot to San Felipe de Austin where he arrived about the end of July. He planned to go at least forty miles farther towards the source of the river, but was unable to reach the mountains, as the Indians on the frontier were troublesome. In the autumn he returned to the coast and wintered on Galveston Bay, again nearly starving while waiting for migrating birds on the uninhabited Galveston Island.

Once more, in the spring of 1834, Drummond made his way, with difficulty, up the Brazos River; transport was hard to get and provisions short, owing to crop failures the year before. He reached San Felipe in April, too late, fortunately, to join a surveying-party bound for the interior under Captain Johnson, all of whom were afterwards massacred by the Indians. Drummond managed to get as far as 'the Garrison', another hundred miles upstream, and was arranging to join a hunting-party of friendly redskins to the source of the Little River, when he heard that letters had arrived at San Felipe, and hastened back to get them, thinking that they might contain instructions from home – and so missed the opportunity.

A drawing of Drummond, made in Glasgow by Sir Daniel Macnee and probably commissioned by Hooker, shows him to have had a sensitive face with a hopeful expression – but he does not look robust; and although he boasted in the spring of 1834 that his health was excellent, in spite of having been 'tried with such fatigue as would have broken down thousands', in the autumn he came down with his third serious illness in four years. This was a 'bilious fever' followed by severe diarrhoea and an outbreak of boils that made him unable to lie down, plus a felon which rendered his hand useless for two months and nearly caused the loss of his thumb. Small wonder that at Christmas-time he was feeling homesick, and longing to see again his family in Scotland; in any case, he would soon have to return, as his stock of clothing and other necessaries was running very low. He had applied for a grant of land in Texas, within reach of the still-unexplored mountains of New Mexico; and he planned to return home before the end of 1835, and then bring over his family to settle in America. Meanwhile, he intended to visit Florida.

In January 1835 he went by sea to Apalachicola, having recovered his health except for an ulcer on one leg 'for which the *Saw Palmetto* is but an indifferent doctor', but finding the place no more than a sandy desert, set sail on 9 February for Cuba, intending to go from there to Key West and then work northwards through Florida. In June, Hooker received three boxes, which were found to contain not plants, but Drummond's scanty personal possessions; they were followed by a letter dated 11 March 1835, from the Consul at Havannah, enclosing Drummond's death certificate. The letter referred to particulars given in an earlier letter, which was never received, so nothing is known of the circumstances of his end. It is obvious, however, that if he was again attacked by illness, he would have had little power of resistance.

Many of Drummond's beautiful Texan plants – *Herbertia drummondiana, Eustoma russellianum, Callirhoe papaver, Sarracenia psittacina* – have proved difficult in cultivation in Britain, and so are not well-known. One of them, however, will keep the name of this cheerful and gallant collector forever alive. *Phlox drummondii* was one of the last items that Drummond sent home, and was named by Hooker in the hope that it would 'serve as a *frequent* memento of its unfortunate discoverer'. In 1842, only seven years after its first introduction to cultivation, this flower was seen by Drummond's brother James, growing in Mrs Molloy's garden in Western Australia.

Douglas and Drummond were connected by their common friendship with Dr Hooker, and by their nationality (their birthplaces were less than thirty miles apart). But the collectors most directly influenced by Douglas were Hartweg, Jeffrey and Lobb, all of whom revisited Douglas's country for the express purpose of following up his finds, and the unfortunate Robert Wallace and Peter Banks, who might have done great things had they not been drowned in the Columbia at the very start of their expedition. The latter were two young gardeners from Chatsworth who were sent out by the Duke of Devonshire in 1838. Their journey from New York to Montreal and Norway House, and thence by the usual Hudson's Bay Company's route across Canada, did not afford much opportunity for plant-collecting, except in regions 'previously rifled of their botanical novelties'. They had crossed the Rockies and descended far down the Columbia when their boat was swamped in the rapids below the Dalles, on 22 October with the loss of twelve lives. Wallace had taken his wife with him, and she, too, was among the dead.

When **Theodor Hartweg** visited California in 1846 on behalf of the Horticultural Society, he had special instructions to get seeds of the evergreen chestnut, *Castanopsis chrysophylla*, a tree discovered by Douglas but not introduced, and which British botanists were particularly anxious to obtain. Douglas had found some of its nuts in the crop of a bird he had shot, more than two hundred miles from the nearest specimen; he ate them, and found them good. 'This fact', he wrote, 'led me first to the knowledge of this tree.'

Hartweg had already spent nearly seven years collecting in Central and South America, and on this, his second expedition, had just completed a season in Mexico. When he came to the port of Mazatlan, intending to sail from there to California, he found to his dismay that no merchant ship had sailed for the past six months, and that none was likely to leave in the near future, the outbreak of war between Mexico and America being hourly expected. He was refused a passage on an American warship, on the grounds that her destination was secret, though it had been known for weeks that she was to leave in a few days for Monterey. Fortunately two British naval vessels arrived soon afterwards; Rear-Admiral Sir George Seymour proved more accommodating, and Hartweg sailed on HMS *Juno* on 12 May 1846, and arrived at Monterey on 7 June. Less than a month later three American warships arrived, and on 7 July the American flag was hoisted in the town, almost without opposition; but other parts of California were less amenable and for the rest of the year Hartweg was unable to venture much beyond Monterey and San Francisco. All available horses had been commandeered by the Mexican commandant, José Castro, to mount his militia; and strangers were regarded with suspicion by the countryfolk, who 'could not be persuaded that a person would come all the way from London to look after weeds'.[1] The missions which Douglas had found so hospitable had been disbanded in 1834–5 and were untenanted and falling to ruin. Although his movements were so restricted, Hartweg was nevertheless the first to find *Cupressus macrocarpa*.

In August a friend offered him a passage by sea to San Francisco Bay, and after stopping for a few days at Santa Cruz, they anchored off Yerba Buena on 2 September. Here Hartweg reclaimed with some difficulty two Wardian cases that had been sent out for him by the Horticultural Society. 'Some miscreant' had informed the Mexican Government that the crates contained contraband, but on examination it was found that 'instead of silk stockings and printed calicos' they held only two small greenhouses and sundries such as nails and kitchen-garden seeds. The Americans, who then took over, demanded papers from Hartweg to prove his ownership, which he did not possess; however, the matter was settled after a great deal of fuss, and in early October he was able to take his cases back to Monterey.

He had been given the option of staying two years in California, or one in California and one in Upper Mexico, and he now decided, in the absence of further instructions, to spend another season in California –

[1] Hartweg in *Journal of the Horticultural Society* II 1847

partly because his first season had been much restricted and partly because Mexico was so unsettled that 'my peaceful occupation might be disturbed, and my personal safety endangered'. Even in California, now apparently quiet, Hartweg thought that 'if these disturbances should break out again during my busy season, it might much affect my plans'.[1] He therefore decided to go to the Sacramento valley, where the settlers being all foreigners, there was less likelihood of trouble. He was invited to make his headquarters at the farm of a friend, situated to the east of the Sacramento, at the junction of the Feather and Yuba rivers, and here he proceeded at the end of March (1847). From this farm he made several excursions – up each of the two rivers to the Sierra Nevada range, and north-west to the Butte mountains between the Feather and the Sacramento. At the end of June he packed his collections, and sending some by water to San Francisco, set out on horseback to return to Monterey. He had been feeling unwell for some time, and on this journey his feverish symptoms developed into a 'Quotidian fever and ague', which reduced him to such a state of debility that he was hardly able to sit on his horse. Back in Monterey (8 July) he soon cured himself 'with the assistance of my little medecine chest', but had further relapses at the end of the month when collecting seed at Santa-Cruz, and in mid-August at Monterey; he did not recover from the latter attack till early September. He then made a tour to the south, but after going as far as Santa Ines and La Purissima, he found little new, and returned to Monterey on 25 October.

Early in November Hartweg was ready to return home; but traffic was so scanty, that it was three months before he was able to get a passage. On 5 February 1848 he boarded a Hawaiian schooner bound for Mazatlan. On arrival twelve days later, he found that the land-crossing of Mexico would be 'extremely hazardous', so proceeded on the same ship to Iztapa in Guatemala. He hired mules and went inland as far as the capital, but was again warned that it would be unsafe to go farther, so he had to return to the coast and continue, still on the same ship, south-east to the port of Realejo in Nicaragua. From here he managed at last to cross the continent – on horseback to Granada with his luggage in an ox-cart, then by canoe along Lake Nicaragua and down the river San Juan to the port of the same name. He was fortunate in finding a ship due to sail for England in three days' time, and reached Southampton on 3 June.

This was Hartweg's last collecting journey; he afterwards retired to his native Germany, where he became Inspector of Gardens to the Duke of Baden at Schwetzingen. He had a number of laurels on which to rest; his introductions from this last trip had included *Cupressus macrocarpa*,[2] two valuable ceanothus (*C. rigidus* and *C. dentatus*) and a number of annuals, among them *Nemophila maculata*, besides the *Castanopsis chrysophylla* and *Zauschneria californica* which he was particularly instructed to procure. He collected a number of conifers, and is credited with the first introduction of the redwood (*Sequoia sempervirens*), but according to Miles Hadfield[3]

[1] *Journal of the Horticultural Society*, III 1848
[2] Not quite the first, but the first considerable introduction
[3] *Quarterly Journal of Forestry* October 1964

there was an earlier introduction, by way of Russia, in 1843. Many of Hartweg's plants were described in the first volumes of the Horticultural Society's *Journal*, along with those of Robert Fortune, just beginning to arrive from China.

Early in 1850 a number of Scottish gentlemen who wanted new conifers for their estates, banded together to form the Oregon Association, with the object of sending a man to collect seed of Douglas's trees, and of anything else he could find. Although they advertised their intentions only at the beginning of February, subscriptions came pouring in, and by the end of the month they were ready to look for a suitable emissary. From several applicants they chose a young man called **John Jeffrey** (1826–1854?), who had been employed for about a year at the Royal Botanic Gardens, Edinburgh. No time was wasted; Jeffrey was given lessons in the method of taking latitude, longitude and altitude, and a grant of £20 to buy books and instruments; the aid of the Hudson's Bay Company was enlisted: and after passing a few days in London, he sailed on one of the Company's ships, the *Princess of Wales*, on 6 June.

He arrived at York Factory on 4 August and thence followed the usual Hudson Bay route across Canada – but with more than the usual delays, for he waited three months at Cumberland House for the party with which he was to continue the journey. When the 'winter express' arrived on 3 January 1851 he was all ready to accompany it, with a team of four dogs to drag his baggage. On they went from post to post, getting a rest, a fresh guide and fresh dogs at each, and reached Jasper House on 21 March – 1200 miles, walked in snowshoes, and on forty-seven nights sleeping 'with no other covering than the friendly pine'; yet suffering nothing worse than slight frostbite 'that was nothing uncommon among us, and little cared for'.[1]

They stayed at Jasper House till the end of April, and Jeffrey sent a long letter home; so far, he had had little opportunity to collect, but he had acquired the name of being 'a most expert and hardy traveller'. The party left for the crossing of the Rockies on 26 April, reached Boat Encampment on the Columbia two days later, and Fort Colville on 12 May. Here Jeffrey made some local excursions and then continued down the Columbia to the mouth of the River Okanagan. After a year of travel, he had reached the country in which he was instructed to collect.

His first season was passed on the borders of what is now British Columbia. He went up the Okanagan and its tributary the Similkameen, and north as far as Thompson's River, and spent the summer and autumn in the neighbourhood of the Fraser River and the Mt Baker range, finding *Abies grandis*, *Pinus albicaulis*, *Rhododendron albiflorum* and *Cladothamnus pyroliflorus*. Early in October he withdrew to Fort Victoria on Vancouver Island for the winter, and from there sent home his seeds and specimens. As the spring of 1852 drew on, he made several short excursions, including one to Fort Rupert at the other end of the island, and towards the end of April got seeds of a new find, *Tsuga heterophylla*. About 21 May he changed his base, going south by way of the Nisqually and Cowlitz

[1] Johnstone *Notes from the Royal Botanic Gardens, Edinburgh* 1950

rivers to Fort Vancouver, and leaving again on 20 June for a journey down the Willamette valley. He crossed the Umptqua River, and travelling at a more favourable season of the year, was able to reach a more southerly point then Douglas had done, visiting Mt Shasta and the Trinity Mountains before returning north to Fort Vancouver by way of Mt. Jefferson in the Cascades. On this journey he harvested seed of *Pinus balfouriana* and *Lilium washingtonianum*. He was back at the Fort in early December, and sorted and packed his collections; but although he prudently sent them by various routes, three of the five packages failed to arrive.

Up to this time Jeffrey had worked well, and was praised by the Chief Factor at the Fort as hard-working, energetic and industrious. He started the following spring as zealously as ever – down the Willamette again to the Umptqua and Rogue River valleys, the Siskyou Range and Mt Shasta; but this time he crossed the neighbouring ranges and worked his way south down the Sierra Nevada, thus making the link between Oregon and California that Douglas had attempted, but failed to achieve. This should have been a rich collecting region, but though he discovered *Penstemon jeffreyanus* and *Cupressus macnabiana*, the collections he sent home were disappointingly small. He reached San Francisco on 7 October, and did some collecting in the vicinity, but was ill for part of the autumn; his last consignment was sent on 25 January 1854.

Jeffrey's three-year contract with the Oregon Association terminated in November 1853, and in view of his unsatisfactory work during the last season, the committee was doubtful whether or not it should be renewed. A committee-member named William Murray was to be in San Francisco in March 1854; he was asked to interview Jeffrey and dismiss or re-engage him as he thought fit – but the collector was not to be found. He had not written to any member of the Association, nor called for letters at the British consulate, and Murray was unable to trace his movements. It was said that he had joined an expedition to explore the Gila and Colorado rivers, and at least three explanations were given for his disappearance – that he had perished of thirst in the Colorado desert, that he had been murdered by a Spanish outcast for the sake of his mules and equipment, or that he was killed when trading with the Indians. It has also been suggested that he was caught up by the California gold-rush; for whatever reason, he was never heard of again.

In some respects Jeffrey succeeded where Douglas failed; but he was travelling twenty years later, and twenty years in a fast-developing country makes a vast difference to the ease of communications. Nevertheless, Jeffrey has not been given quite the credit that is his due; for his Scottish employers being chiefly interested in trees, many of his herbaceous introductions were overlooked, and were afterwards credited to much later collectors. Among the 119 species of seeds he sent home were *Delphinium nudicaule*, *Dodecatheon jeffreyi*, *Fritillaria recurva*, *Lilium washingtonianum*, *Peltiphyllum peltatum*, *Penstemon newberryi* and *Silene hookeri*; but we do not know how many of them were successfully raised. He also introduced *Camassia leichtlinii* – both seeds and bulbs – and *Spiraea densiflora*.

When Jeffrey was in San Francisco (October 1853 to January 1854) he probably just missed meeting **William Lobb**, who left for home that autumn after a four-year visit, and returned again in the following summer. Lobb was another of the collectors sent out chiefly to get more seed of Douglas' conifers, this time for the nursery-firm of Veitch of Exeter, who wanted to introduce the trees to commerce – all previous collections having been made for private subscribers. Lobb had already made two very successful trips to South America, and had proved himself 'quick of observation, ready in resources, and practical in their application'.

It is unfortunate that those who make the most noise, receive the most attention. Lobb spent fourteen years actively collecting and his introductions to horticulture were numerous and important; he was twice as long in California and Oregon as the pioneer Douglas – but all we know of his activities could be engraved on a sixpence. Douglas, working for a learned Society, was obliged by the terms of his contract to keep a journal, and this was preserved and (eventually) printed; both he and Drummond wrote long letters to Hooker, who summarized their contents in some of his interminable but useful publications. Lobb, working for a commercial firm, seems to have been under no such obligations, and nothing written by him was ever printed. All we have are brief references in books written by members of the firm that employed him – in the *Manual of the Coniferae* (1881) published eighteen years after Lobb's death, and *Hortus Veitchii* (1906) another twenty-five years later – by which time any records may already have been lost or destroyed.

Lobb arrived in San Francisco in the summer of 1849 – the year of the Gold Rush,[1] when 80,000 hopeful prospectors came to the town, 500 ships lay in the bay deserted by their crews, and the lawlessness was unprecedented. He seems to have turned his back on all the excitement, and worked his way south to San Diego, finding *Rhododendron occidentale*, and getting seed of *Abies venusta*, which Jeffrey had failed to obtain. He then seems to have made his base at Monterey, and from there visited the Santa Lucia mountains along the coast in the summer of 1850, and San Juan (near San José) to the north, probably early in 1851. In 1852 he went into Oregon and up the Columbia River – how far is not stated, but he found and introduced *Thuya plicata*. He was back in California in 1853, and hearing of the discovery of a new 'Big Tree' in the mountains, he visited the now-famous Calaveras Grove in the late summer or early autumn, and obtained seeds, specimens and living plants of the Wellingtonia, which he successfully transported home to England – by what route is unknown – arriving about the beginning of December. He returned and settled in California the following year; and although his agreement with Messrs Veitch terminated in 1857, he continued to send occasional consignments till his death in San Francisco in 1863. But even less is known of this second sojourn than of the first.

Lobb's Californian introductions included *Dicentra chrysantha*, *Fremontia californica*, *Ceanothus dentatus floribundus*, *C. lobbianus* and *C. veitchianus*,

[1] Douglas had found gold in California, but had ignored it

besides those already mentioned; his plants and Jeffrey's occasionally overlap, and both men are credited with *Thuya plicata*, *Abies concolor*, *Dendromecon rigidum* and *Rhododendron occidentale*. Lobb is chiefly, though erroneously, remembered as the introducer of the Wellingtonia; a small quantity of seed had actually been sent home by J. D. Matthew a month or two before, but it was from Lobb's living and dried specimens that the tree was described by Dr Lindley in the *Gardener's Chronicle* of 24 December 1853, and named by him *Wellingtonia gigantea* (now *Sequoiadendron giganteum*), after 'the greatest of modern heroes', the Duke of Wellington, who had died the previous year.

This section ends, as it began, with the adventures of an explorer. The travels of **J. C. Frémont** (1813–1890) began a little earlier than those of Jeffrey and Lobb; he was primarily a geographer and surveyor, but made many botanical observations and discoveries. His journeys were not undertaken exclusively for scientific ends; they had also open or unavowed political objectives. Frémont, an officer in the US Corps of Topographical Engineers, combined an unusually thorough and comprehensive scientific training with the characteristics of a *Boy's Own Paper* hero. He inspired great devotion in his men, and led them into quite unjustifiable dangers and hardships; he obstinately insisted on dragging a howitzer through the Rockies, but was forced to abandon it in the Sierra Nevada; his impatience of authority led ultimately to his resignation from the army after a trial for mutiny; and the land he subsequently purchased in California turned out to contain one of the richest gold lodes in Mariposa County. His journals record buffalo-hunts, night-raids by Indians, thirst and starvation, blazing deserts and icebound passes; yet he seems to have turned with relief to botany whenever his other preoccupations would permit, and to have had an aesthetic appreciation of plants not shown by many botanists. What other traveller would describe a valley as being 'radiant with flowers'?

This is not the place to recount in detail Frémont's four amazing journeys, which are part of American history. Perhaps his greatest contributions to botany came from the Mojave Desert, where in spite of the aridity he found a flora 'with greater variety than we had been accustomed to see in the most luxuriant prairie countries. . . . Even where no grass would take root, the naked sand would bloom with some rich and rare flower'. He was the first to see and describe in English that fantastic yucca of the Mojave, the Joshua Tree.

He had his share of the botanist's familiar frustrations, and on this (second) journey twice lost part of his collections – once when a mule lost its footing on a track made treacherous by rain and melting snow, and fell down a precipice, and once when a river broke its banks after a sudden storm and flooded their encampment. The plants that remained, like those from Frémont's other expeditions, were sent to the botanist John Torrey for identification; they included some bulbs, and living roots of lewisia. Altogether Frémont discovered about sixteen new species of plants and two new genera – Carpenteria and Fremontia; but it was left to others to introduce them. Frémont played a prominent part in the Mexican War and in the acquisition by the USA of California and Texas; but as Asa Gray declared, these states were 'annexed botanically before they became so politically'.

Canada

Few records have survived of plants observed or introduced by early French travellers in Canada; they were kept too busy trying to stay alive, to have attention to spare for the flora. It is part of an explorer's business, however, to bring home new and strange plants as a proof of his discovery of new and strange lands; and Jacques Cartier, Canada's first investigator, brought at least two new conifers back to France in 1535, and also, it is said, *Lilium canadense*. (Pierre Belon saw one of his trees in the King's garden at Fontainebleau.) Before 1639, Parkinson was growing *Lobelia cardinalis*, which had been found 'neere the river of Canada, where the French plantation in America is seated'; and by 1635 Jacques Cornut was able to describe about forty Canadian plants in his *Historia Canadensium Plantarum*, most of which had been grown by Jean and Vespasien Robin in the Jardin Royal at Paris; but the means by which they were introduced are not recorded.

Almost the first name that is mentioned in connection with Canadian botany is that of N. Diereville, who visited 'Acadia' (Nova Scotia) in 1706 and published an account of the voyage – in verse – in 1708. He brought a number of plant-specimens to Tournefort, who named the Diervilla in his honour; but there is no mention of seeds or living plants. Somewhat later, the resident physicians of the Governor of Quebec, Michel Sarrasin and his successor Jean-François Gaulthier (commemorated in Sarracenia and Gaultheria) sent home specimens and information about Canadian plants, but made no extensive collecting journeys and no recorded introductions.

The line of demarcation between America and Canada was drawn only in 1818, and is a purely artificial boundary; the naturalists ignored it unless it was forced upon them. Most of the plant-hunters who visited the north were travellers in American territory who rocketed into Canada at the apex of their orbits – Kalm and Michaux in the east, Douglas and Jeffrey in the west; and their travels have already been described. The only major collectors to make journeys almost from start to finish within the Canadian boundary were Masson and Drummond.

After his return from his second sojourn in South Africa and the preparation and publication of his monograph on the Stapelias, **Masson** had petitioned George III for re-employment; and his Majesty 'was graciously pleased to order him to explore such parts of North America, under the British Government, as appeared most likely to produce new and valuable plants'.[1] He sailed about the beginning of September 1797, well-furnished with letters of introduction to people of importance in St Johns, Cape

[1] *Annals of Botany* II 1806

Breton Island, Nova Scotia, New Brunswick, Quebec and 'Upper Canada'. But Britain was then in the midst of the French Revolutionary War; the ship was stopped by a French privateer in the Azores, and permitted to proceed, but on 8 November was attacked and taken by a French pirate from San Domingo. Masson and some others were transferred to a German vessel bound for Baltimore, on which they suffered great hardships from bad weather and shortage of food and water; a ship nearing the end of a long voyage would not be in a position to cater for a number of unexpected passengers. Eventually they were taken on board another ship, and arrived in December at New York.

In the spring of 1798 Masson set out once more on his travels; he went by the Hudson and Mohawk rivers to Oswego, and from there made a lakeside journey to Niagara and Fort Erie, returning by much the same route. By early October he was in Montreal, and in December Banks acknowledged the receipt of a large quantity of seeds. And there the record stops; of Masson's seven subsequent years in Canada, almost nothing is known. A study of *Hortus Kewensis* reveals that twenty-four new species were received from him, including the beautiful *Trillium grandiflorum*, and two more are mentioned by Pursh. This was a meagre harvest compared to his South African gatherings; but perhaps in Canada there were not so many new plants to be found – and we do not know how many of his consignments were lost at sea on the way home. Canada at that time was in a much disturbed state, with the Hudson's Bay Company and the rival North-West Company engaged – often quite literally – in cutting each other's throats, and demoralizing the Indians with fire-water; and Masson may have had more than the climate to contend with.

In 1802 Banks wrote to rebuke Masson for the poverty of the collections he had been sending, and to urge him to take any opportunities that might occur for further travels; and this reads strangely when we recall how the same patron reproved him for taking unauthorized journeys in South Africa. Banks wrote again in July 1805, evidently in answer to one or more letters from Masson, granting permission for him to return to England on account of his ill-health. But it was too late; Masson died in Montreal on 23 December of that year – still in harness, at the age of sixty-four. His effects, including journals, were sent to England and became the property of his nephews; some of his drawings and collections were purchased by the younger James Lee, but when enquiry was made about them in 1884, they could not be traced.

Young Lee was indignant at the treatment Masson had received, 'in being exposed to the bitter cold of Canada in the decline of life, after twenty-five years services in a hot climate – and all for a pittance. He has done much for botany and science, and deserves to have some lasting memorial given of his extreme modesty, good temper, generosity and usefulness'. *The Botanist's Repository* speaks of his 'mild, unassuming and universally-allowed amiability of character'.

The year 1825 saw the arrival in Canada of Sir John Franklin's second Arctic Expedition. The naturalist of the party was Dr John Richardson, but he was chiefly interested in zoology and geology, and on this occasion

Thomas Drummond (d. 1835) was appointed to assist him, especially in the department of botany. Of Drummond's early life little is known except that he took over the nursery of Doo Hillock, Forfar, about 1814, when George Don[1] died and his sons were too young to carry on the business. Drummond contributed some information on Scottish plants for Hooker's *Flora Scotica* (1821).

Sir John Franklin's expedition followed the usual Hudson Bay Express route as far as Cumberland House in Saskatchewan, where the party split up, and Drummond was despatched on an independent trip to the Rockies. (There would have been little opportunity for botanizing if he had accompanied either of the main parties, which were to be engaged on coastal surveys beyond the Arctic circle.) He left Cumberland House with the 'Columbia brigade' on 20 August, and the Indians in the neighbourhood of Carlton House being troublesome as usual, they made no long stay there, but pressed on to Edmonton House in Alberta. Drummond usually left the boats when they stopped for breakfast, and proceeded on shore with short excursions inland, joining the boat-party again when they camped for the night. After supper, he 'laid in' his new specimens and changed the papers of the old ones, which usually took till daybreak, when the boats started again; he then slept on board till breakfast-time.

At Edmonton House Drummond left most of his luggage, including his tent, and taking only some linen and a bale of paper, set off (still with the Hudson Bay Brigade) on the 100-mile land-portage north to Fort Assiniboine on the Athabasca. After a few days' rest they started upstream, some in canoes and some, including Drummond, on horseback; it was now October, and a heavy snowfall made the going difficult, and stopped collecting. They reached Jasper House on 11 October, and continued about fifty miles to the last river-station, Upper House, where the Rocky Mountain Portage began. Owing to the difficulty of transporting luggage over the Rockies, Drummond decided to accompany the brigade no farther; he hired an Indian hunter to look after him, and on 18 October said goodbye to the last Europeans he was to see for six months, hardly noticing their departure in his excitement at exploring this unknown botanical region.

He worked slowly northwards up the Snaring River, with the intention of spending the winter at the Company's post on Smoky River; but bad weather and the domestic troubles of his guide caused so much delay that he was unable to get so far. He was obliged to halt at a wintering-station on the Baptiste River, where he arrived on New Year's Day 1826, and where (having all this time no tent) he built himself a brushwood hut. Here his guide left him, for a point a hundred miles downstream, and did not return till the beginning of March. Doubtless this was not as Drummond had planned; but the mind boggles at the thought of his spending two months of his first Canadian winter quite alone, without books, and almost entirely dependent for food on what he could shoot. Moreover, it turned out to be an unusually severe season, and game and birds were

[1] Father of the collector of that name

scarce. He admitted to Hooker that he was lonely; otherwise he seems to have taken it quite as a matter of course.

After his guide returned, Drummond packed up his effects and his few specimens, and set out on 1 April on snowshoes for the comparative civilization of Jasper House, which he reached in six days; here he received his tent, sent up from Edmonton, a fresh supply of paper, and some welcome tea and sugar. Ducks and geese began to reappear, but vegetation that year was a month later than normal. The brigade from over the Rockies arrived on 6 May. Shortly afterwards Drummond had a narrow escape from a grizzly with two cubs; his gun failed to go off, and he was saved only by the timely arrival of some of the men of the brigade – but he found the moss for which he had been looking. Drummond subsequently discovered that 'the best way of getting rid of the bears when attacked by them, was to rattle my vasculum or specimen-box, when they immediately decamp'.[1]

A hunter that he had engaged to take him to the mountains having failed him, Drummond set out on 16 June with the aged Canadian who took the Company's horses to their summer pasture. He was much hampered by the rapid rise of the rivers with snow-water; sometimes a stream crossed in the morning was impassable by the afternoon, and kept him prisoner for days on the farther side. After the bad winter the game was reduced to skin and bone, and the mosquitoes were troublesome; but flowers were abundant, though often very locally distributed. A month was spent in the neighbourhood of Lac-la-Pierre and the Grande Saline, and he was back at Jasper House on 24 July. A visit from a Governor of the Company was expected, and among the Indians assembled to meet him Drummond found one willing to act as a guide to the north; this time he succeeded in reaching Smoky River, where he stayed till late September when he returned by a different route to Edmonton House. In mid-October he again accompanied the Columbia brigade, and this time crossed the Rockies, following a tributary of the Athabasca to a small lake from the other extremity of which flowed a tributary of the Columbia. From the middle of this lake Drummond took a hearty draught, 'pleasing myself with the thought that some of the water I had tasted might have flowed either to the Frozen or Pacific Oceans'.[2] Over the watershed the character of the flora began to change, and he found many new species, in spite of snow and the necessity for keeping up with the party. He went with them as far as Boat Encampment, and then returned to Jasper House with the man in charge of the horses.

At Jasper House he met Finan McDonald (with Douglas's seeds and specimens) and about the middle of November set out with McDonald, his family and some others, to descend to Fort Assiniboine. The river was now very low and full of rocks and rapids; Drummond and McDonald had frequently to leap overboard and haul the boat over the shallows, with the choice of stripping and getting their legs cut and bruised by ice, or keeping on their stockings and suffering the discomfort of having them

[1] and [2] *Botanical Miscellany* I 1830

frozen. After seven days they could go no farther; McDonald and his family camped, while Drummond with one companion went overland to Fort Assiniboine to get help – a journey they expected to accomplish in three days but which actually took five. By mid-December they all got safely to Edmonton House.

In the middle of March 1827 Drummond left with dog-sleighs and two guides to proceed by land to Carlton House; they had to take an unfamiliar route in order to avoid hostile tribes, and losing their way, took longer over the journey than they expected. Owing to snow-blindness, they were unable to hit the game at which they shot, and were driven to eating the skin of a deer that Drummond had intended to preserve – fortunately not too carefully divested of its flesh. Once they had the good fortune to kill a skunk, which (in spite of the flavour) afforded them 'a comfortable meal'. The dogs got so tired that the men had to dismantle the sledges and carry the loads themselves; and the more exhausted they grew, the slower they went. They reached Carlton House at last on 5 April, and were welcomed by Dr Richardson, back from his expedition to the north.

For the rest of his stay in Canada Drummond lived at Carlton House, making only short excursions and finding little new. He left on 19 July to join the remaining members of the Arctic Expedition at Cumberland House, and the journey from there to York Factory on Hudson Bay was too rapid to allow of much collecting. His adventure on the bay with Douglas has already been described; Drummond boasted that he himself was none the worse for it, and on the next day took a five-mile walk through a half-frozen swamp. They sailed a few days later, and got back to England on 11 October. W. J. Hooker's great *Flora Boreali Americana* (1833) was based largely on the collections made by Menzies, Douglas and the members of Franklin's expedition; of the latter, the contributions made by Drummond were by far the most extensive.

Mexico and the Spanish Main

1570	Hernandez	1778	Masson
1674	Willisel	1788	Sessé and Mocino
1687	Sloane	1798	Humboldt and Bonpland
1689	Plumier	1824	Coulter
1690	Harlow	1836	Hartweg
1690	Reed	1838	Linden
1730	Houstoun	1844	Purdie
1734	Millar	1868	Roezl
1754	Jacquin	1938	Balls

'No language . . . can express the emotion which a naturalist feels, when he touches for the first time a land that is not *European*.' A. VON HUMBOLDT *Personal Narrative of Travels c.* 1822

'There is something very exciting in this kind of life to one who makes up his mind to eat the more merrily the scantier or coarser his meal. . . . I felt an enthusiasm and content which I cannot describe.' JOHN HENCHMAN *The Gardener's Magazine* 1835

Mexico and the Spanish Main

It is convenient to treat the lands surrounding the Caribbean Sea, like the Mediterranean countries, as a single unit – and for the same reason; that many naturalists explored two or more regions in the area, and it is preferable not to interrupt the sequence of their travels by recording part in one section and part in another. Most collectors visiting Mexico or Venezuela used the West Indies as stepping-stones.

As the Eastern Mediterranean was dominated for centuries by Turkey, so the Caribbean was dominated by Spain. Europe was just beginning to appreciate the importance of the discoveries of Columbus, when Pope Alexander VI in 1493 kindly gave to Spain all the new-found lands west of a line drawn from pole to pole a hundred leagues to the east of Cape Verde Islands, and to Portugal all the lands to the east of the same line. This was confirmed by the Treaty of Tordesillas in 1506, but the line was moved 270 leagues further to the west, which enabled Portugal to lay claim to Brazil. The countries left out of this convenient arrangement, particularly England and France, disputed the Pope's right to give away half the world and its rumoured, though still undiscovered riches, even though the first landfalls had been made by Spanish and Portuguese navigators. Hence the exploits of Drake, Raleigh and others and the long years of buccaneering and piracy in the so-called Spanish Main. Many Central and South American plants – canna, nasturtium, sunflower, tagetes and yucca, aloe, potato, tobacco and tomato – were cultivated in Europe in the sixteenth century, but the means by which they were introduced are uncertain.

In 1655 Cromwell launched an expedition to the West Indies which resulted in the conquest of Jamaica, the importance of which was hardly appreciated at the time. Kingston became a great mart for pirate booty, and interested persons were able to purchase there plant-specimens that had originally been intended for Spain.

When the Earl of Carbery went to Jamaica as Governor late in 1674, he took with him **Thomas Willisel** as his gardener, on the recommendation of John Aubrey. Willisel had been a soldier in Cromwell's army, under the garden-loving General Lambert; while stationed at St James's he had been taken 'simpling' by some local botanists, and became an enthusiast, teaching himself Latin in order to master the science, though previously 'all the profession he had was to make pegges for shoes'. He collected natural history specimens in various parts of Britain for the Royal Society, helped the botanist Christopher Merrett to compile his *Pinax*[1] (1666) and went on a botanical excursion with John Ray. After the Restoration he was employed by Robert Morison, both at home and in 'several parts' overseas, to obtain plants for the King's physic-garden at Westminster. Willisel was of the true collector type; 'all his cloathes on his back not worth 10 groats', Aubrey describes him, 'an excellent marksman, and would maintain himself with his dog and his gun and his fishing-line . . . if he saw a strange fowl or bird, or a fish, he would have it and case it'.[2] Great things were hoped of him in Jamaica, but he died within a twelve-month of his arrival. Nearly twenty years later, Ray was still lamenting his loss, and praising his skill and hardiness.

It was Dr (later Sir) **Hans Sloane** (1660–1751) who put Jamaica firmly on the botanical map. He had always been 'very much pleased with the Study of Plants and other parts of Nature'. He completed his medical studies in Paris, where he attended Tournefort's lectures on botany (from 6.0 a.m. to 8.0 a.m., in the Jardin des Plantes) and having taken his degree at Orange,[3] continued his botanical studies at Montpellier under Professor Magnol, returning to England in 1684. Three years later, hearing that the Duke of Albemarle was about to go to Jamaica and wanted to take a physician for himself and his family, Sloane thought that this would afford an opportunity for the foreign travel for which he longed: 'I was Young, and could not be so easy if I had not the pleasure to see what I had heard so much of. . . . This voyage seemed likely to be useful to me as a Physician; many of the Antient and best Physicians having travell'd to the places whence their Drugs were brought, to inform themselves concerning them.'

The Duke was already ill before he left England, and died in Jamaica the following autumn; but his new physician can have been of little service on the voyage, for Sloane was a bad sailor, and suffered from 'a very long and tedious Sea-sickness' for a month going out and six weeks coming back 'upon every the least puff of Wind extraordinary'. They left Spithead on 19 September 1687, and anchored at Madeira from 21–23 October. Sloane's medical services were much in demand during this short visit, but he managed nevertheless to record fifty-seven plants, grasses and seaweeds. The next port of call was Bridgetown in Barbados

[1] *Pinax* = a picture: hence a general survey or representation
[2] Aubrey *Natural History of Wiltshire*. 'Case it' = skin it
[3] Where Protestants were not excluded, as they were at Paris

(25 November–5 December, twenty-six plants noted) then Nevis and St Kitts (9–11 December, fourteen plants) and they arrived at Port Royal on 19 December.

In Jamaica Sloane was again much occupied by his profession, but he was able to make at least one botanical excursion from Spanish Town to the north side of the island, with some gentlemen (one of whom 'drew in crayons'), a good guide and a sure-footed horse. They spent a night in a hunter's hut at the foot of Mt Diabolo, where Sloane 'lay on Plantain and Palm Leaves' but was a good deal disturbed by tree-frogs and other night noises. Next day they crossed the mountain and reached St Annes on the north coast; then they went east to the river Nuevo, and back by Archer's Ridge, the Orange River and Guanaboa. During his stay Sloane employed a clergyman, the Rev. Mr Moore, to make drawings for him, 'and carried him with me into several places of the Country, that he might take them on the place'.

After the death of the Duke of Albemarle in the autumn of 1688, his widow decided to return to England, and Sloane, not wishing to desert her, came back also. He left in March 1689, so his sojourn in Jamaica was comparatively short. He brought back a collection of about 800 plant specimens, and possibly some seeds, since the *Caesalpinia* (*Poinciana*) *pulcherrima* (Barbados Pride or Flower Fence) grown at the Chelsea Physic Garden in 1691, is said to have been his introduction. After his return, he quickly became a famous and fashionable physician, and attended at the deathbed of Queen Anne. The first volume of his *Voyage to Jamaica* came out in 1707, the second not till 1725 – thirty-eight years after the voyage took place; the delay being partly due to the pressure of other work, and partly to the care with which the scientific part of the book was prepared. It was the forerunner and model for all the books of naturalists' voyages that were to follow. But long before it appeared, Sloane's journey had had far-reaching effects, for he showed his collections 'very freely to all lovers of such Curiosities'. His former professor, Tournefort, could not himself come to England, but he sent his colleague Dr Gundelscheimer to see them in 1689, with the result that Louis XIV despatched Father Plumier to collect in the Antilles before the year was out.

Sloane was born in Ireland, and among his Irish friends was Sir Arthur Rawdon, an enthusiastic botanist and horticulturist, who was so impressed by Sloane's collections that he immediately despatched a gardener to obtain living plants and seeds. **James Harlow** had been sent to Virginia about 1687 by John Watts, the curator of the Chelsea Physic Garden, but had not done very well there; in Jamaica he was more successful. He sailed in 1690, and for a long time nothing was heard of him. 'I much wonder what is become of James,' wrote Rawdon to Sloane in March 1692, 'I fear he has a design to cheat me, for I cannot hear the least thing from him'; but in May of the same year he returned in triumph, 'He not only brought over a Ship almost laden with cases of Trees and Herbs, planted and growing in Earth, but also a great number of Samples of them, very well preserved in Paper'. He is reported to have brought twenty cases, each containing fifty plants, many of which were distributed to notable gardens of the day. Those kept by Sir Arthur Rawdon at Moira, Co. Down, are reported to have grown well; some were still alive in 1819.

Another collector sent out as a result of Sloane's discoveries was **James Reed,**[1] frequently referred to as 'the Quaker'. He too had previously been sent to Virginia, in this case by the Earl of Portland on behalf of the King's garden at Hampton Court, and was paid £234 11s 9d for collecting 'Foreign Plants'. In 1690 he was sent to Barbados by a syndicate of which Sloane was a member, arriving on 11 May; shortly afterwards he sent a descriptive list of ninety-three plant-species he had found, and in October a consignment of eighty-six trees and shrubs. William Sherard, one of his patrons, raised over sixty Barbados plants from seeds Reed sent home, but in March 1692 Rawdon reported that he had heard that the unfortunate man had been lost on the voyage home. *Hortus Kewensis* credits him with the introduction of *Solanum mammosum* ('Nipple Nightshade, or Bachelor's Pear'), but gives the date as 1699.

Sloane was cautious about establishing new genera, and it was left to Father **Charles Plumier** (1646–1706) to christen many of his plants. Plumier named more than fifty genera after celebrated botanists or patrons of botany, Begonia, Fuchsia and Lobelia among them. He made three voyages to the West Indies, in 1689, 1693 and 1695, and published his important book, *Nova Plantarum Americanarum Genera* (in which the first fuchsia was described) in 1703. He was on his way to embark for Peru in order to investigate the quinine tree when he contracted pleurisy and died in Cadiz. A final ripple from the stone cast by Sloane into the pool washed Mark Catesby to Jamaica in 1714, and to the Bahamas in 1725.

Plumier's practice of naming genera after botanists was followed more than thirty years later by another collector in the region, **William Houstoun** (1695–1733). Of his first visit to the West Indies as a ship's surgeon nothing is known – and indeed very little is recorded of any of his activities. After his return he studied medicine for two years under Boerhaave at Leyden and took his MD in 1728, at the comparatively advanced age of thirty-four. He then entered the service of the Hon. South Sea Company (about ten years after the expansion and ruinous collapse of the famous South Sea Bubble) and returned to the Caribbean in 1730 as surgeon on their snow[2] *Assiento*. He sent specimens to Sloane from Jamaica, Campeachy and Vera Cruz, and quantities of both specimens and seeds to Philip Miller at the Chelsea Physic Garden. On 6 February 1731 the *Assiento* was driven ashore in a gale and wrecked near Vera Cruz, and though most of Houstoun's property was saved, 'the loss of (his) bussiness' obliged him to apply to Sloane for help. He tried to reach the province of Jalapa to enquire after the plant of that name (Jalap, then highly valued as a purgative), but could not get a permit from the Governor; but he sent an Indian, who brought him four small roots.

The dating of Houstoun's movements is obscure, but it seems probable that it was after this that he introduced Jalap roots to Jamaica (where they

[1] Also spelt Read, Rede, Reid, Rheed and Rheede

[2] A small sailing-vessel resembling a brig

'perished by the negligence of the persons entrusted with their cultivation') and brought specimens, seeds and a drawing to London, which enabled the plant to be identified as a member of the Convolvulaceae. (And after all, he had got the wrong species – *Ipomoea jalapa* instead of the true *I. purga*, used by the Mexicans before the Spanish conquest.) This would be about 1732, when he was made an FRS, and probably when he entered into an agreement with Sloane, Lord Petre, the Apothecaries' Company and others, 'for improving botany and agriculture in Georgia' at £200 a year for three years. He never reached Georgia, however; the last seeds he sent to Miller were from Cartagena, and he died of heat in Jamaica on 14 August 1733.

'I met with a great many Plants on the Continent which I could not possibly reduce to any Genus yet described,' Houstoun wrote to Sloane, 'and therefor have made bold to characterise some of them, giving them the names of Botanists, which is a practice now authorised by Custom.' Seven of these names were afterwards adopted by Linnaeus, including that of Buddleia, given by Houstoun to a plant (*B. globosa*) which he probably found in cultivation, as it is a native of Chile and Peru. He made more than seventy-five introductions,[1] but few of them are grown today. Perhaps the most ornamental were *Aphelandra tetragona*, *Passiflora holosericea*, and the orchid *Bletia verecunda*. He is rather inappropriately commemorated in the little North American 'bluet', *Houstonia caerulea*, with which he had nothing to do.

In 1734, Houstoun was succeeded by **Robert Millar,** sent by a syndicate whose members included Sloane, Lord Derby and Lord Petre. He was a surgeon, who had previously done some collecting for Sloane in the Levant. 'Mr. Millar . . . hath done great things,' wrote Sloane in December 1735; 'having sent many seeds from Panama, Cartagena, Portobel and other Places, with Specimens and dry'd plants. . . . We have subscribed afresh to keep him out another year.' He also visited Campeachy in Mexico and 'the hott parts of the West Indies belonging to England and Spaine' but eventually his health broke down and he returned to England in 1740. Philip Miller, to whom he sent seeds and specimens. considered that he 'fell far short of his predecessor', Houstoun.

In 1753 the Emperor Francis I laid out the park and gardens of Schönbrunn near Vienna, and in the following year he sent a party of four to the West Indies, to collect objects of natural history and plants for his new hothouses. The party consisted of his Dutch head-gardener, Richard van der Schott; a young Austrian physician and botanist, **Joseph von Jacquin** (1727–1817) and two Italian zoologists. They sailed in 1754 and visited a number of the islands of the Lesser Antilles – Grenada, Martinique, Guadeloupe, St Kitts, St Eustatius and St Martin. Their first consignment reached Schönbrunn in August 1755, and a second was brought home from Martinique by van Schott early in 1756; another was brought back from St Eustatius by Bonamici, one of the zoologists, in August of the same year, and three more were despatched from Martinique and Curacoa in

[1] They did not include the buddleia

the early part of 1757. But troubles were in store. The Seven Years' War between England and France had broken out in the early months of 1756, and this hampered their movements. In August 1757 Jacquin was attacked by dysentery and was ill for four months; when he recovered he embarked on a ship that was taken by the British, and he was kept prisoner for a time, at first on Montserrat and later on the desert isle of Gonave off Haiti. After his release he seems to have collected in Jamaica and Cuba, and finally sailed from Havana, with the remaining zoologist, Barculli, in the summer of 1759. The plants he took home, together with the previous consignments, constituted the richest collection ever yet brought from the tropics. Jacquin afterwards became one of the premier botanists of Europe, and was made a Baron in 1806.

War and other disasters also hindered **Masson** when he attempted to collect in the West Indies – this time the War of American Independence. He arrived from Madeira late in 1778, and like Jacquin visited many of the Lesser Antilles – Barbados, Grenada, Antigua, St Eustatius, St Christopher, Nevis and St Lucia; but he was unable to get a passport to enter any of the Spanish possessions on the mainland, and his collections were frequently lost through capture at sea, or spoiled by waiting too long for a convoy. 'When the French attacked Grenada he was call'd upon to bear arms in its defence, which he did, & was taken prisoner fighting in the trenches. He was also in the terrible Hurricane of October 14, 1780, at St. Lucie, and lost there all the Collections at that time in his possession, and great part of his Clothes and Papers.'[1] Finding his work unprofitable under such conditions, Masson returned to England in 1781. He introduced a few plants, but nothing of importance.

All this time the botany of Mexico remained practically unknown, though not unexplored, owing to Spain's extraordinary policy of permitting and even encouraging and supporting research, and then suppressing the results. A law was passed in 1556 forbidding the publication of any book containing information about the Spanish colonies, without a licence from the Council of the Indies – and such licences were never granted. The first to suffer from this policy was Dr **Francisco Hernandez** (1515–1578) who sailed to Mexico on a five-year Government-sponsored scientific mission in 1570, and stayed two further years at his own expense – travelling, enquiring, and assembling the material for his great work. He returned to Spain in 1577 with his researches into natural history and ethnography embodied in sixteen folio volumes, complete and ready for the press. Philip II had them beautifully bound and placed in the Escorial library – and that was that. Worn out and disillusioned, Hernandez died a year later, and his books perished in the great fire which gutted the Escorial in 1671; but some parts that had previously been copied, after many vicissitudes, were eventually published in Rome, seventy-three years after the author's death. A similar case was that of the Jesuit father, Barnabas Cobo (1572–1659) who spent forty-five years in Mexico and Peru, and compiled a natural history of the New World in ten books, which never saw the light.

[1] Banks: quoted in *Journal of Botany* 1884

A comparatively enlightened monarch, Charles III, decided to investigate the natural resources of his dominions, and set out various expeditions, including a commission of five, who were to work in Mexico for six years – a period afterwards extended to eight years, and eventually prolonged to fifteen before the last members were recalled. Five scientists arrived at Mexico City in 1788, including Vicente Cervantes, who was to found a chair of botany, and Dr **Martin Sessé y Lacasta** (d. 1809) who was to take charge of a botanic garden; and in Mexico several more members were recruited, including a brilliant young physician and botanist, **José Mariano Mociño** (1757–1819), born in Mexico of Spanish parents, and a draughtsman called Echeverria. Proceedings started literally with a bang – a magnificent firework display, including a set-piece illustrating and illuminating the sexual system of Linnaeus. But within two years the commission had run into difficulties; a new and less sympathetic monarch (Charles IV) had succeeded to the Spanish throne, and considering Mociño's appointment to be unauthorized, ordered him to be dismissed; salaries were in arrears, and Sessé was already paying part of the cost of collecting-expeditions out of his own pocket. The local Viceroy, however, was sympathetic to the scientists, and interpreted the royal commands with all the latitude he could.

When the order for his dismissal arrived, Mociño was already far afield on an expedition with Sessé, Echeverria and another commission member, Castillo, which began in May, 1790. They started north-west to Queretaro, and then went south-west to Morelia and the volcano of Jorullo (which inspired Mociño to compose some verses); by November they had reached a village called Apatzingan in the same province of Michoacan. Here supplies were so short that they were obliged to carry with them 'a commissary nearly equal to what is taken for a voyage at sea, or run the risk of being reduced to the scant and poor provender of the frugal and semi-barbarous Indian'. In June 1791 they reached Guadelajara and sent off their collections, Mociño writing a famous letter advocating the use of camels in Mexico. Then they divided forces, Sessé going north-west to Tépic and along the coast to the Yaqui River, and Mociño north-east to Aguas Calientes.

Sessé now received instructions that Mociño and the 'best of artists' Echeverria were to join a naval expedition which was on the point of leaving to survey the northern borders of California; this was the Viceroy's expedient for the evasion of Charles IV's order for Mociño's dismissal. The botanist hastened to San Blas and embarked on a vessel which arrived at Nootka Sound on 29 April 1792. Here he stayed for five months, studying the local Indian tongue as well as the flora and fauna, and meeting 'the celebrated navigator Wancower'[1] and his surgeon-naturalist Archibald Menzies, who arrived on 28 August. The Spanish pair got back to San Blas on 2 February 1793, and proceeded more or less directly to Mexico City.

A new expedition was now being planned, and Castillo meanwhile

[1] Captain Vancouver. Nootka was then owned by Spain, but was about to be handed over to the British

having died, Mociño, on Sessé's strong recommendation, was accepted in his place as an official member of the commission. For the remainder of the year they were occupied in relatively short excursions; in October Mociño distinguished himself by climbing the active volcano of Tuxtla not once, but twice, the second time accompanied by Echeverria. The volcano had already erupted four times between May and August, once with such force and fury that the panic-stricken inhabitants of Vera Cruz, 200 miles away, barricaded the streets, thinking the English were attacking. The two naturalists crossed ground so hot that they could not keep both feet on it at once, and their legs were scalded to the knees. On the way back they stayed for some weeks at San Andrés.

In February 1794 a consignment of birds and minerals and a few plants was sent home, and shortly afterwards ten cases of living plants for the botanic garden at Madrid; Tehuantepec, San Andrés and Tabasco on the Bay of Compeachy were visited later in the year. But the really big expedition started in June 1795, when Mociño set out with a different draughtsman, Cerda, on a journey that was to last three and a half years. Gradually they worked their way south and east, often staying several months in one place in order to explore it thoroughly – Puebla, Oajaca, Tehuantepec, Chiapas, and after eighteen months, over the border into Guatemala. By 4 March 1797 they had reached San Salvador, where they experienced a violent earthquake, all Mociño's personal possessions being lost in the ruins. They went on, however, and at the end of May reached Leon in Nicaragua, where they lived for nearly a year, exploring the vicinity. Early in 1798 they began their return by the same route, but were detained in Chiapas by an outbreak of leprosy, for which Mociño's services as a physician were required. They got back to Mexico City on 3 February 1799, and were reunited with Sessé, who had meantime been with Echeverria on an expedition to the West Indies, somewhat hampered by the British blockade and their attack on Puerto Rico. Echeverria, his contract at an end, had deserted to a more open-handed employer; for the salaries of Sessé and Mociño had been in arrears since December 1796, and they were reduced to selling or pawning their most necessary possessions in order to live; the overdue account was not settled till 1801. It was probably this financial stringency that put an end to active collecting, and obliged Mociño to return to the practice of medicine and to work on his herbarium.

The return of the diminished commission to Spain was long overdue, but for some reason it did not sail till April 1803, accompanied by a 'gay and jovial' Mociño, all ready to help in compiling a flora of Mexico. They were coldly received in a Spain preoccupied with the Napoleonic wars; their valuable work was ignored and its authors neglected. Sessé died a few years afterwards, and Mociño's subsequent vicissitudes make a pathetic and tragic story, too long to recount here. He died in 1819, old and nearly blind though only sixty-two; what was left of the *Flora Mexicana* was eventually published in 1888, by which time it was long out of date. The expedition that had started with rockets burnt itself out, and nothing was left but the stick.

At the start, however, when everything had been new and hopeful, Cervantes had sent seeds of the dahlia to Cavanilles in Madrid, who in turn some eight years later gave seeds to the Marchioness of Bute,[1] who brought the plant to Britain in 1798. The Marchioness introduced many Spanish colonial species, more than twenty of them Mexican, between 1795 and 1801; most were of botanical interest only, but they included *Cosmos bipinnatus* and *Zinnia elegans*. It is at least possible that these came by the same channel, and that the seed was originally gathered by Sessé or Mociño.

These two unfortunate collectors were followed within a very few years by a far more famous pair – Humboldt and Bonpland. One hears as a rule a great deal about Humboldt, and very little about Bonpland; but it must be admitted that the distinction is deserved. Bonpland had many virtues as a botanist and travelling-companion, to which Humboldt paid ungrudging tribute; but he had not the wide knowledge, the great sweep of imagination and the grasp of underlying principles that enabled Humboldt to make valuable contributions to half a dozen sciences, and when it came to desk-work he was indolent and procrastinating.

The two young men, **Alexander von Humboldt** (1769–1859) a geologist and physicist with considerable mining experience, and **Aimé Bonpland** (1773–1858) a surgeon skilled in botany and comparative anatomy, met in Paris in 1798. Humboldt was very anxious to travel – before it was too late; he had been delicate as a child, and at twenty-six had 'a sad conviction' that he would soon be feeling the effects of age. (He survived triumphantly every danger and hardship, and lived to be ninety.) The previous November, he had signed an agreement to go with the wealthy and eccentric Lord Bristol, Bishop of Derry, on an eight-week expedition to Egypt; but this plan was frustrated first by Napoleon, whose Egyptian campaign was launched from Paris a week or two after Humboldt's arrival, and then by Nelson's activities in the Mediterranean. Humboldt then arranged to join the great exploring expedition to Australia that was being planned, under Captain Baudin; Michaux and Bonpland were appointed as naturalists, but the outbreak of war with Germany and Italy caused the French Government to withdraw the promised funds, and the expedition was indefinitely postponed.[2] The disappointed Humboldt then joined a Swedish Consul on his way to Algiers, who thought he might be able to get permission for Humboldt to botanize in the 'Alps of Barbary' (the Atlas Mountains); but after waiting for two months at Marseilles for the Swedish vessel in which they were to sail, they learned that she had been so badly damaged in a storm that she had had to put back for repairs, and would not arrive till the spring. Humboldt and Bonpland booked a passage on a small local vessel, but heard just in time that the Sultan of Algiers was imprisoning all Frenchmen on arrival and had cancelled the usual caravan to Mecca on the grounds that Egypt was 'polluted by Christians'. Frustrated, the two hired mules, and in January 1799 went to Spain instead. In Aranjuez

[1] Her husband was Ambassador at Madrid from 1795 to 1797
[2] Baudin eventually sailed on 19 October 1800

Humboldt discovered that the ambassador from Saxony to the Spanish court was an old family friend, Baron von Forell; and on 17 March the Baron obtained for him an interview with Charles IV. Such was the impression Humboldt made, that he was given permission to pursue his researches in South America, and provided with a passport and a royal letter of recommendation – an unprecedented concession from a Spanish monarch to a foreigner. The momentous South American journey was initiated, therefore, almost by chance; any quarter of the globe that offered would probably have been accepted by the would-be travellers.

Humboldt gave his hosts little time to change their minds. He visited some of the Spanish museums to learn what had already been collected, saw the drawings sent home by Cervantes and met the returned South American travellers Ruiz and Pavon; his preparations were expedited by the fact that he was paying the expenses out of his own modest patrimony. He and Bonpland left Madrid in May and sailed from Corunna in the corvette *Pizarro* on 5 June 1799.

The *Pizarro* was reputed to be a lucky ship, but a slow one, and by the time she neared South America she had much sickness on board, including a fever 'of a typhoid character', so she docked at Cumana in Venezuela instead of going on to Cuba or Mexico as had been intended. Humboldt – who had escaped both fever and seasickness – and his companion were enraptured by their first sight of the vegetation of the tropics. 'Hitherto we have been running about like a couple of fools,' wrote Humboldt, 'for the first three days we could settle to nothing, as we were always leaving one object to lay hold of another. . . . Aimé Bonpland declares that he should lose his senses if this state of ecstasy were to continue.' They stayed for four months exploring the vicinity, including an excursion south-east over the mountains to Cumanaçoa and thence north-east to Caribe on the coast; and on 18 November they left for Caracas. They went by sea to Nueva Barcelona and there separated – Bonpland, a bad sailor, to take the overland route, collecting by the way, while Humboldt continued by sea to La Guayra with the baggage and valuable instruments. He reached Caracas on 21 November and was rejoined by Bonpland four days later.

Here they stayed for more than two months, for they had planned an ambitious expedition which could not be undertaken till the end of the rainy season. This was to discover and explore the waterway said to connect the Orinoco and the Rio Negro, a tributary of the Amazon. Its existence had been reported by de la Condamine about 1745, but a party sent in search of it in 1754 lost 312 men out of 325, and failed in its object. Since then a number of mission stations had been established along the rivers, but the enterprise undertaken by the two naturalists still required considerable courage.

Leaving Caracas on 7 February 1800, they went at first westward to Cura on the south side of Lake Tacarigua and on to Valencia; then south-east to Calabozo on the river Guarico (where they made some painful but instructive experiments with electric eels), and south again to San Fernando de Apure. Hitherto most of the journey had been across vast barren plains or llanos; but after a few days at San Fernando they transferred to a large canoe with a pilot and four Indians and on 30 March began the descent of the river Apure to its junction with the Orinoco. They reached the Orinoco on 5 April and began to work their way upstream. On the following day a sudden storm half-filled their pirogue with water; the Indians sprang overboard and swam for the shore – but Humboldt, it seems, could not swim. The sturdy and devoted Bonpland offered to swim, across more than a mile of crocodile-infested water, with Humboldt on his back; but fortunately a gust of favourable wind got them out of their predicament. At Urbana they stopped for a change of boat and crew, then continued up the river, in ever-increasing misery from the plagues of insects. The Orinoco was one of the 'white' rivers where mosquitoes abounded; even the natives found them excessive. Above the Maypures rapids the insects were so bad that Bonpland had to crawl inside a large conical native oven, the only place where he could get peace to press his plants. It was a relief when they reached another San Fernando (S. Fernando de Atabapo) and were able to transfer to the 'black' river Atabapo, where there were no mosquitoes. But the harm had been done; from this point Bonpland became increasingly inert and lethargic, already infected with the fever which nearly ended his life.

From San Fernando there were two possible ways to the Rio Negro; south up the Atabapo to Yavita and across a short land portage to the river Guainia, one of the headwaters of the Negro, or a more easterly and longer route by the Upper Orinoco and its extraordinary branch the Casiquiare, which flows the wrong way and eventually joins the Guainia farther down.[1] Humboldt and Bonpland crossed by the first route and descended the Guainia as far as San Carlos; they could not go beyond, without entering Portuguese territory, and the Portuguese were suspicious and hostile. (Humboldt was in danger of being arrested as a spy and sent to Lisbon.) They returned by the Casiquiare (a 'white' river again, notorious for the bloodthirstiness of its insects) carefully surveying and mapping both routes, and then went up the Orinoco as far as Esmeralda, but could not explore the river to its source owing to the hostility of the local tribes. They then started for home down the Orinoco; but at the Apure confluence Bonpland fell ill with severe malaria, and by the time they reached Angostura his life was despaired of. Humboldt was beside himself with anxiety, and to make matters worse, he knew that it was near here that Pehr Loefling had died, at exactly Bonpland's age of twenty-seven. After his sojourn in Spain,[2] Loefling was sent by the Spanish government with two assistants and two young draughtsmen to Venezuela, but he died less than two years later; the drawings brought home by the draughtsmen in 1761 were never published.

Lodged with a local doctor and dosed with quinine and angostura-bark, Bonpland eventually recovered, and a month later was able to make the ten-day journey (13–23 July) across the llanos to the port of Nueva Barcelona. The war then in progress between England and Spain rendered

[1] This was the link of which de la Condamine had heard [2] See p. 38

the movements of shipping more than usually uncertain and precarious. After an interval the two naturalists took a passage on a contraband sloop bound for Trinidad; but she was almost immediately fired on by a privateer from Nova Scotia, and during the parley that followed both ships were overhauled by a British man-of-war. A polite young Englishman came aboard and invited Humboldt to spend the night on the *Hawk*, where Captain Garnier, who had already heard of his exploits, entertained him lavishly, and enabled him next day to continue his voyage to Cumana. Here Humboldt and Bonpland stayed till 16 November, when they took advantage of a neutral American ship to leave the blockaded port for Nueva Barcelona, and thence to Cuba, arriving on 18 December.

It was during their three-month stay in Havana that they encountered the shipwrecked Frasers, father and son.[1] Humboldt treated the forlorn pair with the greatest hospitality, lodged them in his house, supplied them with money and obtained permission for them to travel freely about the island. At this time he was getting anxious about the disposal of his collections, some of which had already been spoiled by insects and damp. The seas were full of pirates; it was reckoned that three out of every four letters sent to Europe failed to arrive, and when he later got to Cartagena he found that no European mail had been received there for a year. He could not send direct to Hamburg as Spanish ships were not allowed to enter neutral ports. He decided therefore to divide the collections into three, one part to be taken by the younger Fraser to London via Philadelphia and sent to Germany from there; one part – Bonpland's share – to France via Cadiz, and one to be stored in Havana to await his own return. The consignment for France was lost by wreck, but that taken in charge by Fraser reached its destination safely.

While in Cuba, Humboldt learnt from the American newspapers that Baudin's expedition to Australia was to sail via Cape Horn, Chile and Peru. He had never quite given up hope of joining it, and now thought he might be able to do so at some South American port. When he left Cuba, therefore, on 15 March 1801, it was to go to Colombia and south as far as Lima in Peru.[2] Disappointed in his expectations – for Baudin after all went via the Cape of Good Hope – he and Bonpland again turned northwards, and sailed from Guayaquil to Mexico in February 1803.

They landed on 22 March at Acapulco, a region then little known, and worked slowly northwards by the hot valleys of Mescala, the high plains of Cilpancingo, the silver mines of Tasco, Cuernavaca with its fine eighteenth-century garden, and Tezcuco, to Mexico City (11 April). Here they met Vicente Cervantes, the last to remain of the Spanish commission of 1788, for Sessé and Mociño had already started their journey home. From the capital Humboldt and Bonpland made several excursions. The first, in May and June, took them north-east to the mining area of Pachuca, Real del Monte and Acropan. In August they started a longer circuit – north to Huehuetoca, northwest to San Juan del Rio and Querétaro, west to Celaya and Salamanca and northwest again to Guana-

[1] About the beginning of February 1801 [2] See p. 284

juato. Here they stayed for some time, visiting the many mines in the vicinity, then in September struck south and west by Valle de Santiago, Cuitzeo, Morelia, Patzcuaro and Ario to the mountain of Jorullo. Between Patzcuaro and Ario they found the wild ancestor of the dahlia, previously known only as a cultivated plant.

Jorullo was a midget among mountains – 4331 ft compared with the 17,881 ft of Popocatepetl and the 21,424 ft of Chimborazo; but it was a volcano and had achieved notoriety by rising some 1680 ft above the surrounding plain in a single night.[1] (Humboldt was an *aficionado* of volcanoes and in South America had climbed every one within reach.) This was their farthest point; they retraced their steps north-east to Lake Cuitzeo and Acambaro, then south-east to Tepetango, Ixtlahuaco and Toluca, where they climbed the Nevada de Toluca (this time a real mountain of nearly 15,000 ft) and so back in the late autumn to Mexico City.

At the turn of the year they made a short local tour in the Valley of Mexico, visiting Santiago, Tezcuco, Guautitlan, San Cristobal and Guadelupe; and in January 1804 set their faces eastward for the first stages of the long trek home. They halted on the way to climb and measure the mountains of Popocatapetl and Ixtaccihuatl, and again at Perote for the ascent of the Cofre de Perote. On 7 March they sailed from Vera Cruz to Havana, to pick up the collections deposited there; but even then they did not return to Europe till they had visited Philadelphia and had stayed for three weeks with President Jefferson at Monticello. They arrived finally at Bordeaux on 3 August in excellent health and spirits, with thirty-two cases of treasures and material for eleven books in twenty-nine volumes – nearly all the work of Humboldt, as Bonpland could not be persuaded to contribute his proper share, and his botanical work had ultimately to be completed by Kunth.

Humboldt claimed to have introduced more than forty species of plants to the gardens of Europe, but the few that have proved of horticultural value were all collected in Mexico. They included *Antirrhinum maurandioides*, *Lobelia fulgens* and *L. splendens*, and 'several varieties of Georgine' (i.e. the dahlia) the seeds of which were raised by the Empress Josephine at Malmaison. His Mexican travels were not nearly so extensive as those of Sessé and Mociño, but his work was of far greater scientific importance – and not only because his results were published and Mociño's were not. But the great value of Humboldt's travels lay less in his own discoveries than in the enormous influence of his writings on others; Darwin, Frémont, Roezl, Schomburgk, Wallace and Webb were among those inspired by his example.

Many of Humboldt's Mexican localities were visited twenty years later by **Thomas Coulter** (1793–1843), who went much farther afield and stayed for ten years instead of ten months. After taking his medical degree in Dublin, this engaging Irishman went to Geneva with a letter of introduction from the great taxonomist, Robert Brown, and studied botany

[1] 20 September 1759

for seventeen months under de Candolle. He then planned an ambitious collecting journey which was to take him from Buenos Aires westward to Mendoza, then up to the western flanks of the Andes through Chile and Bolivia to Lake Titicaca, and finally to Mexico and California. For some reason – perhaps financial – this larger plan was abandoned, and in 1824, Coulter signed a three-year contract to act as physician to the Real del Monte Company, newly-formed in Britain with the object of rehabilitating certain mines which had been left flooded and derelict since Mexico had attained independence in 1821. The manager or commissioner, Captain Vetch, took out a quantity of Cornish mining machinery, and having little confidence in the local labour, recruited a European personnel. It was to care for these that Coulter was appointed, but his energy and benevolence soon led him to supply free treatment and medicines to the impoverished Mexican peons as well.

He arrived at Vera Cruz on 27 January 1825, and after some delay in obtaining mules, reached Real del Monte by way of Jalapa, Apam and Tulancingo on 20 February. He became very friendly with Captain Vetch, who taught him the principles of land-surveying; he also learned to assay ores. At the end of October he was sent to another mine owned by the company – Veta Grande near Zacatecas; a three-week journey by way of Pachuca, Tepetango, San Juan del Rio, Querétaro, Celaya, Salamanca, Guanajuato, Silao, Leon, Lagos and Aguas Calientes. Veta Grande was the only mine that had remained in production throughout the revolutionary wars; the Spanish staff resented British control, and the local Indians were lawless – Coulter reported that in Zacatecas twenty-one murders occurred in a month. In December he made an arduous and urgent midwinter journey to see a sick man at Bolanos, and delayed his return for a day or two, ostensibly to avoid travel on a Sunday, but actually in order to attend a Christmas ball at the priest's house.

Coulter stayed at Veta Grande for more than a year, during the latter part of which he was acting as manager of the mine. (Unfortunately the region was extremely arid, and afforded little scope for the botanist.) Finding his expenditure in this position unduly lavish, Vetch sent him early in 1827 to take charge of the lead-mining and smelting operations at Zimapan in the state of Hidalgo, involving another cross-country journey of nearly three weeks. (Coulter has been accused of lingering unduly on these journeys, in order to botanize by the way.) He stayed at Zimapan till the expiration of his contract, which he did not wish to renew, as Vetch had been succeeded by another commissioner, whom he disliked. Zimapan was in a better botanizing area, and in the spring of 1828 Coulter sent to Europe, at considerable expense, two collections of living cacti – one of seventy species to the botanic garden of Trinity College, Dublin, and one, comprising fifty-seven species, forty-seven of which were new, to de Candolle at Geneva.

We now lose sight of our botanist for more than three years. He seems to have moved over to Western Mexico and engaged in various financial enterprises – profitable at first, 'but four robberies and a shipwreck balanced my accounts without the help of a bookmaker'. We find him again in August 1831 at Guaymas on the Gulf of California; he has wound up his affairs and is now returning to botany and his long-deferred project of a visit to California. The rainy season having set in with unusual severity, he abandoned the idea of going by land, and embarked about 14 August on an American brig for Monterey.

California was still in a very primitive state. Little towns of a few hundred people had begun to arise round the military posts of Monterey, Santa Barbara and San Diego; the only other settlements were the missions, of which there were about twenty, and the only track ran along the coast. Flowers being out of season at the time of Coulter's arrival he 'took a race' down to San Gabriel (now part of Los Angeles) to assess the prospects for the spring. He was back in Monterey before the end of November, and there met David Douglas, returned from his trip to the Sacramento valley. The Scotsman found the Irishman congenial; unfortunately Coulter left no record of his impressions of Douglas. The two worked together till 20 March 1832, when Coulter left for another trip to the south.

He made his way by stages down the coast to San Gabriel again, and beyond to San Louis Rey on the river of the same name, and then upstream to Pala; it was probably hereabouts that he found the magnificent *Romneya coulteri*. Then he went over the mountains and southward along the small stream of San Felipe to the sandplain of Carizal. From there he crossed 100 miles of waterless desert to the Colorado River, which he forded about a mile below its junction with the River Gila; he was probably the first botanist to enter Arizona. Having worked for a time in the neighbourhood of Yuma he returned to Monterey by the same route, after an absence of seventeen weeks.

He planned to return to Mexico about July, but again his movements are obscure. He sailed via Cape San Lucas, and it may have been at this time that he visited Mazatlan, San Blas, Tepic and Guadelajana. By the end of April 1833 he was in Mexico City, engaged on yet another mining project – forming a company with two partners on a commission basis, for the purpose of which he intended to settle in Guanajuato. He was reluctant to return home empty-handed and without the fortune his relations expected him to make – 'I am not avaricious,' he wrote, 'but really I don't like to be always poor'; but owing to the defection of his partners this project, too, was a failure.

Meantime he was detained two months in Guanajuato by a severe outbreak of cholera, for which his medical skill was required, and for which as usual he gave his services free of charge. He was also hampered by 'a very serious-looking revolution', in which the town was besieged, and as his house was directly in line with the small fort that made the most resistance, he experienced the new sensation of having shot and shell whistling round his head. He was still in Guanajuato in November 1833, but got back to Britain some time in 1834, with a collection of some 50,000 plant specimens (counting duplicates) and examples of nearly 1000 species of woods, mostly with their foliage and inflorescences. These he brought safely home, but unfortunately the box containing his journals and botanical manuscripts was lost in transit between London and Dublin, and never recovered. Coulter arranged to hand over his collections[1] to Trinity College on condition that he should be appointed their curator, but his health had suffered severely in his travels, and he died nine years later. The cacti already mentioned were his only recorded introductions, but his name is known to gardeners through the plants he discovered – *Romneya coulteri*, *Philadelphus coulteri* and others. Both he and Douglas found the Californian *Pinus coulteri*, which Douglas named after his friend.

In 1833, when Coulter left Mexico, the fashion – or passion – for orchid-growing was just beginning, and from then onwards orchids figured largely in all collections made in the region. When **Theodore Hartweg** (1812–1871) was sent out by the Horticultural Society in 1836, he was instructed to work in Mexico at altitudes from which hardy or near-hardy plants might be expected; but even in his case, an exception was made for orchids. His personal tastes ran to the lush and tropical, but he was a 'steady, well-informed and zealous young man' and conscientiously carried out his instructions. He was descended from a German gardening family and had worked in the Jardin des Plantes before becoming a clerk at the Horticultural Society's gardens at Chiswick. His visit to Mexico in 1836 was the start of a six and a half year collecting-journey on behalf of the Society.

He arrived at Vera Cruz on 3 December and sent off his first consignment, including sixty-five different orchids, within a fortnight. After ten happy days at Zaquapam he set out on horseback for Jalapa, where he intended to make a short stay; but finding the country unsafe through robbers, he proceeded by the diligence to Mexico City. Here he stayed only long enough to present his letters of introduction, and to make arrangements for continuing his journey to Guanajuato, which he reached on 13 January 1837. It was still too early for flowers, so he filled in time by a short visit to Silao, returning on the 22nd. He stayed in Guanajuato for more than two months; his excursions from there into the mountains involved him in the expense of a horse and a mounted escort, lest he should be stoned or lassoed, 'as has happened to several foreigners'. He moved on to Léon in April and to Lagos in June, but found the country there barren and unrewarding. Continuing northwest, he reached Aguas Calientes in mid-July, and finding this a more productive region, stayed till late September. Then he made nearly a fortnight's journey over bad paths in heavy rains, westward to Bolano, where he hoped to find orchids and pines; but the air was too dry for epiphytes and the pine-seed was not yet ripe. The finest orchid he saw was in the hat of a Quichole Indian; at some risk Hartweg penetrated into the country whence the man came and got the orchid and six others, as well as other new plants. Having left orders with a forester at Bolanos to send him on some pine-seed when it was ready, Hartweg proceeded to Zacatecas, where he hoped to find some

[1] They included plant-specimens previously collected in Europe, and also large numbers of shells

interesting cacti – 'nobody, I believe, having been collecting there yet'; apparently he did not know of Coulter's residence in the neighbourhood twelve years before. He stayed from 14 January 1838 to 22 February, but the cacti that he found were only common kinds. By April he was south-east at San Luis Potosi; but the pine-seeds from Bolanos not having arrived, Hartweg went back there – over 200 miles as the crow flies – to 'look after them' himself. This time he was too late – the seed had all been shed; but he got some other seeds, including those of a white rhododendron.

After a trip to Guadelajara and an unaccountable second visit to Zacatecas, Hartweg went south-east to Morelia about the end of June. This proved to be a good centre, and he stayed for two months, making excursions into the mountains (although the country, especially to the south-west, was exceedingly insecure, with roving bands of robbers up to 300 strong) and finding 'a most beautiful broad-leaved Fuchsia' (*F. fulgens*). His next move was westward to Anganguao, and after a month, north-west to Real del Monte, where again he made a long stay, going on excursions with a resident botanist, Charles Ehrenburg.

Hartweg was more unfortunate than most collectors in the timing of his journeys, and seemed to have a particular propensity for sailing into the most stormy political waters. Before he left England the Admiralty had given authority for his consignments to be conveyed on the official Mexican packets free of charge, 'whenever room could be found for them, without inconvenience to the passengers'. But before he had been long in the country, war began to threaten between Mexico and France; the French blockaded the Mexican ports, and the British packets – the only vessels allowed to leave harbour – were overloaded with cargo and extra passengers, and could not take Hartweg's boxes. Two consignments sent on 2 February and 19 April 1838 reached England together on 27 February 1839, and most of the plants were dead; another, sent in September, never arrived at all. Meantime Mexico was becoming more and more lawless and unsafe; five times Hartweg mentions the prevalence of robbers and twice had narrow escapes. In August 1838 he wrote anxiously to the Horticultural Society to ask for instructions; he was willing, he said, to go anywhere, 'begging only to be provided with a map of the country I am to visit'. The Council of the Society had already become alarmed at the personal risks their collector was running, and by the losses caused by the French blockade, and after careful deliberation decided to send him to Guatemala; but their instructions to that effect did not reach Hartweg till 24 January 1839, when he was at Real del Monte, and the expected hostilities with France had already begun.

He lost no time in getting under way, and left for the south on 2 February; but the situation in Guatemala did not look so good from Mexico as it had done from England several months before. Some sort of revolution was in progress, and Hartweg lingered purposefully by the way, hoping that the trouble would be over before he arrived. He stayed five months in Oajaca, happy to be again in a 'splendid country of constant verdure' after the arid plains and mountains of the north; in

August he moved on to Tehuantepec, and in October he was over the border at Quezaltenango, after a difficult journey with every imaginable setback. He stayed there till the end of the year, climbing an active volcano in the vicinity and finding there 'a most splendid Fuchsia, in flower and seed' – *F. splendens.* He reached the city of Guatemala about the beginning of January 1840, and later extended his journey to South America.

Not all the consignments Hartweg sent home were lost or delayed; 140 species of orchid reached the Horticultural Society safely, a number of cacti, several conifers, and seeds in such large quantities that the Society was able to distribute 7000 packets to its members. Besides the two fuchsias already mentioned, his introductions included *Echeveria fulgens*, of a genus named after Sessé and Mociño's draughtsman; *Lupinus hartwegii* and the valuable *Penstemon hartwegii.*

Hartweg returned to Mexico in 1845, again for the Horticultural Society; but this time he only stayed six months, instead of nearly three years. He arrived at Vera Cruz on 13 November, and after staying for a time with friends near Orizaba, again took the diligence from Jalapa to Mexico City. Then he went like a homing dove to a spot near Angangueo where he had found *Achimenes grandiflora* and *A. heterophylla* seven years before; although not a leaf was now showing, he located and dug up some roots of the former. Back in Mexico City, he went in search of *Abies hirtella* in a locality given by Humboldt; he failed to find it, but found instead *Pinus montezumae* – barely hardy, but one of the most beautiful members of the genus. On 19 December he took the diligence to Guadelajara and from there set out on horseback for Tepic, which was his base for the remainder of his stay; but his outings were hampered by heavy rains, swollen rivers and flooded plains. Early in March 1846, he sent home two consignments of seeds and plants, and at about the same time his luggage arrived from Mexico City, where it had been held up by transport difficulties following a change of government. On 14 March Hartweg left Tepic for San Blas, Mazatlan and California,[1] where this storm-petrel of a collector again ran into a war, this time between Mexico and America.

During Hartweg's first visit to Mexico and Guatemala another collector was working within the same frontiers; but they did not meet until both had entered a third country, Colombia. This was the Belgian horticulturist, **Jean Jules Linden** (1817–1898). He had been one of the first science students in the new University of Brussels, and when only eighteen was sent with an artist and a zoologist, on a Belgian government mission to Brazil. His second journey, again with a Belgian mission, took him to Cuba late in 1837, and to Mexico in March 1838. They seem to have confined their activities to well-known ground – the volcanoes of Popocatepetl, Ixtaccihuatl, Cofre de Perote and Orizaba, and the eastern slopes of the Mexican Cordilleras; Hartweg at this time being north at Zacatecas. Then they went by boat from Vera Cruz to Campeachy, and began to explore Yuca tan; but at Laguna de Terminos Linden nearly died of yellow

[1] See p. 317

fever, and only after three months of painful convalescence was he able to proceed. He then went by sea to the state of Tabasco and across the high regions of Chiapaz to Guatemala, then in revolution, which he entered in the north somewhat later than Hartweg had done in the south. He did not stay long, however, and was back in Belgium via Campeachy, Havana and the USA by February 1841.[1]

Before the end of the same year Linden returned to the Americas,[2] this time accompanied by a stout fellow called Louis Joseph Schlim. They landed at Guayra in Venezuela, and for three months or so explored the flanks of the coastal mountains. Then they began a long journey – westward from Caracas to Valencia, north over the mountains to Puerto Cabello, and through the unhealthy forest of San Felipe to Barquisameto. At Rio Tocuyo the river was in spate, but they forced a passage with the loss of some mules and part of the collections; then they worked southwest down the Eastern Cordilleras – Trujillo, Merida, Socorro and Velez – to Bogotá, which they reached in October 1842. In December they started westward to Ibagué, crossing the River Magdalena, there 325 ft wide, by swimming; and climbing Mt Tolima to the snowline, established a camp at 15,000 ft where they stayed for several weeks, exploring the flora of the high altitudes. Then they continued westward by Cartago and across the Cauca valley to the Pacific, and back by a more northerly route through Honda to Bogotá, where they met Hartweg, who had been as far south as Peru and was now on his way home. He and Linden went together on an excursion north to Pacho, where they found *Odontoglossum crispum*.

After leaving Bogotá, Linden and Schlim returned to Caracas by a similar route, arriving on 17 August 1843, and leaving again on 16 November to explore the 'mysterious' Sierra Nevada of Santa Marta and the Peninsula of Guajira, 'inhabited by ferocious and cannibal Indians'. In March 1844 they sailed from Rio Hacho for a thorough investigation of the little-known south-western corner of Cuba, and were there during the great storm that devastated the island in October of that year. After a visit to North America they got home to Europe, much exhausted, in October 1845.

Besides dried specimens, Linden brought or sent to Europe large numbers of living plants, especially orchids, and though some died in transit he was able successfully to introduce *Masdevallia coccinea*, *Odontoglossum hastilabium*, *O. lindenii*, *O. luteo-purpureum*, *O. triumphans* and many others. On his return he established a nursery specifically for the introduction of new plants, which he removed from Luxemburg to Brussels in 1853 on receiving the appointment of Director of the Brussels Zoological and Botanical Garden, a post that he held till 1861. He retained his interest in commercial horticulture, and eventually became a director of Horticulture Internationale Ltd, which, like the firm of Veitch in England, sent out its own collectors; in 1894 it had no less than seven expeditions in the field. Linden died, full of orchids and honours, in 1898.

[1] His companions had returned the previous autumn

[2] This seems to have been a private enterprise, not a Government mission

After parting from Linden at Bogotá in the spring of 1842, Hartweg worked his way home by way of Jamaica, and at Kingston in May 1843 he met a collector newly arrived from England, **William Purdie** (1817?–1857), with whom he went on excursions during the short time of his stay. Kew had just started on a new period of expansion as a public garden under the direction of Dr William Hooker, and Purdie was the first Kew collector to be sent out since the death of Sir Joseph Banks in 1820. He arrived at Kingston on 14 May and went on his first excursion with Hartweg five days later.

For a year Purdie explored Jamaica; sometimes by boat, sometimes with 'two mules laden with paper, saw, trowels, hampers etc.', and sometimes in forests so dense that the mules had to be left behind; many Wardian cases of exotic plants were despatched to Kew. In May 1844 he left the island and sailed to Colombia, arriving at Santa Marta a few months after Linden had left it. From May to July he collected in the Sierra of Santa Marta, and the following summer moved inland, his main object being to obtain seeds and plants of the ivory-nut palm, *Phytelephas macrocarpa*. In July 1845 he was at Ocana, where he found *Begonia fuchsioides*, and was then planning to go to Bucaramanga, make a side-trip from there to Pamplona and back, and then continue to Bogotá.

In 1846 Purdie was appointed Government Botanist and Superintendent of the Botanic Garden at Port of Spain, Trinidad; and except for two more collecting trips – to Venezuela in 1851 and Puerto Rico in 1854 – he lived there till his death in 1857. Several of the West Indian botanic gardens became the refuges of wearied collectors. Purdie's predecessor at Port of Spain had been David Lockhart, almost the only survivor of Captain Tuckey's ill-fated expedition; the botanic garden at Bath in Jamaica was founded and for a dozen years supervised by James Wiles, who had accompanied as gardener Bligh's second and successful breadfruit voyage; and from 1816–22 the botanic garden on St Vincent was in charge of George Caley, retired from his travels in Australia.

The collectors of hothouse exotics tend to be less well-known than the collectors of hardy plants; this applies to Linden and Purdie, and also to that remarkable man **Benedict Roezl** (1824–1885), though the latter has also a few hardy plants to keep his memory green. He was one of those tiresome collectors who cannot be pinned down to one place – in Colombia at one moment, on the Columbia the next; but Mexico was the starting-point and focus of his activities. The son of a Czech gardener, Roezl was apprenticed at the age of twelve in the gardens of the Count of Thun in Bohemia, and subsequently worked in several important continental gardens, including those of Baron von Hugel[1] at Vienna and Count Lichtenstein in Moravia, and the famous nursery of Van Houtte at Ghent. Having worked much in greenhouses, Roezl in 1854 'could no longer restrain [his] ardent wish to see the tropics', and emigrated to Mexico, where he founded a nursery near the capital. In 1857 he issued a catalogue of the Mexican conifers he had for sale, but his botany was of

[1] Von Hugel had himself travelled and collected in Australia and India

the optimistic kind, and the eighty-two 'new species' listed were all varieties of a few previously-known kinds. In 1861 he relegated the nursery to his partner and took a plantation, where besides the usual crops of sugar, coffee and tobacco, he introduced the cultivation of the Ramie (*Boehmeria tenacissima*) as a textile-plant. He invented a machine for the extraction of the fibres which aroused a good deal of interest in agricultural circles, and when demonstrating this apparatus in Havana in 1868, he got his hand caught in machinery making sixty revolutions a second, and sustained injuries that resulted in the loss of his left arm. It was only then, half-crippled and forty-four years of age, that he embarked on the arduous life of a plant-collector.

He started modestly enough with some travels in Mexico; then in 1869 journeyed by way of Havana and New York to the Sierra Nevada of California, where he collected some of the most valuable of his hardy plants – *Lilium humboldtii*, so-named because he found it on 14 September 1869, the hundredth anniversary of Humboldt's birth, *Lilium roezlii* and *L. parvum*. Then he dashed off to Panama and Colombia; sent home 10,000 orchids and some 500 other species of plants from the neighbourhood of Ocana, and then turned his attention to the Santa Marta range, previously visited by Linden and Purdie. He was there in May and June 1870 – the rainy season, and for twenty days was never dry; but the rare Telepogon orchids that he collected at 11,000 ft died as soon as they were brought down to warmer levels. He obtained, in all, seed of 417 species of Colombian plants and sent about 3000 odontoglossums to Europe, then doubled back via Panama to California again, reaching San Francisco on 1 August; but owing to the outbreak of the Franco-Prussian war,[1] some of the consignments he sent home about this time were delayed in transit and were dead on arrival. After recovering from an illness Roezl embarked on 20 August to go by sea to the Columbia River, visiting Portland – active and bustling with railway-building but still surrounded by uninhabited forests – and Fort Vancouver, the farthest point that could be reached by steamer. Here he found *Lilium columbianum*, and quelled with the power of his eye a black bear whom he encountered on one of his excursions. He also harvested a large quantity of conifer seeds, and travelled 1100 miles back to the Sierra Nevada for those of other species, only to find that the trees, which had cropped heavily the year before, were coneless. After this disappointment he sailed from San Francisco on 15 November for Bonaventura and a journey in north-west Colombia.[2]

In 1862 Roezl spent four months in Europe, during which time he realized his assets, and when he sailed on 3 August from Liverpool to New York he took with him all his savings, which he intended to invest in further journeys. He crossed North America by rail, and during a halt at Denver, Colorado, entrusted almost the whole of his capital to a Danish innkeeper, to look after while he was absent on an excursion to the mountains; he seemed surprised that on his return the innkeeper was nowhere to be found. (Altogether, Roezl reported that he was robbed

[1] On 19 July 1870

[2] See p. 374

seventeen times; one cannot avoid the suspicion that he may have been a little careless.) From Denver he seems to have taken the Atlantic-Pacific Railway by way of New Mexico, where he found *Abies concolor*. Soon he was back in his favourite Sierra Nevada, for further supplies of conifers and lilies; then he shipped from San Francisco to Acapulco in Southern Mexico, and combed the Sierra Madre for orchids, 3500 of which reached London 'in fine condition'. Across the Isthmus of Panama he went, to Guayra and Caracas (eight tons of orchids); back by Cuba and Vera Cruz and across Mexico to the State of Oajaca (the double Poinsettia and ten tons of other plants) and after a slight detour to New York, by Panama to Peru,[1] and back from there to Europe.

After only three months, however, this insatiable collector was back again, and crossed North America via the Cheyenne Territory and Colorado to the Sierra Nevada. Then he went to a new part of Mexico, the vicinity of the volcano of Colina, which he climbed. It was here that the Indians, learning that Roezl paid for orchids at the rate of 10 to 15 francs a hundred, brought him 100,000 plants. (One gets the impression that if parts of Mexico are now desert and devoid of vegetation, it is because Roezl had been there.) Mexico, the first country in which he collected, was also the last; Roezl returned from there to Europe in 1874, and spent his eleven remaining years at Smichow near Prague.

Roezl's vast collections found a ready market in Europe. From the early days he sent seeds and plants to his former employer, Van Houtte; lilies went to Dr A. Wallace in England, orchids to the specialist firm of Henry Sander of St Alban's or to the big London auction-sales, and members of the Bromeliaceae to the botanic garden of St Petersburg, where a study of the order was being made. Many consignments were sent to Herr Ortgies of the Zurich Botanic Gardens, who undertook the distribution of plants and specimens to various subscribers. Nevertheless, Roezl retired with only a modest fortune; few plant-hunters died rich.

In spite of Roezl's depredations, Mexico still had flowers to give. In 1938 **E. K. Balls** and Balfour-Gourlay visited the country, accompanied on this occasion by Mrs Balls. Part of their mission was to collect wild species of potato for the Imperial Agricultural Bureau, and they confined their researches to a few familiar southern mountains, staying, however, at remote and unfamiliar villages away from the beaten track. Even at so late a date, they found quite a number of plants which, if not new to science, were at least new to cultivation.

Their first mountain (at the end of March) was the Nevada de Toluca, where they stayed for some time at Ojos de Agua, at 12,000 ft. On Popocatepetl they found their first potatoes. Cofre de Perote was visited twice, in May and September; it yielded *Agastache mexicana* and *Mahonia atriphylla* var. *saxatilis*. On the Pico de Orizaba they stayed first at the village of Lomagrande, at 9000 ft; the inhabitants had an evil reputation, but received the strangers courteously, and the flora was so good that here, too, they returned later for seeds. At Tesmalaquilla, higher on the same

[1] See p. 375

peak, they lodged in a disused school, and found *Rubus triloba.* The last mountain they climbed was Mt Malinche, the upper slopes of which are now a national park; they ascended it from three different points on two visits. Here, too, the inhabitants, sullen at no longer being allowed to exterminate their forests by resin-gathering and charcoal-burning, had a bad name for turbulence and banditry; but they improved on acquaintance, and the flora was varied and interesting. In November, Balls received a cable from the Imperial Agricultural Bureau, asking him to continue his potato-researches in South America; so he and Balfour-Gourlay sailed directly to Buenaventura, while Mrs Balls, who had been making a series of plant-drawings, returned home.

Some years later Dr Balfour-Gourlay gave some plants of *Rubus triloba* to Captain Collingwood Ingram at Benenden, who crossed them with the related North American *Rubus deliciosus* to produce the hybrid 'Tridel', one of the most excellent flowering shrubs of recent years.

South America

THE AMAZONS

1637	Piso and Marcgrav	1825	Burchell
1814	Cunningham and Bowie	1836	Gardner

THE ARGENTINE

1817	Bonpland	1825	Tweedie

THE ANDES

1709	Feuillée	1840	Bridges
1712	Frézier	1840	Hartweg
1735	De Jussieu	1840	Lobb
1760	Mutis	1859	Pearce
1777	Dombey, Ruiz and Pavon	1869	Roezl
1801	Humboldt and Bonpland	1925	Comber
1826	Cuming	1927	Elliott
1827	Poeppig	1939	Balls
1830	Mathews		

When I was but thirteen or so
I went into a golden land;
  Chimborazo, Cotopaxi
They took me by the hand. . . .

The houses, people, traffic seemed
Thin fading dreams by day;
  Chimborazo, Cotopaxi
They had stolen my soul away.

W. J. TURNER *Romance*

'. . . I went last year to the Andes, and what did I find? Practically nothing but Andean plants. . . .' CLARENCE ELLIOTT *New Flora and Sylva* I 1928–9

South America

The Amazon

For horticultural purposes, South America can conveniently be divided into three regions, the Amazon, the Argentine and the long chain of the Andes, the last by far the most important. The rivers and equatorial forests of the Amazon basin may have seemed a paradise to the explorer-naturalist (especially the entomologist) but it is not from Brazil that our garden-borders are recruited. Some good plants have come from the pampas, but their numbers are few compared to those from Chile and Peru.

In 1637, when the north-east corner of the continent round Pernambuco (Recife) was temporarily owned by the Dutch, the scientifically-minded Count of Nassau-Seigen was sent out to organize the defence of the colony. He took **William Piso** as his personal physician, and two young Germans, **Georg Marcgrav** and H. Cralitz (medical and mathematical students) on purpose to investigate the resources of the country. Piso, whose interest was chiefly in medicinal plants, accompanied the Count in his campaigns against the Portuguese; Cralitz died within twelve months, but for seven years Marcgrav travelled extensively, making astronomical observations and collecting animals and plants. He died of fever in 1644, shortly before the Count returned to Holland with Piso and large natural-history collections, which afterwards afforded material for twelve volumes, three of them on plants. Then the shutters of suspicion and mistrust went up, and no more was heard of Brazilian botany for 170 years, except for such observations as de la Condamine was able to make when he descended and mapped the Amazon from the Andes to the coast in 1743–4. Rio de Janeiro was a port of call for shipping of all nations, but visitors were rarely allowed to go beyond the town.

At the conclusion of the Napoleonic wars a more liberal policy was adopted, and botanists flocked in, as they were later to flock to released Algeria and Japan. Almost the first to arrive were **Allan Cunningham** and **James Bowie**, sent by Banks to Rio in 1814; they were quickly followed by the French St Hilaire and the German von Martius, both of whom published extensive works on the Brazilian flora. Cunningham and Bowie did not go far afield; their longest trip was south-west to Sao Paulo, involving a month of strenuous travel, for which Banks gently chided them, as he thought the results were not worth the hardships endured. After about eighteen months Cunningham was ordered to Australia and Bowie to Cape Town; the former left on 16 September 1816 and the latter twelve days later. They introduced some valuable stove-plants, including the gloxinia (*Sinningia speciosa*) and *Jacaranda ovalifolia*; but their South American sojourn was no more than a curtain-raiser to their more important travels elsewhere.

In exchange, as it were, for Bowie, South Africa lent to Brazil one of her most notable botanical explorers, **William Burchell.** After his return to England in 1815 Burchell spent some years growing his South African bulbs and seeds in his father's garden at Fulham. In 1825, the year of Brazilian independence, he obtained permission to accompany Sir Charles Stuart, who was being sent to Rio de Janeiro on a diplomatic mission. Burchell had made an ambitious plan to cross the continent by Goyaz, Cuiaba and Matto Grosso to Cuzco and Lima in Peru; but he had great difficulty in obtaining the necessary recommendations and passports – having nearly succeeded with one government, there was a change, and he had to start all over again with the next, meanwhile making short local excursions to the Organ and Estrella mountains. On 20 September 1826, fourteen months after his arrival, he set out, armed with eighty-four letters of introduction and the long-awaited passports.

He was in no hurry, however. He went by sea to Santos, and spent three months in a forest hut at the base of the Sierra de Cubateo. In January 1827 he moved to Sao Paulo, where he rented an eight-roomed house for the rainy season at 6s per month. From here he despatched his already large collections before proceeding by leisurely stages north-east to Goyaz, which he reached in November 1827. In Goyaz, where no Englishman had been before, he spent the next rainy season; and having received news during his stay that his father's health was failing, he abandoned his trans-continental project and turned north for the shorter journey to Para. Even this took eighteen months, counting a long wait at Porto Real (Porto Imperiale) for suitable conditions to descend the River Tocantins by boat. At Para he had another eight-month wait before he could obtain a passage home; and by the time he reached England on 24 March 1830, his father was dead. Over the years he had sent home 132 packages, none of which, by a miracle, had been lost; but they seem to have contained only manu-scripts and specimens – there is no report, this time, of the introduction of seeds or living plants.

Burchell travelled almost directly from south to north; the same region, but a little farther to the east, was traversed a few years later from north to south by a more avowed horticulturist, **George Gardner** (1812–1849). He was one of William Hooker's Glasgow students, and sailed for South America (financed by twenty-four subscribers) very soon after taking his MD. He arrived at Rio in July 1836, explored the neighbourhood, and studied the language and customs of the country for a year, in preparation for the three-year journey on which he embarked in July 1837.

He started by sea, and after calling at Pernambuco and exploring some way up the Rio San Francisco, he sailed on northwards to Aracaty, where he began his travels into the interior. He went south-west up the rivers Jaguaribe and Salgado to Crato, then west across the Sierra Cayriris to Oeiras, then steadily south-west again by Parnagua and Santa Maria to Natividade. Here he was on a tributary of the Tocantins River and not far from Burchell's route; but he turned away from the rivers to the mountains, and worked south-east to Arrias, San Domingos and Nostra Senora de Abadir, across diamond-mining country to Diamantina, and thence south to Rio on 2 November 1840. It took him three months to sort and pack his collections, including as they did 'upwards of 60,000 specimens in Botany alone', and when they had been despatched to England he made another excursion to the higher parts of the Organ mountains and the rich country of the Parahyber river. He left for home on 6 May 1841, with 'a large number of the most beautiful plants in a living state' in six large Wardian cases, but owing to a defect in the construction of the cases more than half of the plants died on the voyage home. His successful introductions from this and previous consignments included the orchids *Oncidium forbesii* and *Sophronitis coccinea*, and that magnificent greenhouse climber named like a ballet-dancer, *Tibouchina semidecandra*, which was first discovered by young Linden a year or two before. Gardner afterwards became Superintendent of the Peradeniya Botanic Gardens in Ceylon, and died of an apoplexy when taking off his boots, at the early age of thirty-seven.

Shortly after Gardner's visit the great explorer-naturalists of the Amazon began their voyages of discovery – Wallace and Bates in 1847 and Spruce in 1849. Alfred Russel Wallace retired after four years owing to illness, and his collections were destroyed by fire on the voyage home. Henry Walter Bates stayed for eleven years; he was primarily a zoologist and entomologist, though he collected plant-specimens as well. Richard Spruce explored and collected for fifteen years, ending with a government-sponsored trip to Ecuador to obtain seeds and cuttings of the quinine-tree for transfer to India; in a country largely clothed by towering trees with epiphytes perched almost out of sight on their summits, his principal interest was in liverworts and mosses. None of the three was in any sense a horticulturist, and their stories cannot be told here.

The Argentine

The situation in the Argentine was much the same as that in Brazil; no botanizing was possible so long as the country was under Spanish control. Sir Joseph Banks twice attempted to get permission for a collector to go to Buenos Aires, but neither application was successful. It was not until the Argentine became independent in 1816 that it was possible to enter the country, and for many years afterwards it was difficult to move about in it with safety. There were too many revolutions to be comfortable; fifteen changes of government took place between February and October 1820. Nevertheless, six months after the liberation ceremonies had taken place, the first botanist – **Aimé Bonpland**, no less – landed at Buenos Aires.

After their triumphant return to Paris in 1804[1], Humboldt had obtained a pension for his companion from a grudging Napoleon, and shortly afterwards Bonpland was appointed Superintendent of the Empress Josephine's gardens at Malmaison and Navarre. These were his happiest years. But in 1814 the Empress died; Bonpland grew restless, and decided to return to South America, first making a visit to England to get the latest information on the state of the country. He arrived at Buenos Aires on 29 January 1817 and was received with acclaim and appointed Professor of Natural History to the Medical Faculty. But local jealousies soon made his situation uncomfortable, so he resigned his post and embarked on another journey of exploration.

He planned to cross the Gran Chaco to Bolivia and the Andes, and continue into temperate latitudes the collections he and Humboldt had made in the tropics; but a more immediate object was the investigation of the maté tree (*Ilex paraguariensis*) from whose leaves a tea was made that was considered essential to the welfare of the miners of Peru. The Jesuits had monopolized its manufacture, as the Dutch in the East Indies had monopolized the spice-trade, by establishing enclosed and guarded plantations on the Uruguay and Parana rivers and endeavouring to exterminate the tree in the wild; they exported the tea only in powder-form, so that the leaves could not be identified. They were expelled in 1767 and their plantations went back to the wild.

Bonpland started his travels accordingly up the River Uruguay to the deserted Jesuit station near Itapua, and in doing so approached territory in dispute between the Argentine and Paraguay. He wrote to the Caudillo of Paraguay, Dr José Francia, to ask permission to continue his researches over the border. Francia, as appeared later, himself intended to take over the Jesuit monopoly; he pretended to regard the maté story as a cover for espionage, and sent a force of 400 men to capture and arrest the botanist and his little party. On 3 December 1821 several of Bonpland's servants were killed, the rest and himself wounded; he was taken prisoner, and in spite of all that his influential friends in Europe could do, was kept in captivity for nine years. He was given a certain amount of personal freedom but was not allowed to go beyond the grounds of the mission in which he was housed, where he maintained himself by the practice of medicine and the distillation of brandy from honey. Released at last in February 1830,[2] he settled in the small town of Santa Borja near the Brazilian border, and allied himself to a native woman, by whom he had several

[1] See p. 341 [2] Some authorities give the date as 1831

children. After 1853 he moved to a larger estate on the banks of the Uruguay a few miles south of Restauracion, where he died in 1858, aged eighty-five. He corresponded with Humboldt to the end, and was always talking of returning to France and bringing back his herbarium in person; but he never did. Humboldt, four years his senior, survived him by a year.

James[1] **Tweedie** (1775–1862) was only two years younger than Bonpland, and he, too, lived and travelled in the Argentine till his death, four years after Bonpland's; the two corresponded, but so far as is recorded, never met. Tweedie was born in Lanarkshire and trained in horticulture, rising to be head-gardener at Edinburgh Botanic Garden and later working as a landscape-gardener on several Scottish estates. He was already fifty years of age when he emigrated to the Argentine in 1825, but like Drummond in Australia, he accomplished more collecting in his declining years than most men in a lifetime. He seems to have settled at Santa Catalina near Buenos Aires and resumed the practice of landscape gardening, but his principal support was a store in the city, looked after by his family during his absences.

His first recorded journey took place about 1832, in the company of H. S. Fox, the British envoy at Rio de Janeiro. They went by boat about sixty miles up the River Uruguay, then returned to the sea and followed the coast northwards to Rio, 'not passing a single port or point where the ship could go, without landing and strictly searching every hill and valley where anything was to be found'.[2] Next came a disastrous voyage to the south, when he embarked on a small vessel for Patagonia provisioned for five days; it grounded on a sandbank and took nineteen days to reach Bahia Blanca, after the sacrifice of half its cargo. The hard and brackish water of Bahia Blanca made Tweedie ill, and he dared not venture far from the fort, because of numerous and hostile Indians.

For his third journey Tweedie joined a *tropa* – a caravan of seventeen waggons, 240 cattle, forty-four horses, thirty-five mules and thirty-two persons, bound for Tucuman in the interior; a journey that was expected to last forty to fifty days but actually took some time longer. They left on 2 March 1835 and slowly wound their way north-westwards, covering 120 miles in rather more than a fortnight, during which time Tweedie found only seven new plants. They then entered the province of Santa Fé and pressed rapidly (by their standards) across a belt of arid country inhabited only by roving Indians. When they reached the River Corcouneon the botanizing was better; they followed its course upstream for about a hundred miles, through the settlement of Frayle Muerto, above which the river was called Rio Tercero, as it was the third to be crossed between Cordova and Buenos Aires. (The others were the rivers Primero, Segundo, Quarto and Quinto – a useful if unimaginative system of nomenclature). The Segundo, 400 yards wide and four feet deep, was crossed on 6 April; on their return journey three or four months later it was completely dried up. Two days later they crossed the Cordova and

[1] His name is sometimes given as John
[2] *Journal of Botany* 1834

again entered desert country where water was scarce and bad; and when, on 15 April, they came to a well-wooded area, it was uninhabited because so infested with 'tigers' – though Tweedie took what he called a 'cautious stroll' in the forest in the evening.

On 17 April they reached the River Saladillo, but it was in spate and they had to wait a fortnight for it to subside enough to be fordable. Several other *tropas* assembled on its banks, and the place was like a fair. When the waters went down, passengers and baggage were loaded into crude canoes each made of a single ox-hide with the corners tied together, which were towed across the river by stalwart Indian girls, who swam across the stream holding the tow-ropes in their teeth. Meantime the men unloaded the waggons, manhandled them across, and reloaded them on the other side, the combined operation taking eight days. This gave Tweedie time to botanize, but it was already the dry season and he found little. They got off at last on 8 May by a road so narrow that if two waggons met it was necessary to cut down some trees to enable them to pass. To make up time after the long delay they travelled all night, Tweedie on a waggon loaded with bales of cloth; towards morning the driver fell asleep and the waggon overturned, and our traveller fell from a height with the cloth on top of him, but fortunately was little hurt. A week later the discovery of some new plants could not altogether prevent him from 'falling into a sort of melancholy fit' as it was his sixtieth birthday, 9000 miles from home and among people worse than savages. On approaching Santiago the country improved, and at last the Andes came in sight. Tucuman was now only forty leagues away, and Tweedie, whose patience with the slow pace of the caravan was 'quite exhausted', went ahead with a party of five men and some mules bound for the mines of Peru, reaching the town on 24 May.

Tweedie does not say how long he stayed in Tucuman. He made at least one two-day trip to 'a branch of the Snowy Cordilleras', but found little of interest in the parts he was able to reach, and he was two months too late for the seed-harvest in general. He managed, however, to send a box of seeds to Dr W. Hooker in Glasgow – 'being from a strange country, they may be in request for your Botanic Garden' – which included *Blumenbachia* (*Loasa*) *lateritia* and *Passiflora tucumanensis*. He returned by a similar, though not identical route, and was back in Buenos Aires by the beginning of October, 'so much disfigured by the effects of weather and sun, to say nothing of dirty and tattered garments, that several of my old acquaintances did not know me'.[1] He had accomplished 'this pleasure-trip', as he called it, of nearly 2000 miles, almost entirely on foot.

After his return Tweedie planned another visit to Brazil, to explore 'the hilly province of St Paul's'; Gardner reported that he was in Rio at the time of his own arrival in July 1836, although they did not meet. On 12 April 1837 he took an opportunity of accompanying a rancher (a fellow Scot) to his estate in the south; he stayed for about ten days, exploring for thirty miles in all directions, and then went on to the Sierra de Tandil,

[1] Tweedie, *Annals of Natural History* VOL IV

about 300 miles from Buenos Aires, across barren and almost uninhabited country. The mountains were so stony that he thought the whole pampas must have been cleared of stones to supply them; they bore no shrubs and trees, and few new plants. He returned to Buenos Aires after an unrewarding absence of nearly a month; it was at Tandil, however, that he collected seeds of *Verbena platensis* (=*teucrioides*).

This was not Tweedie's last journey. He refers to having been at some time four hundred miles up the Uruguay, and there suffering shipwreck in a gale, when the skies at noon were as black as midnight; and about 1845, when he was over seventy, he started with a companion for Patagonia. This time they were held up by unexpectedly-flooded rivers; their provisions began to run out and starvation threatened, but a dark patch on the distant plain turned out on investigation to be a group of dwarf pines, and on the seeds of these they lived till the waters subsided. Needless to say, cones of this life-saver were sent home, and the species successfully introduced.

Tweedie corresponded with Dr William Hooker, and many of his plants were grown at the Glasgow Botanic Garden; but he also sent seeds to the botanic gardens of Liverpool and Dublin, and to various Scottish individuals. He introduced a number of handsome stove-plants and is commemorated in the beautiful *Tweedia caerulea*; but his most valuable gifts to the general gardener were the silvery Pampas Grass, *Cortaderia selloana*, several verbenas, *Tritelia*[1] *uniflora* and *Petunia violacea*, the ancestor of our bedding petunias. He also introduced *Cyphomandra betacea*, the Tree-Tomato, which combines fragrant flowers and edible fruit. He never returned to England, but died in Santa Catalina, aged eighty-six.

The Andes

The first published accounts of the natural history of the Andes were produced by Frenchmen. On the death of the childless Charles II of Spain in 1700, a grandson of Louis XIV succeeded to the Spanish throne, and France and Spain subsequently became allies in the War of the Spanish Succession; hence the temporary chink in the Spanish armour which allowed Father **Louis Feuillée** (1660–1732) to make scientific researches at Lima and elsewhere in Chile and Peru.

Feuillée, like Plumier, was a member of the Order of Minims, and his principal function was to make astronomical, hydrographical and geographical observations, to measure altitudes and latitudes and draw up maps and charts;[2] he was anxious to determine the exact position of the ports of Chile and Peru because of the 'immense treasures' daily brought from there for the enrichment of Europe. He was supported by the Abbé Bignon and the Secretary of State, the Comte de Pontchartrain, under whose auspices Tournefort had made his journey to the Levant; and was directed to make all natural history observations, with drawings and des-

[1] Also known as *Milla*, *Brodiaea* or *Ipheion*

[2] He had previously made voyages for this purpose to the Levant and the Antilles

criptions of plants and their uses, but was not instructed to send specimens or seeds.

His voyage to South America must almost have established a record for setbacks and delays; he left Marseilles on 14 December 1707 and reached Concepcion in Chile on 21 January 1709 – including, however, a three-month stay at Buenos Aires. After visiting Valparaiso he arrived at Callao on 9 April, and presented his letter of introduction to the Viceroy, who gave him permission to proceed to Lima. Here his time was very fully occupied, making astronomical observations when the weather allowed (in seven months he had only one clear night); coaching a local doctor in astronomy; mapping the city, at the request of the Viceroy, and conducting religious services – but he still contrived to draw a plant or an animal daily, which he got an Indian to bring, as he had no time to collect for himself. He sailed from Callao to Concepcion on 15 January 1710, and after a stay of rather more than a month, worked northwards again up the coast, visiting several ports – Valparaiso, Coquimbo, Cobija, Arica and Ylo – before returning to Europe. The first part of his *Journal des Observations* was published in 1714, and it concluded in 1725 with an appendix on medicinal plants containing fifty plates, among them a fuchsia and species of alstroemeria, mimulus and tropaeolum.

About the time that Feuillée returned to France, **Amedée François Frézier** (1682–1773) was leaving it. The Spaniards might well have been justified in regarding him with suspicion, for he frankly admits that he went to South America as a spy. He was an experienced military engineer, and was sent by Louis XIV on a merchant-ship 'to pass as a Trader only, the better to insinuate himself with the *Spanish governors*, and to have all Opportunities of learning their Strength, and whatever else he went to be inform'd of'. His voyage, too, was delayed by several false starts, but once fairly launched on 14 January 1712 he made a relatively quick passage and reached Concepcion by 18 June. He visited much the same ports as Feuillée, between Concepcion in Chile and Callao in Peru, with excursions inland to Santiago and Lima. In his book (*A Voyage to the South Sea* . . . 1717) he records a number of observations on economic, ornamental or poisonous plants, and he brought home from Concepcion plants of a large-fruited strawberry (*Fragaria chiloensis*) which was cultivated there. Fresh water was short on the six-month voyage, and only five plants reached Marseilles alive; but one of these, crossed with the Virginian strawberry, became the ancestor of our garden strawberries of today. Frézier may have taken a special interest in these 'fraises' as he was descended from Scottish ancestors whose name was originally Frazer and whose blazon was three strawberry-flowers ('Frazirs' in Scots heraldry) on a blue field.

The next botanist of importance was also French. In 1735 the French Academy of Science planned two expeditions, to take measurements of a degree of latitude at the equator and in the extreme north, in order to test Newton's theory that the world was an oblate spheroid. The most convenient place to take the equatorial measurement was in Ecuador,[1] where

[1] The northern measurement was made in Lapland

the city of Quito was situated very near the line; and the Spanish government gave a grudging assent, on condition that two Spanish scientists were included in the team. A party of ten accordingly set out, the principal geodesists being Godin, Bouguer, and the leader Charles Marie de la Condamine, with **Joseph de Jussieu** (1704–1779) as physician-botanist. The two Spanish members, Juan and Ulloa, joined the expedition at Cartagena, and at Cape Pasado de la Condamine co-opted a brilliant young scientist called Maldonado – making thirteen in all. The expedition certainly had its share of bad luck; but it is only the exploits of the botanist de Jussieu that can be recounted here.

Joseph was the volatile younger brother of the celebrated botanists Bernard and Antoine de Jussieu, and the fourth son in a family of sixteen. He began with the study of medicine, gave it up in favour of engineering and mathematics, returned to medicine at the urgency of his brothers and took degrees in Rheims and Paris; then eagerly accepted the opportunity of going to South America, and spent the remaining months before the departure of the expedition in the intensive study of botany.

The party sailed from La Rochelle on 16 May 1735, and as usual their progress was leisurely. A three-month voyage took them to Martinique, where de Jussieu botanized ardently and with de la Condamine made the ascent of Mt Pelée; they then proceeded by way of several of the Lesser Antilles to San Domingo, and from there across the Caribbean to Cartagena. After being joined by Juan and Ulloa they sailed in November to the notoriously unhealthy Porto Bello in Panama; de Jussieu was stricken by fever, but nevertheless made large collections. At Panama there was a long delay while a dilatory Governor arranged their further passage, which enabled de Jussieu to explore the Rio Chagres. They embarked on the Pacific in March 1736; then the party divided, de la Condamine landing at Manta and de Jussieu and some others at Guayaquil, to proceed by different routes to Quito. By the end of April 1736, nearly a year after leaving France, they were assembled and ready to begin. De Jussieu's mathematical training enabled him to take a share in the geodesic work Bouguer considered him the most valuable of his helpers), and he also made a thorough investigation of the plants yielding quinine and coca – the products were known in Europe but not their sources. He stayed with the expedition at their encampment in a high Andes valley till 1739, when he went on a journey to Loja to examine the quinine-trees on the spot.

After seven years, the expedition began to disintegrate; one by one its members dropped off. The mathematician Godin, his part of the work done, took a teaching post in Lima; a physician had been murdered by the natives, a draughtsman killed in an accident; a chain-bearer had married a local girl and settled down. In 1743 the remaining members left for home – Bouguer by the shortest way, through Colombia and by the Rio Magdalena to Cartagena; de la Condamine and Maldonado by the longest – down the whole length of the Amazon, which they mapped as they went. De Jussieu remained behind; it is said that he had not enough money to accompany de la Condamine, though one would have expected the Academy, which had sent out the expedition, to have born the cost of the return. Be that as it may, de Jussieu resumed the practice of medicine in Quito, and a year later, when he was once more in funds and preparing to depart, an epidemic of smallpox broke out, and the authorities found his services so valuable that they refused to let him go, and threatened with severe penalties anyone who helped him to depart. This crisis over, he received orders from France to retrieve some instruments that had been left in the hands of Godin at Lima. On the way there he made a detour eastwards to Los Canelos, to see the (so-called) cinnamon-trees that grew there – a very hard journey through steep and desert mountains; then crossed back into the central valley of the Cordilleras, where he found *Heliotropium peruvianum* – a discovery by which he was 'intoxicated with delight'. South, then to Lima; and finding Godin about to return home via Buenos Aires, de Jussieu decided to go with him.

They set off accordingly south-east by Huancavalica and Cuzco to Lake Titicaca; but when (after nine months) they got to La Paz, de Jussieu found the vegetation so new and interesting that he stopped to examine it, and left Godin to go on by himself. Following more slowly in the same direction, and making a detour to investigate the cultivation of the coca shrub, he reached Chuquisaca – and that was the nearest he ever got to Buenos Aires. Ulloa reported that 'on the way to Buenos Ayres' he was robbed of his notebooks and plants by an absconding servant, and that despairing of replacing them he 'withdrew into a shell' at Lima; but it was five years or more before he returned to that city. We find him first north-east in Santa Cruz de la Sierra, and then south-west again at the mining-centre of Potosi. His funds had apparently run out, and he again resorted to the practice of medicine; and the local Governor pressed his engineering talents into service, to aid in the making of roads.

In 1755 he set out for Lima, once more intending to return to France, but was discouraged by hearing that his mother and some of his brothers had died, and that some collections he had sent home had been lost at sea. He fell into a depression and wrote to his brother Bernard that he was turning to mathematics for consolation, since botany had caused him so many vexations. He resumed medical practice, but latterly his mental powers began to fail, and when he at last got back to Paris in 1771, he left most of his wits behind him. He was most kindly cherished by the surviving members of his family, but never regained his lucidity. A sad story to lie behind the fragrance of the heliotrope, which he had successfully introduced.

The Spanish botanist **José Celestino Mutis** (1732–1808) is too important to be omitted, although he was not a traveller or collector to any great extent. He came to Bogotá in 1760 as physician to the new Viceroy, the Marquis della Vega; and for the first twenty years of his stay in Colombia he was permitted, rather than encouraged, to botanize. His time was much occupied by a number of other pursuits. He founded the chair of philosophy, mathematics and natural history at the university of Bogotá and became its first professor in these subjects; and when it was discovered that he could measure the distance of the sun and moon and foretell eclipses, only the influence of the Viceroy saved him from the Inquisition. He spent a great deal of time and money in a fruitless attempt to rediscover the lost pre-Conquest silver-mines, and in 1776 was appointed superintendent of the mines of Sapo near Ybague; there were also short periods when he 'embraced the clerical profession', and acted as professor of medicine. Twice he appealed to Charles III for a commission to conduct a botanical expedition, but it was not till 1779 that he was officially appointed botanist and astronomer, with a sufficient salary to enable him to devote all his time to botany, and not till 1783 that he was put in charge of an *Expedicion Botanica*, or, as Smith called it, 'a tribe of botanical adventurers'. His pupils and associates ranged the country, but Mutis himself does not seem to have travelled very far. He set up a headquarters at Maraquito, but about 1790 ill-health obliged him to return to the capital, there to supervise a botanic garden and a team of draughtsmen – twelve in 1791, but the number is said to have risen ultimately to thirty – who executed some six or seven thousand plant-drawings under his direction. Those who have followed Spain's botanical record so far, will not be surprised that the first of these drawings to see the light were published in 1954.

The next scientific expedition to South America, like those of Feuillée and de la Condamine, was of French inception, but again it was stipulated that Spanish botanists should be included, and eventually the Spanish element ousted the French, as effectively and as fatally as a young cuckoo ousts the legitimate fledgelings from the nest. In 1775 the French Controller-General, Mons. A. R. J. Turgot, decided to send a botanist to South America, partly to replace the many shipments from Joseph de Jussieu that had failed to arrive. He asked the advice of the ageing Bernard de Jussieu, who recommended a pupil, **Joseph Dombey** – a name we associate with Dickensian pomposity and gloom, but this Dombey was gay, charming and extravagant; when appointed to the South American expedition in August 1775 he had to mortgage half his salary in advance, to satisfy his creditors.

The scales were weighted against Dombey from the start. The Spanish authorities, having consented to the project, deliberately set out to exploit him in every possible way. He was to present two specimens of every plant he found, with notes, to the professors of the botanic garden at Madrid, for them to select the best and return the worst; if only one specimen was found, he might keep it, but must substitute a drawing and description. The associates chosen for him, **Hipolito Ruiz Lopez** (1754–1816) and **Antonio Pavon y Jiminez** (1754–1840) were young pharmacists with little previous knowledge of botany; he was to instruct them in the science and in the preparation of herbaria. Two young draughtsmen were also appointed, whose qualifications were to be 'bachelorhood, skill and a gentle disposition' – but they refused to supply drawings to Dombey. These four received salaries of 10,000 livres when in the field; Dombey's original 3000 livres from the French government was doubled to 6000 livres, but he frequently had to lend money to his better-paid companions. He applied to the Spanish authorities for material and equipment, but did not receive them, and was obliged to purchase at his own expense paper and instruments that he could have bought more cheaply in Paris; for he was kept kicking his heels, first in Paris and then in Madrid, for two years and two months, before the dilatory Spaniards were ready to depart. Dombey was twelve years older than either Ruiz or Pavon, and far more experienced; yet it was Ruiz who was nominated the leader of the party. It says much for Dombey's enthusiasm and good nature that he ever embarked at all.

They sailed at last about the end of October 1777, and arrived at Callao on 8 April 1778, proceeding to Lima the following day; this was to be their base for nearly four years. The South American winter was approaching, and there was little to be done, but Dombey collected some seeds from the repositories of ants in the sand, where the want of moisture had prevented them from germinating. When spring began at the end of July they set off up the coast to the 'delightful plains' of Chancay, about forty miles to the north, continuing to Huaura in the sugar-cane district and returning to Lima by 22 October. Then they made a short excursion southward to Lurin, Pachacamba and Casa Blanca; so far, they had hardly been out of sight of the sea, and had found little new.

In 1778 Dombey sent his first duplicate collections to France and Spain, and in March 1779 a big consignment was packed on board the *Buen Consejo* – seven cases for France from Dombey, ten for Spain from Ruiz and Pavon, four containers of living plants and some seeds and bulbs. Dombey's consignment contained some Peruvian antiquities, including an Inca robe which had cost him nearly seven months' salary, and which he destined for Louis XVI. But Spain that year joined France and the American colonies in their war against England, and the *Buen Consejo* was captured by the British. Her cargo was sold by auction at Lisbon, and the South American collections were ransomed by Spain. The French consul tried in vain to get Dombey's boxes; their less important contents were eventually restored to France – but not the valuable Inca robe. It was two years before the travellers heard of this loss, but they had already decided to store all further collections at Lima till the war was over.

Early in March 1779 Dombey was asked by the local Spanish authorities to analyse the mineral waters of Chuchiu – a five-month trip which he performed at his own expense, as he regarded himself as an envoy of France, and not entitled to accept money from Spain. Two months later Ruiz and Pavon made a difficult crossing of the Andes to Tarma, about 120 miles to the north-east, where Dombey rejoined them at the end of September. At Tarma there was a rich flora, including many orchids, and a further excursion eastward brought them to a tributary of the Amazon and their first experience of a tropical forest.

From January to April 1780 they were back in Lima; then they returned to Tarma and beyond to Huanuco, over 100 miles to the north-west. This was the farthest Spanish settlement, and they used it as an advance headquarters, to which they returned several times on subsequent occasions. Their excursions eastward were limited by local unrest, which later in the year flared into a serious uprising under Tupac Amaru, a descendant of the Incas, which cost thousands of lives. Dombey went to Lima in

October to store collections and replenish funds and supplies; on his return to Huanuco in December he found the inhabitants in great distress, threatened by the rebellion and without food or money. Dombey offered the President of the Council 2000 piastres and twenty loads of corn, and to raise two regiments at his own expense; heartened by this offer, the town organized its own defences and refused the money, which Dombey thereupon gave to a hospital for the poor.

The question arises of how Dombey came to possess so much cash. He kept two servants, whose wages and equipment cost him more than his salary; he made generous loans to his companions; he dispensed free medicines and medical services, he bought the Inca robe and other antiquities, and the seventy-three cases in which he packed his final collections cost as much as he earned in three years. One authority frankly states that the money was obtained by gambling: Dombey was very popular with the ladies of Lima and attended all their card-parties. If this was the explanation, his luck must have been phenomenal.

The expedition had originally been intended to last for four years, but the time was extended, perhaps owing to the difficulties of the return journey in wartime; and after a further nine months in Lima (March–December 1781) with local forays to replace plants lost in the *Buen Consejo* (of which they had just heard) the party left by sea for Talcahuano and Concepcion in Chile. At the time of their arrival (March 1782) Concepcion was in the grip of an epidemic of plague; Dombey temporarily abandoned botany, settled in the city and dedicated his whole time to visiting the poor, and furnishing them with food, medicines, vinegar, sugar, beds and even nurses at 5 livres a day, until the epidemic was over. At Lima, in spite of his free doctoring and good behaviour, he had been regarded (as a heretic foreigner) with hostility; at Concepcion he was hailed as a 'messenger from heaven' and offered a position as physician at 10,000 livres a year, and the hand of a beautiful heiress, to whom he was much attracted. He resisted the temptation, however, and late in March left with his party for San Fernando and Santiago. At this time great interest was being taken in the newly-discovered araucaria as a possible source of timber for shipping; Pavon, accompanied by a naval officer, was the first European to penetrate the forests where it grew and obtain botanical specimens, in spite of the danger from hostile Indians. From Santiago the versatile Dombey was sent to examine mines and to look for new sources of quicksilver; he found one at Xarilla near the port of Serena, and a new gold-mine into the bargain. This trip cost 15,000 livres, but again he refused to be reimbursed by Spain. On 15 October the party left Chile and returned to Lima.

The war ended that year (1783) and it was now possible for the botanists to return. Dombey was anxious to go home; his health had been deteriorating, and he was much troubled by scurvy. He received permission and a free passage – 'at which he never ceased to marvel' – and sailed for Cadiz on 4 April 1784, with the French share of his collections – seventy-three cases of minerals, manuscripts, dried plants, antiquities, wood and bark samples, fish, birds, insects, reptiles and shells. The Spanish share he handed over to Ruiz and Pavon, who despatched it with their own collection by the *San Pedro de Alacantara* – fifty-five cases and thirty-one tubs of living economic plants, with some bulbs and seeds. The Spanish botanists, however, wished to remain for another season or two, and this they were permitted to do. A month after Dombey had left they set out again for Huanuco, and then worked for several weeks from a new base at Pazuzo. From October till the following June (1785) they were at Huanuco again, and then moved on to the hacienda of Macora, in the tropical forest to the north-east.

On 7 August 1785 Ruiz returned after an excursion, to find the encampment a furnace. A fire lighted to burn brushwood cleared from the fields had spread to the botanists' hut. Pavon, at home ill, had rescued all that he could, entering the hut three times before the roof fell in, but very little was saved. Tents, equipment, clothes, two months' provisions, presses, drying paper and collections, books, three years' journals, four years' revised plant descriptions – everything went; a nearby garden where rare plants were being grown was also destroyed. To cap all, it rained heavily on the unsheltered party all night. There was nothing for it but to return to Huanuco.

A new order from Spain had authorized the expedition to be continued indefinitely; but now it was Ruiz, shaken by his losses and suffering from the cumulative effects of five attacks of fever, who asked to be recalled. It was March 1787 before his letter was received in Spain, and October of that year before the desired permission arrived. Meantime the two botanists continued to work in the regions east and north-east of Huanuco, and sent another large consignment home by way of Lima early in 1787. After three further months at Huanuco to clear up, and nearly two at Lima, they left for good on 1 April 1788, again with large collections, including 102 live plants in twenty-four tubs. They had been in South America ten years all but a week.

Ruiz afterwards wrote that he had suffered 'heat, fatigue, hunger, thirst, nakedness, want, storms, earthquakes, plagues of mosquitoes and other insects, continuous danger of being devoured by jaguars, bears and other wild beasts, traps of thieves and disloyal Indians, treason of slaves, falls from precipices and the branches of towering trees, fording of rivers and torrents, the fire at Macora . . . the separation from Dombey, the death of the artist Brunete, and the most touching of all, the loss of manuscripts'. Well might he count the loss of Dombey among the disasters, considering how much the expedition had profited from his botanical knowledge, his medical skill and his financial generosity; but until the fire at Macora at least, Ruiz could hardly complain of nakedness and want. His wardrobe then contained five suits (one of silk), three pairs of velvet breeches and seven of plain white; two dressing-gowns, sixteen pairs of stockings, fourteen shirts, twelve pairs of shoes, three cloaks, and etceteras such as hair-nets and sleeping-caps; his camping equipment included four table-cloths, many pieces of plate, a chintz bedspread and a silver chamber-pot. The contrast with the austerities accepted by other collectors – for example Tournefort and Douglas – is very striking.

Meanwhile, Dombey had had a troubled voyage. The weather was bad round Cape Horn; seventy-two of those on board were ill with scurvy and thirty-two died; Dombey himself had dysentery and the ship's steering broke. They limped into Rio on 4 August and stayed for more than three months to recuperate and repair. Dombey was received with honour by the Portuguese authorities, and despite the rainy season explored the vicinity, adding about 200 plants and some minerals and butterflies to his collections. He eventually reached Cadiz on 22 February 1785.

Dombey had obtained from Ruiz and Pavon receipts for the Spanish share of his collections, and in view of this and his gratuitous services in South America he expected to pass through the customs with little trouble. But the *San Pedro de Alacantara* was overdue, and presently it was learnt that she had been wrecked off the Portuguese coast, with the loss of thirty-nine lives and all the Dombey, Ruiz and Pavon collections. The Spanish government impounded Dombey's whole cargo – including five chests of his own personal property, his Brazilian collections, and the presents made to him by the Portuguese viceroy at Rio – while they wrote to France and demanded a half-share, in place of the consignment that had been lost; and this the French government conceded. Division began on 13 June – nearly four months after his arrival – and went on till 4 August. Meantime, Dombey's expensive cases – they were double, made with great care, cemented together and covered with skins – were roughly broken open and many of their contents damaged; they were stored in damp warehouses and Dombey was not allowed access to them, or even to sow the seeds; nearly all the living plants died. Spain acquired thirty-seven boxes out of his seventy-three; an exact copy was taken of all descriptions and field-notes, and his own property was not released nor his departure allowed until he had promised not to publish anything before Ruiz and Pavon's return. He reached Paris at last with the sorry remnants on 13 October 1785.

It is hardly surprising that the once happy and confident Dombey became a misanthrope and a recluse. The French Revolution four years later completed his downfall. He decided to quit a country that had become the scene of so much carnage, and obtained a permit to go to America to buy corn for France and other commodities. He landed at Guadelupe in the West Indies, but this, too, was in a state of revolution; he was imprisoned, released, fell into a river, had acute fever, and was told to leave the island; his ship was attacked by two corsairs, he was detected 'disguised as a Spanish sailor' and imprisoned by the British on Montserrat, where he died in 1796.

The Spanish government, having so effectively robbed Peter, were in no great hurry to pay Paul. Ruiz and Pavon got back in 1788; it was January 1793 before suitable accommodation was found for them, where their cases could at last be unpacked. The first volume of their *Flora Peruviana et Chilensis* appeared ten years after their return, in 1798; eight volumes and an appendix were planned, but only three were published. (The book was largely based on Dombey's work; his name appeared in the preface, but not on the title page.) Ruiz died in 1816, and Pavon, impoverished, was reduced to hawking round his spare herbarium specimens. Some of them were bought by the British botanist A. B. Lambert, who succeeded in raising several Peruvian plants from seeds thirty years old.

The South Americans believed that every German was a miner, and every Frenchman a doctor; this was certainly true of their next scientific visitors, **Humboldt** and **Bonpland**. After their explorations on the Orinoco and their stay in Cuba,[1] this devoted pair set sail from Batabano on 15 March 1801, and on the 24th reached the mouth of the River Sinu (west of Cartagena) where they landed to observe an eclipse of the moon, but were attacked by a band of *cimarrons* (escaped negro slaves) whom they beat off with difficulty. They reached Cartagena by 30 March and were told that the season of trade-winds was over and the passage by sea from Panama to Guayaquil was likely to be slow. Humboldt was very anxious to visit Mutis at Bogotá, so they decided to make the journey by land – if it can be called a land-journey when it started with a fifty-five-day river passage up the Rio Magdalena (21 April–13 June) progress being delayed by the violence of the current. They left the river at Honda and went by mule and horseback over incredibly bad mountain roads to the plain of Santa Fé and the city of Bogotá. Here they were welcomed by Mutis with great cordiality, and lodged in a house adjacent to his own. Their encounter would have been worth witnessing – Humboldt, small, dapper, at thirty-two already going a little bald at the temples, but with something engagingly youthful in his expression, and the tall, aged, long-nosed Spaniard, jowled like a Great Dane, who must nevertheless have been less formidable than his portraits suggest, since he was loved as well as revered in Bogotá.

The travellers stayed three months in the capital, partly to allow Bonpland to recuperate after another bout of fever, contracted on the journey. When they resumed their march on 8 September they took with them at Mutis's request a young scientist called Francisco José de Caldas, the inventor of a method of determining altitudes by the temperature at which water boils. They started almost due west by Pandi and Ibagué, and over the central Cordillera by the difficult pass of Quindiu. Part of the way led through dense uninhabited forest and provisions for more than a month had to be carried by oxen, the road being impossible for horses or mules; the men went on foot, in the later stages through dense bamboo scrub where the 'pucks on the roots' ruined their boots. They reached Cartago with feet sore and bleeding, but enriched with many new plants – one of which, *Melastoma mutisii*, they never saw elsewhere. Then they went south to Caldas's home-town of Popayan, where the botanizing was good and the climate delightful; they stayed for the month of November, and climbed the volcano of Purace. There followed another gruelling journey across the *paramos* of Pasta – a bitterly cold, desert mountain region, mostly above the tree-line and in the rainy season; no wonder Humboldt afterwards declared with such emphasis that he did not find South America enervating! Still bearing south-west, by Ypiales, Tulcan and Ibarro, they reached Quito in Ecuador on 6 January 1802.

[1] See p. 340

Humboldt and Bonpland stayed seven months at Quito, climbing with enthusiasm every available volcano. Humboldt ascended Pinchincha three times, and the natives attributed signs of renewed volcanic activity to the heretic foreigners having thrown gunpowder into the crater. He also climbed Cotopaxi, but his greatest mountaineering feat was the ascent of Chimborazo. He did not quite reach the top – he had to give up when he found a deep crevasse still separated him from the ultimate summit – but he reached 19,286 ft, at that time the greatest height ever attained by man. On the way down he collected some chips of rock as souvenirs, knowing that in Europe he was sure to be asked for 'a fragment of Chimborazo'. At Quito, Humboldt heard that Baudin, whose expedition he had so long hoped to join, had, indeed, sailed, but had gone to Australia the other way, eastward round the Cape of Good Hope, and would not be calling at South America. He took the disappointment philosophically, and in July the two friends resumed their journey to the south.

They went by Cuenca to Loja, to see the quinine trees described by de Jussieu, and then by the valley of the Huancabamba (which they crossed twenty-seven times) to Jaen, where de la Condamine had started his journey down the Amazon. They stayed at Jaen for seventeen days, checking Condamine's map, then continued southwards to Cajamarca and across to Truxillo on the Peruvian coast. Here at last they left the mountains, and for a change followed the coast six hundred miles to Callao, taking frequent sea and air temperatures. Humboldt had been puzzled by the existence of a coastal strip on which rain never fell, although the air was moist. He found it was due to a cold current (now called the Humboldt current) setting northwards along the coast; the air chilled by this current rolling inland, and encountering the warmer land temperatures, absorbed instead of depositing moisture. The existence of the current had been known for centuries, but only Humboldt deduced its effect on the climate; it was Humboldt, too, who introduced to Europe the guano from the Peruvian bird-islands, which had been used by the Indians as a fertilizer before the Spanish conquest.

Some two months were spent in Lima, during which they observed the transit of Mercury; then they left by sea on Christmas Day for the port of Guayaquil. On 5 January 1803, Cotopaxi, which they had climbed a few months before, erupted with such violence that they heard the explosion when still six leagues at sea. During their six-week stay at Guayaquil they went a short excursion up the River Guayas to watch the volcano in action. They sailed for Acapulco in Mexico on 15 February.

Humboldt has been called the greatest of scientific travellers. He made original and important observations on plant-geography, metereology, terrestrial magnetism, topography, hydrography and ethnology, and invented the use of isothermal lines on maps. He was also a good linguist, an excellent descriptive writer and a good draughtsman, and had great personal charm. But this giant of the intellect was physically unimpressive; young Joseph Hooker met him in Paris in 1845 and was disillusioned to find him 'a paunchy little German. . . . I expected to see a fine fellow 6 feet without his boots, who would make as few steps to get up Cimborazo, as thoughts to solve a problem. I cannot now at all fancy his trotting along the Cordillera, as I once supposed he would have *stalked*'. Humboldt by then was seventy-six, so perhaps this judgment was hardly fair.

For a quarter of a century after the departure of Humboldt and Bonpland, political disturbances during the liberation of the colonies from Spain kept the Andes more or less unvisited; but in the late 1820s there was what almost amounted to a botany-rush. In Valparaiso, Cuming took up full-time collecting in 1826, the German Poeppig and the Italian Bertero arrived in 1827 and Bridges in 1828, while north in Peru, Mathews came to Lima in 1830. The most important of these collectors was the second.

No sooner had **Eduard Frederich Poeppig** (1798–1868) taken his medical degree in Leipzig than he sailed for Cuba, and he had already spent five years in travelling and collecting in the West Indies and North America when he landed at Valparaiso in mid-March 1827. He did not go far afield during his first season; but having shipped his first collections at the end of September, moved on to Santiago, and then to a small hut near San Felipe de Acongua, where he made a long stay. Early in 1828 he started for Mendoza, but unfortunately lost a great part of his equipment in crossing a river, and had to return to Valparaiso. Unable to embark on a long expedition until the lost books and instruments were replaced, he spent the Chilean winter at the coastal town of Talcahuano, and in October moved eastward into the mountains, being the first scientist to climb the volcano of Antuco and to report the existence of glaciers in the Andes. He found orchids in the valleys and alpines on the heights; the zones of vegetation were distinct, but even on the highest levels 'the violet *Amaryllis* and variously-tinted *Alstroemerias*' dispelled any illusion that he might be in Europe. Such were the treasures he discovered that he was 'often compelled to relieve his full heart by uttering loud shouts of joy, to which his faithful dog, the sole companion and witness of his delight, responds by many a yelp of exultation',[1] but the presence of the dog provoked some alarming attacks by condors. In spite of the unsettled state of the country and the difficulty of the terrain, he made an excursion to the araucaria forests and obtained fresh seed, which he sent to Germany; he claimed that the trees raised from this seed were the first in Europe, but there were at least two earlier introductions to Britain.

In May 1829, his new equipment having at last arrived, Poeppig sailed from Valparaiso for Callao and Lima. A journey to the north having been frustrated by an outbreak of hostilities between Peru and Colombia, he turned north-east to Huanuco, formerly the hunting-ground of Ruiz and Pavon. From July 1829 till April 1830 he lived at Pampayoco, where he made some of his richest collections. Early in May he embarked with a servant, his baggage and some porters, on three light rafts on the River Huallaga, and for two years the virgin forests of the Amazons closed over his head.

He went by slow stages down the stream, staying several months at the

[1] *Companion to the Botanical Magazine* I 1835

deserted village of Tocache, and then through Ibitos, Sion and Juanguy to Yurimaguas, the first village in the practically unknown Mayas country. Here he made his solitary home for eight months (December 1830–July 1831) the only European in a vast province, 'without shoes and without clothes, often without a monkey to dine on',[1] and accomplished an enormous amount of work, largely because there was nothing else to do. At the end of this time he set out in a small boat on the homeward journey – down the Huallaga to its junction with the Amazon, and then on the breast of the great river to Para, which he reached on 22 April 1832. By October he was back in Leipzig, with a very large harvest of plant and animal specimens and with manuscript materials for a number of botanical works; like Humboldt he wrote an influential book on his travels, which he treated separately from his scientific publications. He brought or sent home a number of seeds and plants, and introduced to Germany species of francoa, puya and escallonia, but perhaps his most important find was that spectacular water-lily, *Victoria* (*Euryale*) *amazonica*, which had previously been seen, but not named or botanically described.

About this time there was an apparently insatiable demand for natural history collections, and especially for 'sets' of dried plants, and though prices were not high, many naturalists were largely, if not entirely supported by this means. Cuming, Bridges and Mathews were self-employed and not backed by any particular firm, institution or patron. **Hugh Cuming** (1791–1865) was born near the Devonshire port of Kingsbridge and apprenticed to a sailmaker, and his primary interest was in shells and corallines. He settled in Valparaiso in 1819, and embarked on a business that must have been profitable, since in 1826 he was able to build himself a yacht especially equipped for making and stowing natural history collections, and cruised the Pacific for more than a year, looking for shells and plants. His interest in botany was stimulated by one of those chance encounters that make it seem as if naturalists in the field are drawn together by some force of magnetism. In early March 1830 Captain King, accompanied by James Anderson as botanist, was engaged on a survey ot the coasts of South America. 'Anderson went out one day looking for plants', reports the *Gardener's Magazine*,[2] 'and met Cuming among the rocks at Concepcion looking for shells etc. They were strangers to each other. but felt the greatest delight when they found they were of the same country, and almost on the same pursuit, on this savage and inhospitable coast. Ever since this circumstance, they look on each other as two brothers, and Cuming learned from Anderson how to dry plants, and the other duties of a collector.'

Cuming explored the western coasts from Chiloe to Nicaragua, with occasional excursions inland, till 1831, when he returned for a time to England; he afterwards spent four years collecting in the East Indies. He introduced a number of plants, but none of them seems to have been of great garden importance.

Thomas Bridges (1807–1865) was one of those who chose a collecting

[1] Poeppig, in *Journal of Botany* 1834
[2] VOL XVI 1840

career, without waiting to be chosen, or until an opportunity happened to be offered. A 'very enterprising young man' of Wrexham in Norfolk, who had 'spent some years among the plants already in England', he 'solicited some gentlemen to whom he was known'[1] for advances to enable him to travel, and after a boisterous passage of nearly twenty weeks reached Valparaiso on 8 August 1828, where he set up as a brewer of small beer, to maintain himself till his collecting began to produce returns. He gradually extended his botanical researches farther and farther afield, but broke off in order to spend a year at Quillota helping to lay out a farm for a friend. He began his first long collecting journey in October 1832, when he accompanied the party of the Commissary of the Indians, who was going into the little-known interior of Valdivia to stop one of the passes to the Pampas and prevent the troublesome Petahuatche tribe from 'intruding' on the western side of the Andes. After visiting the lake of Runco and its islands Bridges returned to Valparaiso in February 1833 with specimens of nearly 300 species of plants, many of them new. 'We understand', said Hooker in his lofty botanical way, 'that he has also sent numerous seeds and roots for cultivation';[2] but of what kind is not stated. Later in the same year, Bridges visited Chiloe and southern Chile; then he again took a post as superintendent of an estate, and had no time for collecting for a further two years, resuming botanical work in 1841 with a journey northward along the coast to Coquimbo and Copiago.

Bridges went home to England in 1842, and returned to South America some two years later, arriving this time at the port of Cobija, on 13 September 1844. He then went by Calama and Potosi to the Bolivian capital of Chuquisaca, and through the British consul obtained an interview with the President, who granted permits for further travel in the interior. On Christmas Eve he came to Cochabamba, where he found a fine climate and a good flora, and stayed for more than three months (mostly engaged on entomology and ornithology) before setting out eastward, early in April 1845, to cross 'an enormous ridge' and 'fall into the tropical forests' of the river Mamore, which he proposed to follow downstream to Trinidad. North of Trinidad the Mamore receives a tributary, the Yacuma, and it was on this river that Bridges found and collected seed of *Victoria amazonica.* When he returned to London in June 1846 he brought these seeds with him, carefully packed in wet clay; they sold at 2s apiece, and dried specimens of the plant for 30s.[3] Bridges afterwards complained that he received no credit for this introduction; but only two of his seeds germinated and the resulting plants died before flowering. The giant water-lily was successfully grown four years later from seeds obtained from Demerara.

During this visit to England Bridges married Cuming's niece,[4] but it is not recorded whether he met her uncle in Chile or, later, in England. He brought his wife to Valparaiso in September 1851, and started a nursery (in both senses) but after another visit to England in 1855 he settled in

[1] *Gardener's Magazine* VI 1831 [2] *Journal of Botany* I 1834
[3] The average price for plant specimens was £2 a hundred
[4] In January 1847

California, and from there made a two-year collecting and exploring trip to British Columbia. In 1865 he visited Nicaragua, but on the way back to San Francisco he contracted malaria and died at sea.

Much less is known about **Andrew Mathews**, who had been employed by the Horticultural Society as a gardener before he came to Peru, about 1830, and made his headquarters at Lima. He made a number of journeys, mostly in the regions explored by Ruiz and Pavon; he had not only studied their works, but saw 'many of their unpublished drawings and their original named species, which he was so fortunate as to meet with in the country'. A few details of his earlier travels have survived, but nothing between the end of 1834 and his unexplained death at Cachapoyas in 1841.

In South America, as elsewhere, the botanists are much better served than the horticulturists; comparatively detailed accounts are to be found of the exploits of Humboldt and Bonpland or Ruiz and Pavon, but very little has been printed on the travels of Lobb, Hartweg, Pearce or Roezl. Information fails just where it is most wanted, since it is to these four, and especially to Lobb and Pearce, that we are indebted for our most valuable South American plants.

William Lobb (1809–1863) was the first of the twenty-two collectors sent out by the firm of Veitch, and South America was his first assignment. He was a Cornishman, and brought up as a gardener, but attained 'considerable proficiency' in botany in his spare time, and also 'cherished an ardent desire for travel and adventure'.[1] He sailed from Plymouth to Rio de Janeiro in 1840, and like every previous botanical visitor made first for the Organ mountains, sending home *Begonia coccinea*,[2] *Passiflora actinea*, and several orchids, including *Cycnochis pentadactylon*. Veitch announces casually[3] that he then went overland from Rio to Chile; even on a modern map this does not seem a journey to be undertaken lightly. Of his movements in Chile we only know that he went south to the araucaria forests and procured a large quantity of seed. This was not by any means the first introduction, but the tree was still very rare in Britain, and it was Lobb's consignment that established it in general cultivation. He also found *Desfontainea spinosa* and *Tropaeolum azureum*.

Lobb returned to England in 1844, renewed his contract and sailed again for Rio in April 1845. He revisited the Organ mountains and obtained *Hindsia violacea* and *Escallonia organensis*, and then went on to Valparaiso – perhaps by sea this time, but we are not told. We only know that he went as far south as Valdivia, northern Patagonia, and the island of Chiloe; but what a harvest he brought or sent home! *Nothofagus obliqua*, the hardiest of the beautiful antarctic beeches; *Berberis darwinii*, *Embothrium coccineum*, *Escallonia macrantha*, *Tricuspidaria hookerianum*, and others in the first rank of hardy or near-hardy shrubs; the brilliant *Tropaeolum speciosum*, and for the greenhouse *Ruellia spectabilis*, *Streptosolen jamesonii* and the lovely climbing *Lapageria rosea*. Conifers were then very popular, and Lobb introduced several fine species, including *Fitzroya cupressoides*

[1] and [3] *Hortus Veitchii*
[2] Found, but not introduced, by Gardner

and the botanically-interesting *Saxe-Gothaea conspicua*, named after Prince Albert. Lobb returned to England in 1848, and in the following year embarked on the first of his two trips to California.[1]

Still less is known about the other Veitch collector, **Richard Pearce**. He was a Devon man and had been employed by the firm as a gardener; like Lobb, he made two journeys to South America, in 1859–62 and (approximately) 1863–6 – even the dates are uncertain. His orders were comprehensive; he was to look for timber trees and conifers; hardy flowers, trees and shrubs; stove and greenhouse plants, and orchids; and on his first trip he responded with *Mutisia decurrens*, *Eucryphia glutinosa* and *Berberidopsis corallina*. On his second trip he is known to have visited Lima, Cuzco, La Paz and Tucuman; he got three important begonias (*B. boliviensis*, *B. pearcei* and *B. veitchii*, parents of the tuberous begonias of today) and an ancestral hippeastrum (*H. pardinum*) besides *Escallonia virgata* and *Nierembergia repens*. Back in London in 1866, Pearce transferred his services to the firm of William Bull, almost next door to Veitch's Chelsea branch, and undertook another visit to South America on behalf of his new employer; but he took yellow fever in Panama, and died on 17 July 1868 after an illness of only four days. He was probably little more than thirty.

The South American travels of Hartweg and Roezl were chiefly confined to the equatorial regions, and the plants they introduced were not hardy. **Hartweg** sailed from Realejo in Guatemala early in 1841, and after waiting in Guayaquil till the end of the rains, proceeded in mid-May to his most southerly point, Loja in Ecuador. From there he worked gradually north by a number of well-known centres, staying for an average of four months at each place; Cuenca, Riobamba (a shorter visit), Quito, Popayan, Bogotá (where he had fever and met Linden), Honda, and down the Rio Magdalena to Cartagena. From there he crossed to Jamaica, and sailed for home on 3 June 1843. He introduced a number of handsome stove-plants – *Achimenes longiflora*, *Isoloma bogotense*, *Monochaetum hartwegianum*, *Stenomesson incarnatum* and *S. aurantiacum* (=*hartwegii*) – but except for *Cattleya maxima*, few orchids; those he sent home from Loja nearly all perished on the homeward journey round Cape Horn.

Thirty years later, in 1871, **Benedict Roezl** made a six-month journey in north-western Colombia – from the port of Bonaventura to Buga on the River Cauca; down the Cauca valley to Cartago; to Sonson and Amalfi in the Central Cordillera; then probably by the River Nechi to the Magdalena, and down that highway to the coast. He then circled back by Colon and Panama to Payta, on the coast of North Peru, and made a quick but profitable excursion inland, finding several new orchids; then back to Bonaventura, and so home to Europe.

Improved transport, including the occasional railway, had by this time made it easier to get quickly from place to place, and Roezl skated over the Americas like a water-beetle on the surface of a pond. After four months in Europe, during which he visited England, Germany, Belgium

[1] See p. 321

and Switzerland, he sailed from Liverpool on 3 August 1872, and within two years had been to New York, Colorado, California, Mexico, Venezuela, Cuba, Mexico again, New York again, Panama, Peru, Bolivia, Ecuador and Colombia; to Europe for three months, then back to North America, Colorado, California and Mexico once more. During this second South American trip Roezl covered a good deal of ground. From Callao he was able to go to Lima and Oroza by train, and thence to Tarma and back – a trip that yielded 10,000 bulbs; then he went south-east by Puno, Lake Titicaca and La Paz to the mountains of Illimani and back by way of Tacna and Arica on the coast to Lima again. He revisited his former hunting-grounds in the vicinity of Payta and Bonaventura on the way home. Besides orchids and other stove-plants, he sent or brought to Europe calceolarias, fuchsias, mutisias, begonias and tropaeolums in wholesale quantities. *Begonia froebelii* (named after a nurseryman of Zurich) was among his discoveries.

The resources of the Andes were by no means exhausted, and new plants continued to be found during the present century. In 1925, **H. F. Comber** was offered the post of collector for the Andes Syndicate, chiefly promoted by the Hon. H. D. Maclaren[1] of Bodnant. He worked well to the south, on the Chile-Argentine border in Chillan and Valdivia, and besides some useful introductions (*Fabiana violacea*, *Azara integrifolia* and *Alstroemeria ligtu* var. *angustifolia*) made a number of discoveries, including *Hydrangea integerrima*, *Pernettya leucocarpa*, and seven or eight excellent berberis species, among them *B. linearifolia*, *B. montana* and *B. comberi*. Perhaps his greatest gift to horticulture was the hardier Norquinco Valley form of *Embothrium coccineum*, which has enabled this spectacular shrub to be grown in a much wider range of gardens than was formerly the case.

Comber was very soon followed by **Clarence Elliott** (1881–1969). The latter had made many journeys to collect plants for the nursery that he had founded near Stevenage in 1907, but except for one visit to the Falkland Islands and a spell of fruit-farming in South Africa in early life, his travels had not hitherto gone beyond Europe. His South American trip was undertaken in company with Dr W. Balfour-Gourlay, who was collecting plant-specimens for Kew and Edinburgh, and who later went on several expeditions with E. K. Balls.

They sailed on 4 August 1927 on RMS *Orbita*, and made short stops at Hamilton, Havana, Panama, Callao (whence they went on a brief visit to Lima), Arica and Antofagusto; but collecting did not really begin till they reached Valparaiso on 7 September, and were at once confronted with the plant-hunters' dilemma – what, in this exciting new flora, should be collected and what passed over – 'So fine a thing must have been sent home years ago; and yet, on the other hand, why has one never seen it?' It was still rather early for the Andean spring, but bunches of exquisite leucocorynes were being sold in the streets. On enquiry, they learned that this flower grew wild at Coquimbo, farther to the north; so after visiting Santiago to meet the local botanical authorities and consult their herbaria,

[1] Later Lord Aberconway

they proceeded to Coquimbo by train. (It chanced to be the anniversary of Independence Day, 18 September; the train was garlanded with flowers, and Elliott was able to do some useful botanizing on the engine.) At Coquimbo they were invited to stay at the farm of a local magnate, where the lovely, fragrant Glory of the Sun (*Leucocoryne ixioides*) grew in sheets in the meadows – a sight which, they felt, justified the whole journey to Chile. Local labour was recruited to dig up quantities of bulbs, but unfortunately the plant has not proved sufficiently hardy for cultivation out-of-doors in Britain. In a gorge near Coquimbo they found *Alstroemeria violacea* in bloom; a few of the previous year's seeds were still scattered round the clump, and these Elliott laboriously gathered, one by one, from the steep rough scree, with cactus thorns 'not only among stones I scrabbled in, but among stones I sat and kept slipping on!'

From Coquimbo they worked their way gradually back to Valparaiso, with side-trips into the mountains, and then took the Trans-Andean railway to Rio Blanco (*c.* 9000 ft) where they found *Alstroemeria ligtu*, and climbed a further 3000 ft to the Argentine frontier before returning to Valparaiso again. Then south to Temuco and the 'delicious forest settlement' of Pucon, an oasis of green in the parched Chilean summer. Afterwards they took a small coastal steamer to Punto Arenas in the Magellan Straits, and from there visited desolate sheep-stations on Tierra del Fuego, and the complicated shores of Last Hope Inlet to the north. They sailed for home from Punto Arenas via the Falklands early in 1928.

The same two gardener-botanists made another South American journey in 1929–30, but this time their itinerary has not, apparently, been recorded. More important, perhaps, are the names of the plants that they introduced, or re-introduced. They included the fantastic little *Calceolaria darwinii* from Magellan, and the sea-green *Puya alpestris* (first found by Poeppig); the hardy *Verbena corymbosa* and the elegant 'white' form of *Fuchsia magellanica*. Like Comber, Elliott also re-introduced *Alstroemeria ligtu* and the hardier *A. haemantha*; and these two species became the parents of the lovely race of ligtu hybrids, one of the most valuable recent additions to our hardy herbaceous flowers.

On returning from his first South American journey Clarence Elliott brought home, besides plants, what he called 'arkloads' of living animals for the London Zoo – a Galapagos turtle, a pair of rare pygmy deer, birds of various sorts, including two hens of the Chilean breed that lays blue eggs, and giant edible frogs, one of which had retained its tadpole tail and had to be doctored daily with iodine. It is unlikely, however, that he kept them *all* in his cabin, as John Nash's misleading drawing would seem to imply.

Some ten years later Balfour-Gourlay returned to South America, this time with **E. K. Balls**. They had been working together in Mexico[1] when Balls received a request from the Imperial Agricultural Bureau to investigate the wild potato species of the Andes, material of which was wanted for research. They sailed from Mexico to Buenaventura in

[1] See p. 350

November 1938, and crossing to Bogotá, began from there to work their way south. At Quito, where they spent Christmas, they were joined by a potato-expert, Dr Jack Hawkes; and then proceeded, now by car, now by train, and from Callao to Mollendo by boat, to Jujuy and Salta in Argentina, their farthest points south. From Jujuy they went camping for several days in the mountains, and had bad weather, but found good plants; then they retraced their steps northwards. At La Paz Balfour-Gourlay was obliged to quit the party and return to England, but Balls continued to Quito, and had an unnerving experience when attempting to climb Mt Cintisana. He approached it from the wrong point and the ascent took four days instead of the expected two; he spent the first night at 14,000 ft rolled up with his guide in the man's poncho, the next in some sheds at 16,000 ft where he had, indeed, a bed of sorts, but felt 'lonely and unprotected' without his previous night's companion. He had obtained permission from the Agricultural Bureau to collect ornamental plants for himself, besides fulfilling their requirements, and he found many attractive species; but with the outbreak that year of the Second World War, they never reached general cultivation. His wild potatoes fared better and are still in use for experimental and breeding work.

Epilogue

Is plant-hunting at an end? Have we by now exhausted all the floral resources of our limited planet? In every developing country a stage is reached when it is no longer necessary for travellers to come from afar to investigate the flora – when the task of discovery and distribution can be taken over by native-born botanists and nurserymen. Except in a few remote and backward areas, therefore, plant-collecting as hitherto understood tends to diminish, if not to cease.

E. H. Wilson, writing in 1927, thought that the day of the plant-hunter was done, and that the world's flora was 'almost an open book'; yet since then the Asiatic expeditions of Kingdon-Ward and Ludlow and Sherriff have produced many fine plants, and our gardens have been enriched by contributions from several other parts of the world. The amount of ground that can be covered by a single collector is limited by the strength of his legs and the length of his eyesight, and even when these are augmented by field-glasses and mechanical transport, there is always a hope that some undiscovered beauty may lurk in the next valley.

True, the pioneering days are over. No longer may the bearded 'Grass Man' astonish the natives in regions where the white man has never before been seen. But many of the plants so laboriously gathered have been lost to cultivation, and many would be worth re-introduction. The advent of the aeroplane and the land-rover, the refrigerator and the polythene bag, has changed beyond recognition the conditions of the collector's work; but fashions in gardening also change, and the demand for a different type of plant is a constant stimulus to fresh endeavour. There will probably be many future expeditions in search of plants, which will still demand the qualities of hardiness, courage and enterprise that make the stories of the plant-collectors such inspiring reading.

Bibliography

> For out of the old feldis, as men saieth
> Cometh all this newe corne fro yere to yere
> And out of olde bokis, in gode faieth
> Comith all this newe science that men lere.
>
> CHAUCER

ABEL, C. *Narrative of a Journey in the Interior of China* 1819

ALLEN, MEA *The Tradescants* 1964

AITON, W. *Hortus Kewensis (2nd ed.)* 1810

ANDERSON, A. W. *The Coming of the Flowers* 1950

(ANON.) *Histoire des Découvertes faites . . . dans plusieurs contrées de la Russie et de la Perse* 1779–81

ARCHER, M. *Natural History Drawings in the India Office Library* 1962

BACKHOUSE, J. *Narrative of a Visit to the Australian Colonies* 1843

BARTRAM, J., ed. HARPER, F. 'Diary of a Journey' (in *Transactions of the American Philosophical Society*, N.S. xxxiii pt. 1) 1942

BARTRAM, W. *Travels through North and South Carolina* 1792

DE BEER, G. R. *Sir Hans Sloane and the British Museum* 1953

BELON, P. *Les Observations de Plusieurs Singularitez . . .* 1554

BERKELEY, E. and D. S. *John Clayton, Pioneer of American Botany* 1963

BERNARD-MAITRE, H. 'Le Père le Chéron d'Incarville (in *Archives Internationales d'Histoire des Sciences* II) 1948–9

BIDWILL, J. C. *Rambles in New Zealand* 1841

BLANFORD, W. T. 'Account of a Visit to . . . Independent Sikkim' (in *Journal of the Asiatic Society of Bengal*) 1871

BRACKENRIDGE, H. M. *Journal of a Voyage up the River Missouri* (2nd ed.) 1816
BRADBURY, J. *Travels into the Interior of America* 1819
BRETSCHNEIDER, E. *A History of European Botanical Discoveries in China* 1898
BRITTEN, J. and DANDY, J. E. *The Sloane Herbarium* 1958
BROOKS, E. ST J. *Sir Hans Sloane* 1954
BURBIDGE, F. W. *The Gardens of the Sun* 1880
BURCHELL, W. J. *Travels in the Interior of Southern Africa* 1822
BURKILL, I. H. 'Chapters on the History of Botany in India' (in the *Journal of the Bombay Natural History Society*) 1956

CATESBY, M. *The Natural History of Carolina* 1730–48
CHEESEMAN, T. F. *Manual of the New Zealand Flora* (2nd ed.) 1925
COCKAYNE, L. *New Zealand Plants and Their Story* 1919
COMBER, H. F. 'Plant-Collecting in Tasmania' (in *New Flora and Silva* III) 1931
COOPER, R. E. 'Botanical Tours in Bhutan' (in *Notes from the Royal Botanic Garden, Edinburgh* No. 87) 1933
COOPER, R. E., ed. 'George Forrest, V.M.H.' (*The Scottish Rock-Garden Club*) 1935
COULTER, T. 'Notes on Upper California' (in the *Journal of the Royal Geographical Society V*) 1835
COWAN, C. F. 'Sir Hugh Low' (in the *Journal of the Society for the Bibliography of Natural History IV*) 1968
COWAN, J. M. *The Journeys and Plant Introductions of George Forrest* 1954
COX, E. H. M. *Farrer's Last Journey* 1926
Plant Hunting in China 1945
The Plant Introductions of Reginald Farrer 1930
CRAWFORD, D. G. *A History of the Indian Medical Service* 1914
CRAWFORD, J. *Journal of an Embassy . . . to the Court of Ava* (2nd ed.) 1834

DARLINGTON, W. *Memorials of John Bartram and Humphrey Marshall* 1849
DAVY VIRVILLE, A. DE. *Histoire de la Botanique en France* 1954
DAWSON, W. R. (compiler) *The Banks Letters* 1958
DIEFFENBACH, E. *Travels in New Zealand* 1843
DOUGLAS, D. *Journal kept by David Douglas during his Travels* 1914

ELLIOTT, C. 'Chile and the Andes' (in *The Gardener's Chronicle*) 1927–9
'The Road by Which I have Come' (in *The Countryman*) 1947
ELWES, H. J., ed. HAWKE, E. G. *Memoirs of Sport, Travel and Natural History* 1930
EWAN, J. AND N. 'John Lyon . . . and his Journal' (in *Transactions of the American Philosophical Society*, N.S. vol. 53 pt. 2) 1963

FAIRCHILD, D. 'Frank Meyer' (in the *National Geographical Magazine*) 1919
FARRER, R. *On the Eaves of the World* 1917
The Rainbow Bridge 1921
FARRINGTON, E. I. *Ernest H. Wilson, Plant Hunter* 1931

FEUILLÉE, R. P. L. *Journal des Observations* 1714
FORTUNE, R. *Three Years' Wanderings in the Northern Provinces of China* 1847
A Journey to the Tea-Countries of China 1852
A Residence among the Chinese 1857
Yedo and Peking 1863
FOX, H. M. *Abbé David's Diary* 1949
FRASER, J. *A Short History of the Agrostis Cornucopiae* 1789
FRÉMONT, J. C., ed. NEVINS, A. *Narratives of Exploration and Adventure* 1956
FRÉZIER, A. F. *A Voyage to the South Sea* 1717
FRICK, G. F. and STEARNS, R.P. *Mark Catesby* 1961

GARDNER, G. *Travels in the Interior of Brazil* 1846
GARSIDE, S. 'The Schönbrunn Gardens' (in *Journal of South African Botany*) 1941
GMELIN, J. G., trs. KERALIO, M. DE *Voyage en Sibérie* 1767
GOURLIE, N. *The Prince of Botanists* 1953
GRAUSTEIN, J. E. *Thomas Nuttall, Naturalist* 1967
GRIFFITH, W., ed. MCLELLAND, J. *Journals of Travels in Assam, Bootan, etc.* 1847

HADFIELD, M. *Pioneers of Gardening* 1955
Gardening in Britain 1961
HAGEN, V. W. von *South America Called Them* 1949
HAMY, E. T. 'Voyage d'André Michaux en Syrie et en Perse' (in *Comte Rendu . . . du Neuvième Congrès Internationale de Géographie*) 1911
HARPER, F. *The Travels of William Bartram* 1958
HARSHBERGER, J. W. *The Botanists of Philadelphia and their Work* 1899
HARTWEG, T. 'Journal and Notes' (in *Journal of the Horticultural Society* vols. 1 and 2, 1846–7, and *Transactions of the H.S.* new series, vols. 2 and 3) 1842–8
HASLUCK, A. *Portrait with Background* 1955
HASSELQUIST, F. *Voyages and Travels in the Levant* 1766
HOOKER, J. D. *The Rhododendrons of the Sikkim Himalayas* 1849
Himalayan Journals (2nd ed.) 1855
HOOKER, J. D. *Introductory Essay to the Flora of Tasmania* 1859
HOOKER, J. D. and BALL, J. *Journal of a Tour in Morocco* 1878
HOOKER, W. J. *Flora Borealis Americana* 1840
Niger Flora 1849
HOVE, A., ed. GIBSON *Tours for Scientific and Economical Research*, (1787–8) 1855
HUMBOLDT, A. von, trs. WILLIAMS, H. A. *Personal Narrative of Travels* 1818–19
HUTCHINSON, J. *A Botanist in Southern Africa* 1946
HUXLEY, L. *The Life and Letters of Sir J. D. Hooker* 1918

INGRAM, C. *Isles of the Seven Seas* 1936

JOHNSTONE, J. T. 'John Jeffrey and the Oregon Expedition' (in *Notes from the Royal Botanic Garden, Edinburgh*) 1950

KAEMPFER, E., trs. SCHEUZER *History of Japan* 1728
KALM, P., trs., FORSTER *Travels in North America* (2nd ed.) 1772
'Letter to John Bartram' (in the *Gentleman's Magazine*) 1751
KARSTEN, M. C. *The Old Company's Garden at the Cape* 1951
'Andreas Sparrman and Francis Masson' (in *Journal of South African Botany*, vols. XXIII–XXVII) 1957–61
KAULBECK, R. *Tibetan Trek* 1934
KINGDON-WARD, F. *The Land of the Blue Poppy* 1913
In Farthest Burma 1921
The Mystery Rivers of Tibet 1923
From China to Hkamti Long 1924
The Romance of Plant Hunting 1924
Plant-Hunting on the Edge of the World 1930
Plant-Hunting in the Wilds 1931
A Plant-Hunter in Tibet 1934
Plant-Hunter's Paradise 1937
Assam Adventure 1941
Burma's Icy Mountains 1949
KINGDON-WARD, F., ed. STEARN, W. T. *Pilgrimage for Plants* 1960

LACROIX, A. 'Notice Historique sur les cinq de Jussieu' (in *Mémoires de l'Académie des Sciences* series 2 vol. 63) 1941
LEE, I. *Early Explorers in Australia* 1925
LEIGHTON, C. *Cape Floral Kingdom* 1960
LEMMON, K. *The Golden Age of Plant Hunters* 1968
LEWIS, M. and others *A History of the Expedition under the Command of Captains Lewis and Clark*, 1804–6 (*Great American Explorers* series) 1905
LINNAEUS, C., ed. SMITH, J. E. *Tour in Lapland* 1811
LOUDON, J. C. *Arboretum et Fruticetum Britannicum* 1838
LÖWENBERG, J., trs. LASSEL *Life of Alexander von Humboldt* 1873
LOW, H. *Sarawak* 1848

MAIDEN, J. H. *Sir Joseph Banks, the Father of Australia* 1909
MAIN, J. 'Reminiscences of a Voyage to and from China' (in *Paxton's Horticultural Register, V*) 1836
MCKAY, H. M. 'William John Burchell' (in *Journal of South African Botany*) 1941
MCKELVEY, S. D. *Botanical Exploration of the Trans-Mississippi West* 1955
MARIES, C. 'Rambles of a Plant Collector' (in *The Garden*) 1881
MASSON, F. 'An Account of Three Journeys from the Cape Town' (in *Philosophical Transactions of the Royal Society* LXVI) 1776
Stapeliae Novae 1796
MICHAUX, A. *Journal of Travels into Kentucky*, 1793–6 (Thwaites, *Early Western Travels* III) 1904
MICHAUX, F. A. *Travels to the Westward of the Allegany Mountains* 1805
MISCHENKO, P. 'To the Memory of L. F. Mlokosevich' (in *Transactions of the Botanic Garden of the University of Yuriev*, XII) 1911

NUTTALL, T. 'Travels into the Old North-West, 1810' (*Chronica Botanica* 14) 1951
A Journal of Travels . . . in the Year 1819 (*Early Western Travels* XIII) 1905

PALLAS, P. S., trs. GAUTHIER DE LA PEYRONIE *Voyages de M.P.S. Pallas* 1788–93
Travels through the Southern Provinces of the Russian Empire 1802–3
PIM, S. *The Wood and the Trees* 1966
PULTENEY, R. *Historical and Biographical Sketches of the Progress of Botany in England* 1790
PURSH, F. C. *Flora Americae Septentrionalis* 1814

RAY, J. *Collection of Travels* (2nd ed.) 1738
RHOADS, S. N. *Botanica Neglecta: William Young Jr.* 1916
RICKETT, H. W. 'The Royal Botanical Expedition to New Spain' (in *Chronica Botanica* 11) 1947
RITCHIE, R. T. *Lord Amherst* 1894
RODGERS, A. D. *John Torrey* 1942
ROGERS, C. *Trodden Glory* 1949

SELIGMAN, R. 'June in the Great Atlas' (in the *Bulletin of the Alpine Garden Society* V) 1937
SKOTTSBERG, C. 'Pedr Kalm' (in *Colloques Internationaux du Centre National de la Recherche Scientifique*) 1957
SLOANE, H. *A Voyage to . . . Jamaica* 1707–25
SMITH, J. E. *The Correspondence of Linnaeus and other Naturalists* 1821
SMITH, LADY *Memoir and Correspondence of Sir J. E. Smith* 1832
STANDLEY, P. C. *Trees and Shrubs of Mexico* (*U.S. National Herbarium*, vol. 23) 1920–6
STAUNTON, G. *An Authentic Account of an Embassy . . . to the Emperor of China* 1797
STEARN, W. T. 'Botanical Exploration to the Time of Linnaeus' (in *Proceedings of the Linnean Society*) 1958
'Botanical Gardens and Botanical Literature in the Eighteenth Century' (*Catalogue*, Hunt Botanical Library II) 1961
'Grisebach's Flora of the British West Indian Islands' (in *Journal of the Arnold Arboretum* 46 No. 3) 1965
STEARN, W. T. ed. *Humboldt, Bonpland and Kunth and Tropical American Botany* 1968
STEELE, A. R. *Flowers for the King* 1964
STEJNEGER, L. *Georg Wilhelm Steller* 1936
SYNGE, P. M. *Mountains of the Moon* 1937

TAYLOR, N. M. *Early Travellers in New Zealand* 1959
THOMSON, T. *Western Himalaya and Tibet* 1852
THUNBERG, C. P. *Travels in Europe, Africa and Asia*, (1770–79) 1795
TOURNEFORT, J. P. DE *A Voyage into the Levant* 1718

TOWNSEND, J. K. *Narrative of a Journey Across the Rocky Mountains* (*Early Western Travels* XXI) 1905

TRUE, R. H. 'Some Neglected Botanical Results of the Lewis and Clark Expedition' (in *Proceedings of the American Philosophical Society* No. 67) 1928

'John Bradbury' (*ibid.*, No. 68) 1929

TUCKEY, J. K. *Narrative of an Expedition to Explore the River Zaire* 1818

TURNER, D. *Extracts from the Correspondence of Richard Richardson* 1835

TWEEDIE, J. 'Buenos Ayres' (in *Annals of Natural History* I & IV) 1838–42

VEITCH, H. *Hortus Veitchii* 1906

VEITCH, J. G. 'Travels in Japan' (in the *Gardener's Chronicle*) 1860–1

'Extracts from Journal . . . during a trip to the Australian Colonies' (in the *Gardener's Chronicle*) 1866

VEITCH, J. H. *A Traveller's Notes* 1896

A Manual of the Coniferae (2nd ed.) 1900

WHELER, G. *Journey into Greece* 1682

WILSON, E. H. *A Naturalist in Western China* 1913

Plant Hunting 1927

WOOLLS, W. *Lectures on the Vegetable Kingdom* 1879

Periodicals and works of reference:

Annals of Botany

Botanical Magazine

Botanical Miscellany

Companion to the Botanical Magazine

Gardener's Chronicle

Gardener's Magazine

Journal of Botany

Journal of The Royal Horticultural Society

Journal of the Linnean Society

Kew Bulletin

Philosophical Transactions

Proceedings of the Linnean Society

Transactions of the Horticultural Society

Transactions of the Linnean Society

Allgemeine Deutsche Biographie

BRITTEN, J. and BOULGER, G. S. *A Biographical Index of Deceased British and Irish Botanists* 2nd ed. 1931

BARNHART, J. H. *Biographical Notes upon Botanists* 1965

Biographie Universelle

Dictionary of National Biography

R.H.S. Dictionary of Gardening

Rees' Cyclopaedia

Unpublished letters from the Kew collection and other sources

List of illustrations

Plates between pages 176 and 193

Figures in the text

Index

Index of Persons (collectors' names and main references in bold)

Index of Plants